T0213444

LONDON MATHEMATICAL SOCIETY LECTURE NOTE SERIES

Managing Editor: Professor Miles Reid, Mathematics Institute, University of Warwick, UK

All the titles listed below can be obtained from good booksellers or from Cambridge University Press. For a complete series listing visit www.cambridge.org/mathematics

London Mathematical Society Lecture Note Series: 363

Foundations of Computational Mathematics, Hong Kong 2008

Edited by

FELIPE CUCKER

City University of Hong Kong

ALLAN PINKUS

Technion

MICHAEL J. TODD

Cornell University

CAMBRIDGE
UNIVERSITY PRESS

CAMBRIDGE
UNIVERSITY PRESS

University Printing House, Cambridge CB2 8BS, United Kingdom

Cambridge University Press is part of the University of Cambridge.

It furthers the University's mission by disseminating knowledge in the pursuit of education, learning and research at the highest international levels of excellence.

www.cambridge.org
Information on this title: www.cambridge.org/9780521739702

© Cambridge University Press 2009

First published 2009

A catalogue record for this publication is available from the British Library

ISBN 978-0-521-73970-2 Paperback

Contents

Preface

The Society for the Foundations of Computational Mathematics supports and promotes fundamental research in computational mathematics and its applications, interpreted in the broadest sense. It fosters interaction among mathematics, computer science and other areas of computational science through its conferences, workshops and publications. As part of this endeavour to promote research across a wide spectrum of subjects concerned with computation, the Society brings together leading researchers working in diverse fields. Major conferences of the Society have been held in Park City (1995), Rio de Janeiro (1997), Oxford (1999), Minneapolis (2002), Santander (2005), and Hong Kong (2008). The next conference is expected to be held in 2011. More information about FoCM is available at its website http://www.focm.net.

The conference in Hong Kong on June 16 – 26, 2008, was attended by several hundred scientists. FoCM conferences follow a set pattern: mornings are devoted to plenary talks, while in the afternoon the conference divides into a number of workshops, each devoted to a different theme within the broad theme of foundations of computational mathematics. This structure allows for a very high standard of presentation, while affording endless opportunities for cross-fertilization and communication across subject boundaries. Workshops at the Hong Kong conference were held in the following nineteen fields:

- Approximation theory
- Asymptotic analysis
- Computational algebraic geometry
- Computational dynamics
- Computational number theory
- Foundations of numerical PDEs
- Geometric integration and computational mechanics
- Image and signal processing
- Information-based complexity
- Learning theory
- Multiresolution and adaptivity in numerical PDEs
- Numerical linear algebra

- Optimization
- Real-number complexity
- Relations with computer science
- Special functions and orthogonal polynomials
- Stochastic computation
- Stochastic eigenanalysis
- Symbolic analysis

In addition to the workshops, eighteen plenary lectures, covering a broad spectrum of topics connected to computational mathematics, were delivered by some of the world's foremost researchers. This volume is a collection of articles based on the plenary talks presented at FoCM 2008. The topics covered in the lectures and in this volume reflect the breadth of research within computational mathematics as well as the richness and fertility of interactions between seemingly unrelated branches of pure and applied mathematics.

We hope that this volume will be of interest to researchers in the field of computational mathematics and also to non-experts who wish to gain some insight into the state of the art in this active and significant field.

Like previous FoCM conferences, the Hong Kong gathering proved itself to be a rather uncommon but very stimulating meeting place of researchers in computational mathematics and of theoreticians in mathematics and computer science. While presenting plenary talks by foremost world authorities and maintaining the highest technical level in the workshops, the conference, like previous meetings, laid emphasis on multidisciplinary interaction across subjects and disciplines in an informal and friendly atmosphere.

We wish to express our gratitude to the local organizers and administrative staff of our hosts at the City University of Hong Kong, and wish to thank the Croucher Foundation, the Department of Mathematics at City University of Hong Kong, the Liu Bie Ju Centre for Mathematical Sciences at City University of Hong Kong, and the French General Consulate of Hong Kong and Macau for their financial assistance and for making FoCM 2008 such an outstanding success. We would like to thank the authors of the articles in this volume for producing in short order such excellent contributions. Above all, however, we wish to express our gratitude to all the participants of FoCM 2008 for attending the meeting and making it such an exciting, productive and scientifically stimulating event.

Contributors

Peter Bürgisser Institute of Mathematics, University of Paderborn,
D-33098 Paderborn, Germany
pbuerg@upb.de

Alicia Dickenstein Departamento de Matemática, FCEN,
Universidad de Buenos Aires, Ciudad Universitaria, Pab. I,
C1428EGA Buenos Aires, Argentina
alidick@dm.uba.ar

Nira Dyn School of Mathematical Sciences, Tel Aviv University, Tel
Aviv, Israel
niradyn@post.tau.ac.il

Ulrik S. Fjordholm Department of Mathematics, University of Oslo,
P.O. Box 1053, Blindern, N–0316 Oslo, Norway
ulriksf@ulrik.uio.no

A. Jentzen Institut für Mathematik, Johann Wolfgang Goethe
Universität, D-60054 Frankfurt am Main, Germany
jentzen@math.uni-frankfurt.de

P.E. Kloeden Institut für Mathematik, Johann Wolfgang Goethe
Universität, D-60054 Frankfurt am Main, Germany
kloeden@math.uni-frankfurt.de

Siddhartha Mishra Centre of Mathematics for Applications (CMA),
University of Oslo, P.O. Box 1053, Blindern, N–0316 Oslo, Norway
siddharm@cma.uio.no

A. Neuenkirch Institut für Mathematik, Johann Wolfgang Goethe
Universität, D-60054 Frankfurt am Main, Germany
neuenkirch@math.uni-frankfurt.de

Carles Simó Departament de Matemàtica Aplicada i Anàlisi,
Universitat de Barcelona, Gran Via, 585, Barcelona 08007, Spain
`carles@maia.ub.es`

Eitan Tadmor Department of Mathematics, Institute for Physical
Sciences and Technology (IPST), and Center of Scientific
Computation and Mathematical Modeling (CSCAMM), University
of Maryland, MD 20742-4015, USA
`tadmor@cscamm.umd.edu`

R. Wong Department of Mathematics, City University of Hong
Kong, Tat Chee Avenue, Kowloon, Hong Kong
`mawong@cityu.edu.hk`

Henryk Woźniakowski Department of Computer Science, Columbia
University, New York, NY 10027, USA; *and*
Institute of Applied Mathematics, University of Warsaw, ul.
Banacha 2, 02-097 Warsaw, Poland
`henryk@cs.columbia.edu`

1

Smoothed Analysis of Condition Numbers

Peter Bürgisser

Institute of Mathematics
University of Paderborn
D-33098 Paderborn, Germany
email: pbuerg@upb.de

Abstract

The running time of many iterative numerical algorithms is dominated by the condition number of the input, a quantity measuring the sensitivity of the solution with regard to small perturbations of the input. Examples are iterative methods of linear algebra, interior-point methods of linear and convex optimization, as well as homotopy methods for solving systems of polynomial equations. Thus a probabilistic analysis of these algorithms can be reduced to the analysis of the distribution of the condition number for a random input. This approach was elaborated upon for average-case complexity by many researchers.

The goal of this survey is to explain how average-case analysis can be naturally refined in the sense of smoothed analysis. The latter concept, introduced by Spielman and Teng in 2001, aims at showing that for all real inputs (even ill-posed ones), and all slight random perturbations of that input, it is unlikely that the running time will be large. A recent general result of Bürgisser, Cucker and Lotz (2008) gives smoothed analysis estimates for a variety of applications. Its proof boils down to local bounds on the volume of tubes around a real algebraic hypersurface in a sphere. This is achieved by bounding the integrals of absolute curvature of smooth hypersurfaces in terms of their degree via the principal kinematic formula of integral geometry and Bézout's theorem.

1.1 Introduction

In computer science, the most common theoretical approach to understanding the behaviour of algorithms is *worst-case analysis*. This means proving a bound on the worst possible performance an algorithm can

have. In many situations this gives satisfactory answers. However, there
are cases of algorithms that perform exceedingly well in practice and
still have a provably bad worst-case behaviour. A famous example is
Dantzig's simplex algorithm. In an attempt to rectify this discrepancy,
researchers have introduced the concept of *average-case analysis*, which
means bounding the expected performance of an algorithm on random
inputs. For the simplex algorithm, average-case analyses have been first
given by Borgwardt (1982) and Smale (1983). However, while a proof of
good average performance yields an indication of a good performance in
practice, it can rarely explain it convincingly. The problem is that the
results of an average-case analysis strongly depend on the distribution
of the inputs, which is unknown, and usually assumed to be Gaussian
for rendering the mathematical analysis feasible.

Spielman and Teng suggested in 2001 the concept of *smoothed analysis
of algorithms*, which is a new form of analysis of algorithms that arguably
blends the best of both worst-case and average-case. They used this new
framework to give a more compelling explanation of the simplex method
(for the shadow vertex pivot rule). For this work they were recently
awarded the 2008 Gödel prize. See Spielman and Teng (2004) for the
full paper.

The general idea of smoothed analysis is easy to explain. Let $T\colon \mathbb{R}^p \to \mathbb{R}_+ \cup \{\infty\}$ be any function (measuring running time etc). Instead of
showing "it is unlikely that $T(a)$ will be large," one shows that "for
all $\overline{a}$ and all slight random perturbations $\overline{a} + \delta a$, it is unlikely that
$T(\overline{a}+\delta a)$ will be large." If we assume that the perturbation δa is centered
(multivariate) standard normal with variance σ^2, in short $\delta a \in N(0, \sigma^2)$,
then the goal of a smoothed analysis of T is to give good estimates of

$$\sup_{\overline{a}\in\mathbb{R}^p} \operatorname{Prob}_{\delta a \in N(0,\sigma^2)}\{T(\overline{a} + \delta a) \geq \epsilon^{-1}\}.$$

In a first approach, one may focus on expectations, that is on bounding

$$\sup_{\overline{a}\in\mathbb{R}^p} \mathbb{E}_{\delta a \in N(0,\sigma^2)} T(\overline{a} + \delta a).$$

Figure 1.1 succinctly summarizes the three types of analysis of algorithms.

Smoothed analysis is not only useful for analyzing the simplex algorithm, but can be applied to a wide variety of numerical algorithms. For
doing so, understanding the concept of condition numbers is an important intermediate step.

A distinctive feature of the computations considered in numerical anal-

Worst-case analysis	Average-case analysis	Smoothed analysis
$\sup\limits_{a \in \mathbb{R}^p} T(a)$	$\mathbb{E}_{a \in \mathcal{D}} T(a)$	$\sup\limits_{\overline{a} \in \mathbb{R}^p} \mathbb{E}_{a \in N(\overline{a}, \sigma^2)} T(a)$

Fig. 1.1. Three types of analysis of algorithms. $\mathcal{D}$ denotes a probability distribution on $\mathbb{R}^p$.

ysis is that they are affected by errors. A main character in the understanding of the effects of these errors is the *condition number* of the input. This is a positive number which, roughly speaking, quantifies the errors when computations are performed with infinite precision but the input has been modified by a small perturbation. The condition number depends only on the data and the problem at hand. The best known condition number is that for matrix inversion and linear equation solving. For a square matrix A it takes the form $\kappa(A) = \|A\|\|A^{-1}\|$ and was independently introduced by Goldstine and von Neumann (1947) and Turing (1948).

Condition numbers are omnipresent in round-off analysis. They also appear as a parameter in complexity bounds for a variety of efficient iterative algorithms in linear algebra, linear and convex optimization, as well as homotopy methods for solving systems of polynomial equations. The running time $T(x, \epsilon)$ of these algorithms, measured as the number of arithmetic operations, can often be bounded in the form

$$T(x, \epsilon) \leq \big(\text{size}(x) + \mu(x) + \log \epsilon^{-1}\big)^c, \tag{1.1}$$

with some universal constant $c > 0$. Here the input is a vector $x \in \mathbb{R}^n$ of real numbers, $\text{size}(x) = n$ is the dimension of x, the positive parameter ϵ measures the required accuracy, and $\mu(x)$ is some measure of conditioning of x. (Depending on the situation, $\mu(x)$ may be either a condition number or its logarithm. Moreover, $\log \epsilon^{-1}$ might be replaced by $\log \log \epsilon^{-1}$.)

We discuss the issue of condition-based analysis of algorithms in Sections 1.2–1.4, by elaborating a bit on the case of convex optimization and putting special focus on generalizations of Renegar's (1995a, 1995b) condition number for linear programming. We also discuss Shub and Smale's (1993a) condition number for polynomial equation solving.

Let us mention that L. Blum (1990) suggested to extend the complexity theory of real computation due to Blum, Shub, Smale (1989) by

measuring the performance of algorithms in terms of the size and the
condition of inputs. However, up to now, no complexity theory over
the reals has been developed that incorporates the concepts of approximation and conditioning and allows to speak about lower bounds or
completeness results in that context.

Smale (1997) proposed a two-part scheme for dealing with complexity
upper bounds in numerical analysis. The first part consists of establishing
bounds of the form (1.1). The second part of the scheme is to analyze the
distribution of $\mu(x)$ under the assumption that the inputs x are random
with respect to some probability distribution. More specifically, we aim
at tail estimates of the form

$$\text{Prob}\left\{\mu(x) \geq \epsilon^{-1}\right\} \leq \text{size}(x)^c \epsilon^\alpha \quad (\epsilon > 0)$$

with universal constants $c, \alpha > 0$. In a first attempt, one may try to
show upper bounds on the expectation of $\mu(x)$ (or $\log \mu(x)$, depending
on the situation). Combining the two parts of the scheme, we arrive
at upper bounds for the average running time of our specific numerical
algorithms considered. So if we content ourselves with statements about
the probabilistic average-case, we can eliminate the dependence on $\mu(x)$
in (1.1). This approach was elaborated upon for average-case complexity by Blum and Shub (1986), Renegar (1987), Demmel (1988), Kostlan
(1988), Edelman (1988, 1992), Shub and Smale (1993b, 1994, 1996),
Cheung and Cucker (2002), Cucker and Wschebor (2003), Cheung et
al. (2005), Beltrán and Pardo (2007), and others. We only briefly discuss a few of these results in Section 1.5. Instead, we put emphasis on the
analysis of the GCC-condition number $\mathcal{C}(A)$ of linear programming introduced by Goffin (1980) and Cheung and Cucker (2001), see (1.11). This
is a variation of the condition number introduced by Renegar (1995a,
1995b). We discuss a recently found connection between the average-
case analysis of the GCC-condition number and covering processes on
spheres, and we present a sharp result on the probability distribution of
$\mathcal{C}(A)$ for feasible inputs due to Bürgisser et al. (2007).

The main goal of this survey is to show that part two of Smale's
scheme can be naturally refined by performing a smoothed analysis of
the condition number $\mu(x)$ involved. This was already suggested by
Spielman and Teng in their ICM 2002 paper. For the matrix condition number, results in this direction were obtained by Wschebor (2004)
and Sankar et al. (2006). A recent paper by Tao and Vu (2007) deals
with the matrix condition number under random discrete perturbations.
Dunagan et al. (2003) gave a smoothed analysis of Renegar's condition

number of linear programming, thereby obtaining for the first time a smoothed analysis for the running time of interior-point methods, see also Spielman and Teng (2003).

A paper by Demmel (1988) has the remarkable feature that the probabilistic average-case analysis performed there for a variety of problems is not done with ad-hoc arguments adapted to the problem considered. Instead, these applications are all derived from a single result bounding the tail of the distribution of a conic condition number in terms of geometric invariants of the corresponding set of ill-posed inputs. Bürgisser et al. (2006, 2008) recently extended Demmel's result from average-case analysis to a natural geometric framework of smoothed analysis of conic condition numbers, called *uniform smoothed analysis*. This result will be presented in Section 1.6. The critical parameter entering these estimates turned out to be the degree of the defining equations of the set of ill-posed inputs. This result has a wide range of applications to linear and polynomial equation solving, as explained in Section 1.6.1. In particular, it easily gives a smoothed analysis of the condition number of a matrix. Moreover, Amelunxen and Bürgisser (2008) showed that this result, after suitable modification to a spherical convex setting, also allows a smoothed analysis of the GCC-condition number of linear programming.

The mathematical setting of uniform smoothed analysis has a clean and simple description. The set of ill-posed inputs to a computational problem is modelled as a subset Σ_S of a sphere S^p, which is considered the data space. In most of our applications, Σ_S is an algebraic hypersurface, but for optimization problems Σ_S will be semialgebraic. The corresponding conic condition number $\mathcal{C}(a)$ of an input $a \in S^p$ is defined as

$$\mathcal{C}(a) = \frac{1}{\sin d_S(a, \Sigma_S)},$$

where d_S refers to the angular distance on S^p. For $0 \leq \sigma \leq 1$ let $B(\overline{a}, \sigma)$ denote the spherical cap in the sphere S^p centered at $\overline{a} \in S^p$ and having angular radius $\arcsin \sigma$. Moreover, we define for $0 < \epsilon \leq 1$ the ϵ-*neighborhood* of Σ_S as

$$T(\Sigma_S, \epsilon) := \{a \in S^p \mid d_S(a, \Sigma_S) < \arcsin \epsilon\}.$$

The task of a uniform smoothed analysis of $\mathcal{C}$ consists of providing good upper bounds on

$$\sup_{\overline{a} \in S^p} \mathrm{Prob}_{a \in B(\overline{a}, \sigma)}\{\mathcal{C}(a) \geq \epsilon^{-1}\},$$

Fig. 1.2. Neighborhood of the curve Σ_S intersected with a spherical disk.

where a is assumed to be chosen uniformly at random in $B(\overline{a}, \sigma)$. The probability occurring here has an immediate geometric meaning:

$$\mathrm{Prob}_{a \in B(\overline{a}, \sigma)}\{\mathcal{C}(a) \geq \epsilon^{-1}\} = \frac{\mathrm{vol}\,(T(\Sigma_S, \epsilon) \cap B(\overline{a}, \sigma))}{\mathrm{vol}\,(B(\overline{a}, \sigma))}. \tag{1.2}$$

Thus uniform smoothed analysis means to provide bounds on the relative volume of the intersection of ϵ-neighborhoods of Σ_S with small spherical disks, see Figure 1.2. We note that uniform smoothed analysis interpolates transparently between worst-case and average-case analysis. Indeed, when $\sigma = 0$ we get worst-case analysis, while for $\sigma = 1$ we obtain average-case analysis. (Note that $S^p = B(\overline{a}, 1) \cup B(-\overline{a}, 1)$ for any $\overline{a}$.)

In Section 1.7 we explain the rich mathematical background behind our uniform smoothed analysis estimates. We first review classical results on the volume of tubes and then state the principal kinematic formula of integral geometry for spheres. Finally, in Section 1.7.3, we outline the proof of the main Theorem 1.2, which proceeds by estimating the integrals of absolute curvature arising in Weyl's tube formula (1939) with the help of Chern's (1966) principal kinematic formula and Bézout's theorem.

1.2 Condition Numbers for Linear Algebra

A numerical computation problem can often be formalized by a mapping $f\colon U \to Y$ between finite-dimensional real or complex vector spaces X and Y, where U is an open subset of X. The space X is interpreted as the set of inputs to the problem, Y is the set of solutions, and f is the solution map. Small perturbations δx of an input x result in a perturbation δy of the output $y = f(x)$. In order to quantify this effect with regard to small relative errors, we choose norms on the spaces X and Y and define the *relative condition number* of f at x by

$$\kappa(f, x) := \lim_{\epsilon \to 0} \sup_{\|\delta x\| \leq \epsilon \|x\|} \frac{\|f(x + \delta x) - f(x)\| / \|f(x)\|}{\|\delta x\| / \|x\|}.$$

If f is differentiable at x, this can be expressed in terms of the operator norm of the Jacobian $Df(x)$ with respect to the chosen norms:

$$\kappa(f, x) = \|Df(x)\| \frac{\|x\|}{\|f(x)\|}.$$

In the case $X = Y = \mathbb{R}$, the logarithm of the condition number measures the *loss of precision* when evaluating f: if we know x up to ℓ decimal digits, then we know $f(x)$ roughly up to $\ell - \log_{10} \kappa(f, x)$ decimal digits.

Consider matrix inversion $f \colon \mathrm{GL}(m, \mathbb{R}) \to \mathbb{R}^{m \times m}, A \mapsto A^{-1}$, measuring errors with respect to the L_2-operator norm. A perturbation argument shows that the condition number of f at A equals the *classical matrix condition number*

$$\kappa(A) := \kappa(f, A) = \|A\| \, \|A^{-1}\|$$

of the matrix A. It is easy to see that $\kappa(A)$ also equals the condition number of the map $\mathrm{GL}(m, \mathbb{R}) \to \mathbb{R}^m, A \mapsto A^{-1}b$ for fixed nonzero $b \in \mathbb{R}^m$. In fact, $\kappa(A)$ determines the condition number for solving a quadratic linear system of equations. It is also known that $\kappa(A)$ dominates the condition number of several other problems of numerical linear algebra, like the Cholesky and QR decomposition of matrices, see Amodei and Dedieu (2008). Moreover, the condition number $\kappa(A)$ appears in Wilkinson's round-off analysis of Gaussian elimination with partial pivoting (together with the so-called growth factor), see Wilkinson (1963) and Higham (1996).

Let us return to the problem of matrix inversion. We can interpret the set of singular matrices $\Sigma := \{A \in \mathbb{R}^{m \times m} \mid \det A = 0\}$ as its *set of ill-posed instances*. Let $\mathrm{dist}(A, \Sigma)$ denote the distance of the matrix A to Σ, measured with respect to the L_2-operator norm. The distance of A to Σ with respect to the Frobenius norm $\|A\|_F := (\sum_{ij} a_{ij}^2)^{1/2}$ shall be denoted by $\mathrm{dist}_F(A, \Sigma)$. The theorem of Eckart and Young (1936) states that $\mathrm{dist}(A, \Sigma) = \mathrm{dist}_F(A, \Sigma) = \|A^{-1}\|^{-1}$. As the right-hand side equals the smallest singular value of A, this is just a special case of the well-known fact that the kth largest singular value of A equals the distance of A to the set of matrices of rank less than k (with respect to the L_2-operator norm). We rephrase Eckart and Young's result as the following *condition number theorem*

$$\kappa(A) = \frac{\|A\|}{\text{dist}(A, \Sigma)} = \frac{\|A\|}{\text{dist}_F(A, \Sigma)}. \tag{1.3}$$

It is remarkable that the condition number $\kappa(A)$, which was defined using local properties, can be characterized in this global geometric way.

Demmel (1987) realized that this observation for the classical matrix condition number actually holds in much larger generality. For numerous computation problems, the condition number of an input x of norm one, say, can be bounded up to a constant factor by the inverse distance of x to a corresponding set of ill-posed inputs Σ. It is this key insight that allows to perform probabilistic analyses of condition numbers by geometric tools.

To further illustrate this connection, consider *eigenvalue computations*. Let $\lambda \in \mathbb{C}$ be a simple eigenvalue of $A \in \mathbb{C}^{m \times m}$. The sensitivity to compute λ from A is captured by a condition number $\kappa(A, \lambda)$, see (1.22). Wilkinson (1972) proved that

$$\kappa(A, \lambda) \le \frac{\sqrt{2} \, \|A\|_F}{\text{dist}_F(A, \Sigma_{\text{eigen}})},$$

where Σ_{eigen} is the set of matrices in $\mathbb{C}^{m \times m}$ having a multiple eigenvalue.

Clearly, condition numbers are a crucial issue when dealing with finite precision computations and round-off errors. When considering iterative methods (instead of direct methods), it turns out that, even when assuming infinite precision arithmetic, the condition of an input often affects the number of iterations required to achieve a certain precision. A famous example for this phenomenon is the *conjugate gradient method* of Hestenes and Stiefel (1952). For a given linear system $Ax = b$, A a symmetric positive definite matrix, the conjugate gradient method starts with an initial value $x_0 \in \mathbb{R}^n$ and produces a sequence of iterates $x_0, x_1, \ldots, x_n = x^*$ satisfying

$$\|x_k - x^*\|_A \le 2 \left(\frac{\sqrt{\kappa(A)} - 1}{\sqrt{\kappa(A)} + 1} \right)^k \|x_0 - x^*\|_A,$$

where the A-norm of a vector v is defined as $\|v\|_A := (v^T A v)^{1/2}$. Therefore, roughly $\frac{1}{2} \sqrt{\kappa(A)} \ln \frac{1}{\epsilon}$ iterations are sufficient in order to achieve $\|x_k - x^*\|_A \le \epsilon \|x_0 - x^*\|_A$.

1.3 Condition Numbers for Convex Optimization

We restrict our discussion to feasibility problems in convex conic form. Let X and Y be real finite-dimensional vector spaces endowed with norms. Further, let $K \subseteq X$ be a closed convex cone that is assumed to be regular, that is $K \cap (-K) = \{0\}$ and K has nonempty interior. We denote by $L(Y, X)$ the space of linear maps from Y to X endowed with the operator norm. Given $A \in L(Y, X)$, consider the feasibility problem of deciding

$$\exists y \in Y \setminus \{0\} \quad Ay \in K. \tag{1.4}$$

Two special cases of this general framework should be kept in mind. For $K = \mathbb{R}^n_+$, the nonnegative orthant in $\mathbb{R}^n$, one obtains the homogeneous *linear programming feasibility problem*. The feasibility version of homogeneous *semidefinite programming* corresponds to the cone $K = \mathcal{S}^n_+$ consisting of the positive semidefinite matrices in $\mathbb{R}^{n \times n}$.

The feasibility problem dual to (1.4) is

$$\exists x^* \in X^* \setminus \{0\} \quad A^* x^* = 0, \; x^* \in K^*. \tag{1.5}$$

Here X^*, Y^* are the dual spaces of X, Y, respectively, $A^* \in L(X^*, Y^*)$ denotes the map adjoint to A, and $K^* := \{y^* \in Y^* \mid \forall x \in K \; \langle y^*, x \rangle \geq 0\}$ denotes the cone dual to K.

We denote by $\mathcal{D}$ the set of instances $A \in L(Y, X)$ for which the problem (1.4) is strictly feasible, i.e., there exists $y \in Y$ such that $Ay \in \text{int}(K)$. Likewise, we denote by $\mathcal{P}$ the set of $A \in L(Y, X)$ such that (1.5) is strictly feasible, i.e., there exists $x^* \in \text{int}(K^*)$ with $A^* x^* = 0$.

Both $\mathcal{D}$ and $\mathcal{P}$ are disjoint open subsets of $L(Y, X)$ and duality in convex optimization implies that $\mathcal{P}$ is the complement of the closure of $\overline{\mathcal{D}}$ in $L(Y, X)$, cf. Boyd and Vandenberghe (2004). The *conic feasibility problem* is to decide for given $A \in L(Y, X)$ whether (1.4) or (1.5) holds. The common boundary $\Sigma := \partial\mathcal{D} = \partial\mathcal{P}$ of the sets $\mathcal{D}$ and $\mathcal{P}$ can be considered as the set of ill-posed instances. Indeed, for given $A \in \Sigma$, arbitrarily small perturbations of A may yield instances in both $\mathcal{D}$ and $\mathcal{P}$.

Renegar (1995a) defined the *condition number* of the conic feasibility problem by

$$C(A) := \frac{\|A\|}{\text{dist}(A, \Sigma)}. \tag{1.6}$$

He observed that the number of steps of interior-point algorithms solving the conic feasibility problem can be effectively bounded in terms

of $C(A)$. Before elaborating on this important issue, let us character-
ize the condition number $C(A)$ in a different way. Suppose there exists
$e \in \text{int}(K)$ such that the unit ball $B(e, 1)$ centered at e is contained
in K. We define $\lambda_{\min} \colon X \to \mathbb{R}$ by $\lambda_{\min}(x) := \max\{t \in \mathbb{R} \mid x - te \in K\}$
and note that $x \in K \Leftrightarrow \lambda_{\min}(x) \geq 0$. For $K = \mathbb{R}^n_+$ and $e = (1, \ldots, 1)$ we
have $\lambda_{\min}(x) = \min_i x_i$, while in the case $K = \mathcal{S}^n_+$ and e being the unit
matrix, $\lambda_{\min}(x)$ equals the minimum eigenvalue of x.

The problem (1.4) is feasible iff there exists $y \in Y$ of norm one such
that $\lambda_{\min}(Ay) \geq 0$. A vector y maximizing $\lambda_{\min}(Ay)/\|y\|$ may be in-
terpreted as a best-conditioned solution, due to the following max-min
characterization in Cheung et al. (2008):

$$\text{dist}(A, \Sigma) = \left| \max_{\|y\|=1} \lambda_{\min}(Ay) \right|. \tag{1.7}$$

Actually, in that paper a more general result is shown. Suppose we have
a multifold conic structure: $X = X_1 \times \cdots \times X_r$, where X_i is a normed
vector space, $K = K_1 \times \cdots \times K_r$ with regular closed convex cones K_i
in X_i, and $e_i \in \text{int}(K_i)$ such that the unit ball centered at e_i is contained
in K_i. We have a corresponding function $\lambda^i_{\min} \colon X_i \to \mathbb{R}$. Then (1.4) can
be written as

$$\exists y \in Y \setminus \{0\} \quad A_1 y_1 \in K_1, \ldots, A_r y_r \in K_1,$$

where $A_i \in L(Y, X_i)$ is the composition of A with the projection onto X_i.
Generalizing (1.6), we define the corresponding *multifold condition number* $\mathcal{C}(A)$ by

$$\mathcal{C}(A) := \left(\min_{B \in \Sigma} \max_i \frac{\|A_i - B_i\|}{\|A_i\|} \right)^{-1}.$$

It is easy to see that $\mathcal{C}(A) \leq C(A)$ when taking $\|A\| = \max_i \|A_i\|$. Note
that in the case $r = 1$ of just one factor, we retrieve $C(A) = \mathcal{C}(A)$.
The condition number $\mathcal{C}(A)$ seems a more natural measure of conditioning in the multifold setting, when allowing component normalization as
preconditioning. Cheung et al. (2008) proved the following *condition
number theorem*, extending (1.7),

$$\frac{1}{\mathcal{C}(A)} = \left| \max_{\|y\|=1} \min_i \frac{\lambda^i_{\min}(A_i y)}{\|A_i\|} \right|. \tag{1.8}$$

Let us now have a closer look at the important special case of $X_i = \mathbb{R}$,
$K_i = \mathbb{R}_+$, $e_i = 1$ for $i = 1, \ldots, n$. We endow $X = \mathbb{R}^n$ with the L_∞-norm
and $Y := \mathbb{R}^{m+1}$ with the L_2-norm. The problem (1.4) now reads as the

linear programming feasibility problem

$$\exists y \in \mathbb{R}^{m+1} \setminus \{0\} \quad a_1 y \geq 0, \ldots, a_n y \geq 0, \qquad (1.9)$$

where $a_i \in \mathbb{R}^{m+1}$ denote the rows of the given matrix $A \in \mathbb{R}^{n \times (m+1)}$, which we may assume to be scaled to Euclidean length one. We can therefore interpret the input A as a sequence of n points $a_1, \ldots, a_n$ on the unit sphere $S^m := \{y \in \mathbb{R}^{m+1} \mid \|y\| = 1\}$. By self-duality of the nonnegative orthant, the feasibility problem (1.5) translates to

$$\exists x \in \mathbb{R}^n \setminus \{0\} \quad A^T y = 0, \; x \in \mathbb{R}^n_+. \qquad (1.10)$$

The multifold condition number $\mathcal{C}(A)$ corresponding to this setting has been introduced and investigated by Goffin (1980), and Cheung and Cucker (2001). We will refer to it as the *GCC-condition number*.

There is a nice geometric characterization of $\mathcal{C}(A)$: Fix an input A, interpreted as a sequence of points $a_1, \ldots, a_n \in S^m$. For $y \in S^m$ we have $a_i y = \cos \theta_i(y)$, where $\theta_i(y) \in [0, \pi]$ denotes the angle between y and a_i. Put $\theta(y) := \max_i \theta_i(y)$. Then $\rho(A) := \min_{y \in S^m} \theta(y)$ is the angular radius of a smallest spherical cap enclosing all the points a_i. This quantity captures the GCC-condition number. Indeed, using $\lambda^i_{\min}(x_i) = x_i$, the condition number theorem (1.8) translates to

$$\mathcal{C}(A)^{-1} = |\cos \rho(A)|. \qquad (1.11)$$

Moreover, we note that (1.9) is feasible iff $\rho(A) \leq \pi/2$ and hence $A \in \Sigma$ iff $\rho(A) = \pi/2$.

1.3.1 Condition based analysis

We turn now to the relation of conditioning to complexity. Freund and Vera (1999) gave a condition based analysis of Khachyian's (1979) ellipsoid method. The essence of their argument is rather simple so that we are going to sketch it briefly.

Suppose we are in the general conic setting and A is a feasible instance of (1.4). We define the *width* $\tau(A)$ of the cone of solutions $A^{-1}(K)$ as the maximum ratio $r/\|y\|$ over all balls $B(y, r)$ contained in $A^{-1}(K)$. Let $y_0 \in Y$ be a best conditioned solution of norm 1, that is, maximizing the right hand side of (1.7). Then it is not hard to see that $B(y_0, C(A)^{-1}) \subseteq A^{-1}(K)$, hence $C(A)^{-1} \leq \tau(A)$. (In the case $K = \mathbb{R}^n_+$ we even have $\mathcal{C}(A)^{-1} \leq \tau(A)$.)

Suppose now that $Y = \mathbb{R}^m$ is endowed with the L_2-norm and consider the compact convex set $\tilde{K}_A$ obtained by homogenizing with an additional variable t and intersecting with the unit ball B_{m+1}:

$$\tilde{K}_A := \{(y, t) \in Y \times \mathbb{R}_+ \mid Ay \in K, \; \|y\|^2 + t^2 \leq 1\}.$$

Freund and Vera (1999) showed that

$$\ln\left(\frac{\text{vol } B_{m+1}}{\text{vol } \tilde{K}_A}\right) \leq (m+1) \ln\left(2 + \frac{6}{\tau(A)}\right),$$

where vol denotes the $(m+1)$-dimensional volume.

We assume now $X = \mathbb{R}^n$ and that the convex cone $K \subseteq \mathbb{R}^n$ is given by a separation oracle, i.e., for a given $x_0 \in \mathbb{R}^n$ the oracle either answers $x_0 \in K$ or provides a hyperplane separating x_0 from K. (Note that for $K = \mathbb{R}^n$ the separation oracle is trivial.) Running the ellipsoid method (see Grötschel, Lovász and Schrijver (1988)) on the convex set $\tilde{K}_A$, starting with the enclosing ball B_{m+1}, we arrive at the following:

Theorem 1.1 *The ellipsoid method, applied to the homogenized convex set $\tilde{K}_A$, either finds a feasible point $y \in A^{-1}(K)$ or decides $A^{-1}(K) = \emptyset$ with a number of iterations bounded by $2(m+1)^2 \ln(2 + 6\,C(A))$. Each iteration step involves one call of the separation oracle plus $\mathcal{O}(m^2)$ arithmetic operations and one square root for the computation of the next ellipsoid.*

This general result is impractical, but it has the beauty of showing by a simple argument that the complexity of rather general conic feasibility problems is polynomially bounded in the dimensions m, n and $\ln C(A)$. (Of course we assume that the cost of one call to the separation oracle is polynomially bounded in n, m.)

A great deal of motivation for the work described so far in this section comes from the major open problem of whether the linear programming feasibility problem LPF (1.9) can be algorithmically solved with a number of arithmetic operations polynomial in m and n. In fact this problem is listed as one of Smale's problems (2000) for the next century. Motivated by this question, Renegar (1995a, 1995b) introduced the condition number $C(A)$ and proved by interior-point methods that the complexity of LPF is polynomially bounded in m, n and $\ln C(A)$. This considerably added to our understanding of the complexity of LPF. The well-known fact that LPF for rational inputs is solvable in polynomial time in the Turing model is a simple consequence of this. Indeed, it is sufficient to note that for rational matrices $A \notin \Sigma$, $\ln C(A)$ is polynomially bounded in the bitsize of A. (One also has to check that there is no explosion of bit size in the computations, which is straightforward.)

The most efficient known algorithms for solving convex optimization problems in theory and practice are interior-point methods, cf. Nesterov and Nemirovskii (1994). We do not want to enter this vast field and just mention that Cucker and Peña (2002) gave a condition based analysis of a finite-precision primal-dual interior-point method for solving the linear programming feasibility problem LPF (1.9) with a number of iterations bounded by

$$\mathcal{O}\big(\sqrt{m+n}\,(\log(m+n) + \log \mathcal{C}(A))\big).$$

Hereby, each iteration costs at most $\mathcal{O}((m+n)^3)$ arithmetic operations. In that paper, for the first time, a round-off analysis of an interior-point algorithm was performed, and it was shown that the amount of precision required can be bounded in terms of $\mathcal{C}(A)$. For an early condition based analysis of LPF (in terms of another condition number) we refer to Vavasis and Ye (1995).

A solution to LPF (1.9) can be found by the perceptron method with a number of iterations bounded by $\mathcal{O}(1/\tau(A)^2)$, see Rosenblatt (1962). A more efficient, re-scaled version of the perceptron algorithm has been developed by Dunagan and Vempala (2004), which uses only $\mathcal{O}(n\ln(1/\tau(A)))$ iterations. Recently, this result was extended to conic systems by Belloni, Freund and Vempala (2007).

1.4 Condition Numbers for Polynomial Equation Solving

Condition numbers for solving systems of complex polynomial equations were introduced and studied by Shub and Smale (1993a). The geometric viewpoint of looking for roots of homogeneous equations in complex projective space adds a lot to the elegance and mathematical feasibility of the theory.

We briefly review the setting, for more details and a simplified treatment see Blum, Cucker, Shub, and Smale (1998). Fix $d_1, \ldots, d_n \in \mathbb{N} \setminus \{0\}$ and denote by $\mathcal{H}_d$ the vector space of polynomial systems $f = (f_1, \ldots, f_n)$ with $f_i \in \mathbb{C}[X_0, \ldots, X_n]$ homogeneous of degree d_i. For $f, g \in \mathcal{H}_d$ we write

$$f_i(x) = \sum_\alpha a_\alpha^i X^\alpha, \quad g_i(x) = \sum_\alpha b_\alpha^i X^\alpha,$$

where $\alpha = (\alpha_0, \ldots, \alpha_n)$ is assumed to range over all multi-indices such that $|\alpha| = \sum_{k=0}^{n} \alpha_k = d_i$ and $X^\alpha := X_0^{\alpha_0} X_1^{\alpha_1} \cdots X_n^{\alpha_n}$. The space $\mathcal{H}_d$ is endowed with a Hermitian inner product $\langle f, g \rangle = \sum_{i=1}^{n} \langle f_i, g_i \rangle$, where

$$\langle f_i, g_i \rangle = \sum_{|\alpha|=d_i} a_\alpha^i \, \overline{b_\alpha^i} \binom{d_i}{\alpha}^{-1}.$$

Here, the bar denotes complex conjugate and $\binom{d}{\alpha}$ denotes the multinomial coefficients. The reason for choosing this inner product is that it is invariant under the natural action of the unitary group $U(n+1)$ on $\mathcal{H}_d$. This property is crucial for the whole development. We denote by $\|f\|$ the corresponding norm of $f \in \mathcal{H}_d$.

Let $\mathbb{P}^n := \mathbb{P}(\mathbb{C}^{n+1})$ and $\mathbb{P}(\mathcal{H}_d)$ denote the complex projective spaces associated to $\mathbb{C}^{n+1}$ and $\mathcal{H}_d$, respectively. These are complex manifolds that naturally carry the structure of a Riemannian manifold. The solution variety defined as $V := \{(f, \zeta) \in \mathbb{P}(\mathcal{H}_d) \times \mathbb{P}^n \mid f(\zeta) = 0\}$ is a smooth submanifold of $\mathbb{P}(\mathcal{H}_d) \times \mathbb{P}^n$ and hence also carries a Riemannian structure. (We identify $f \in \mathcal{H}_d$ and its corresponding element in $\mathbb{P}(\mathcal{H}_d)$.)

The computational problem under investigation is now the following: given $f \in \mathbb{P}(\mathcal{H}_d)$, find $\zeta \in \mathbb{P}^n$ such that $f(\zeta) = 0$. Suppose that ζ is a simple solution of f. By the implicit function theorem, the projection map $V \to \mathbb{P}(\mathcal{H}_d), (f', \zeta') \mapsto f'$ can be locally inverted around (f, ζ). The solution map G is the local inverse of this projection. Following the scheme of Section 1.2, it is natural to define the condition number at (f, ζ) as the operator norm of the derivative of G at ζ: $\mu(f, \zeta) := \|DG(\zeta)\|$. A calculation shows

$$\mu(f, \zeta) = \|f\| \, \left\| (Df(\zeta)_{|T_\zeta})^{-1} \mathsf{diag}(\|\zeta\|^{d_1-1}, \ldots, \|\zeta\|^{d_n-1}) \right\|,$$

where $Df(\zeta)_{|T_\zeta}$ denotes restriction of the derivative of $f \colon \mathbb{C}^{n+1} \to \mathbb{C}^n$ at ζ to the tangent space $T_\zeta := \{v \in \mathbb{C}^{n+1} \mid \langle v, \zeta \rangle = 0\}$ of $\mathbb{P}^n$ at ζ. Note that $\mu(f, \zeta)$ is homogeneous of degree 0 in both arguments and hence defined for $(f, \zeta) \in V$ outside the *subvariety of ill-posed pairs*

$$\Sigma' := \{(f, \zeta) \in V \mid \operatorname{rank} Df(\zeta)_{|T_\zeta} < n\}. \tag{1.12}$$

We remark that $(f, \zeta) \in \Sigma'$ means that ζ is a multiple root of f.

In order to simplify the statement of the condition number theorem below, one considers the *(normalized) condition number* defined as

$$\mu_{\mathrm{norm}}(f, \zeta) := \|f\| \, \left\| (Df(\zeta)_{|T_\zeta})^{-1} \mathsf{diag}(\sqrt{d_1} \|\zeta\|^{d_1-1}, \ldots, \sqrt{d_n} \|\zeta\|^{d_n-1}) \right\|.$$

The *condition number theorem* in Shub and Smale (1993a) gives a characterization of μ_{norm} in terms of the inverse distance to the nearest ill-posed input. It states that for $(f, \zeta) \in V \setminus \Sigma'$

$$\mu_{\text{norm}}(f, \zeta) = \frac{1}{\sin d_\zeta(f, \Sigma_\zeta)}. \qquad (1.13)$$

Here $d_\zeta(f, \Sigma_\zeta)$ denotes the distance of f to $\Sigma_\zeta := \{f' \in \mathbb{P}(\mathcal{H}_d) \mid (f', \zeta) \in \Sigma'\}$ measured in the Riemannian metric of the "fiber" $\{f' \in \mathbb{P}(\mathcal{H}_d) \mid f'(\zeta) = 0\}$.

If $f \in \mathcal{H}_d$ has only simple zeros $\zeta_1, \ldots, \zeta_q$ we define the *condition number of f for approximating roots* as

$$\mu_{\text{norm}}(f) := \max_{i \le q} \mu_{\text{norm}}(f, \zeta_i),$$

otherwise we set $\mu_{\text{norm}}(f) := \infty$.

In closing this discussion we mention that Wilkinson (1963), Wozniakowski (1977), and Demmel (1987) studied condition numbers for finding the roots of polynomials in one variable.

We now briefly discuss how to compute an approximate zero of a system of polynomial equations by homotopy continuation and how condition numbers enter the complexity estimates.

By an *approximate zero* of f (in the strict sense) associated with a zero ζ of f we understand a point z such that the sequence of Newton iterates (adapted to projective space)

$$z_{i+1} := N_f(z_i) := z_i - (Df(z_i)_{|T_{z_i}})^{-1} f(z_i)$$

with initial point $z_0 := z$ converges immediately quadratically to ζ, i.e.,

$$d_R(z_i, \zeta) \le \left(\frac{1}{2}\right)^{2^i - 1} d_R(z_0, \zeta),$$

for all $i \in \mathbb{N}$, where d_R refers to the Riemannian distance on $\mathbb{P}^n$.

Suppose we are looking for a root of $f \in \mathbb{P}(\mathcal{H}_d)$. We use a "start system" $(f_0, \zeta_0) \in V$ and define $f_t := tf + (1-t)f_0$ for $t \in [0, 1]$. If the line segment $\{f_t \mid t \in [0, 1]\}$ does not meet the *discriminant variety*

$$\Sigma := \{f' \in \mathbb{P}(\mathcal{H}_d) \mid \exists \zeta' \ (f', \zeta') \in \Sigma'\}, \qquad (1.14)$$

then there exists a unique lifting to a "solution curve" $[0, 1] \to V, t \mapsto (f_t, \zeta_t)$. Since $f_1 = f$, ζ_1 is the root of f we are looking for. The idea is now to partition $[0, 1]$ into k parts by $t_i = i/k$ for $i = 0, \ldots, k$ and to successively compute approximations z_i of ζ_{t_i} by Newton's method. More specifically, we set $z_0 := \zeta_0$ and for $1 \le i \le k$

$$z_i := N_{f_{t_i}}(z_{i-1}).$$

Put $D := \max_i d_i$ and let L denote the length of the curve $(f_t)_{0 \leq t \leq 1}$ in $\mathbb{P}(\mathcal{H}_d)$. The main result in Shub and Smale (1993a) states that

$$k = \mathcal{O}\left(L D^2 \max_{t \in [0,1]} \mu_{\mathrm{norm}}(f_t, \zeta_t)^2 \right) \tag{1.15}$$

iterations are sufficient to achieve that z_i is an approximate zero of f_{t_i}, for all $1 \leq i \leq k$. In particular, z_k is an approximate zero of f.

So by the condition number theorem (1.13), the number k of Newton iterations depends on how close the solution curve $(f_t, \zeta_t)_{t \in [0,1]}$ approaches the variety Σ' of ill-posed pairs. As a suitable start system it has been proposed to take $g = (X_0^{d_1-1} X_1, \ldots, X_0^{d_n-1} X_n)$ together with its zero $e = (1, 0, \ldots, 0)$.

Let us mention some recent improvements. Shub (to appear) introduced the *condition metric* on the solution variety V by scaling the Riemannian metric on V with μ_{norm}^2. He showed that for a given smooth curve $\gamma \colon [0,1] \to V$ in V, the number of Newton steps sufficient to follow a homotopy along γ is bounded by $\mathcal{O}(D^{3/2} \operatorname{Length}(\gamma))$, where $\operatorname{Length}(\gamma) := \int_0^1 \mu_{\mathrm{norm}}(\gamma(t)) \|\dot\gamma\| \, dt$ is the length of γ in the condition metric. Beltrán and Shub (to appear) proved that any $(f, \zeta) \in V$ can be connected to (g, e) by a curve γ with

$$\operatorname{Length}(\gamma) \leq 9nD^{3/2} + 2\sqrt{n} \ln \left(\mu_{\mathrm{norm}}(f, \zeta)/\sqrt{n} \right).$$

Note that this is a much better bound than (1.15), as the condition number $\mu_{\mathrm{norm}}(f, \zeta)$ has been replaced by its logarithm. Unfortunately the above result is not algorithmic, so that it only suggests, but does not imply a considerable complexity improvement.

1.5 Average-Case Probabilistic Analysis

Recall Smale's two part scheme for analyzing iterative numerical algorithms from the introduction. In the previous three sections, we illustrated the first part of this scheme in several important examples. We continue now the discussion of these examples with regard to the second part of the scheme.

The first example is the classical condition number $\kappa(A) = \|A\| \cdot \|A^{-1}\|$ of a random matrix $A \in \mathbb{R}^{m \times m}$. Suppose that the entries of A are independent standard normally distributed. Edelman (1988) derived sharp estimates on the distribution of $\kappa(A)$ by analyzing the distribution of the smallest and largest singular value of random matrices. In particular, he showed that

$$\mathbb{E}\left(\ln \kappa(A)\right) = \ln m + c + o(1), \quad m \to \infty$$

where $c \approx 1.537$. Edelman (1992) also gave closed formula expressions for the distribution of the related quantity $\kappa_F(A) := \|A\|_F \cdot \|A^{-1}\|$. In the case where the entries of A are complex numbers with independent standard normally distributed real and imaginary part, the resulting closed form is so amazingly simple, that we state it here:

$$\text{Prob}\{\kappa_F(A) \geq \epsilon^{-1}\} = 1 - (1 - \min\{1, m\epsilon^2\})^{m-1}.$$

We move now to the probabilistic analysis of the GCC-condition number $\mathcal{C}(A)$ for the linear programming feasibility problem (1.9). The input $A = (a_1, \ldots, a_n)$ is assumed to be uniformly distributed in the product $(S^m)^n$ of spheres and our goal is to determine the induced probability distribution of $\mathcal{C}(A)$. Let $\rho(A)$ denote the angular radius of a smallest spherical cap enclosing all the points a_i. The geometric characterization (1.11) states that $\mathcal{C}(A)^{-1} = |\cos \rho(A)|$. Moreover, (1.9) is feasible iff $\rho(A) \leq \pi/2$. We denote by $\mathcal{F}_{n,m}$ the set of all feasible instances.

Our problem can be restated as one concerning coverage processes on spheres. For $\alpha \in [0, \pi]$ let $p(n, m, \alpha)$ denote the probability that randomly chosen spherical caps with centers $a_1, \ldots, a_n$ and angular radius α do *not* cover the sphere S^m. Of course, we assume a_i to be uniformly and independently chosen from the uniform distribution of S^m.

We claim that for $0 < \epsilon \leq 1$

$$\text{Prob}\left\{A \in \mathcal{F}_{n,m} \text{ and } \mathcal{C}(A) \geq \epsilon^{-1}\right\} = p(n, m, \pi/2) - p(n, m, \alpha_f(\epsilon))$$
$$\text{Prob}\left\{A \notin \mathcal{F}_{n,m} \text{ and } \mathcal{C}(A) \geq \epsilon^{-1}\right\} = p(n, m, \alpha_i(\epsilon)) - p(n, m, \pi/2),$$

where $\alpha_i(\epsilon) := \arccos \epsilon \leq \pi/2$ and $\alpha_f(\epsilon) := \arccos(-\epsilon) \geq \pi/2$. Indeed, the caps of radius α with center $a_1, \ldots, a_n$ do not cover S^m iff there exists $y \in S^m$ having distance greater than α from all a_i. The latter means that the cap of radius $\pi - \alpha$ centered at $-y$ contains all the a_i. Hence $p(n, m, \alpha) = \text{Prob}\{\rho(A) \leq \pi - \alpha\}$. This implies

$$p(n, m, \pi/2) - p(n, m, \alpha_f(\epsilon)) = \text{Prob}\left\{\pi - \alpha_f(\epsilon) \leq \rho(A) \leq \pi/2\right\}.$$

This equals the probability that $A \in \mathcal{F}_{n,m}$ and $\cos \rho(A) \leq \epsilon$, which can be rewritten as $A \in \mathcal{F}_{n,m}$ and $\mathcal{C}(A) \geq \epsilon^{-1}$. Hence the first assertion follows. The second one is shown similarly.

The problem to determine $p(n, m, \alpha)$ is classic, see Solomon (1978). It has been completely solved for $m = 1, 2$, but little is known for $m \geq 3$. If $n \leq m + 1$ and $\alpha \leq \pi/2$ it is not hard to see that $p(n, m, \alpha) = 1$.

We therefore focus on the more interesting case $n > m$. Wendel (1962) showed that

$$\text{Prob}\{A \in \mathcal{F}_{n,m}\} = p(n, m, \pi/2) = 2^{1-n} \sum_{k=0}^{m} \binom{n-1}{k}. \qquad (1.16)$$

Further, a general result by Janson (1986) implies an asymptotic estimate of $p(n, m, \alpha)$ for $\alpha \to 0$.

Motivated by the probabilistic analysis of the linear programming feasibility problem, Bürgisser et al. (2007) recently discovered a closed formula for $p(n, m, \alpha)$ in the case $\alpha \geq \pi/2$, and an upper bound for $p(n, m, \alpha)$ in the case $\alpha \leq \pi/2$ that is asymptotically sharp for $\alpha \to \pi/2$. To state this result, let $\lambda_m(t)$ denote the relative volume of a spherical cap of radius $\arccos t \in [0, \pi/2]$ in S^m. It is well known that for $t \in [0, 1]$

$$\lambda_m(t) = \frac{\mathcal{O}_{m-1}}{\mathcal{O}_m} \int_0^{\arccos t} (\sin \theta)^{m-1} \, d\theta,$$

where $\mathcal{O}_m := \text{vol}\,(S^m) = 2\pi^{\frac{m+1}{2}}/\Gamma(\frac{m+1}{2})$ denotes the m-dimensional volume of the sphere S^m. Put $\epsilon := |\cos(\alpha)|$. Bürgisser at al. (2007) proved that for $\alpha \geq \pi/2$,

$$p(n, m, \alpha) = \sum_{k=1}^{m} \binom{n}{k+1} C(m, k) \int_0^{\epsilon} t^{m-k}(1 - t^2)^{\frac{1}{2}km-1} \lambda_m(t)^{n-k-1} \, dt.$$

Moreover for $\alpha \leq \pi/2$, $p(n, m, \alpha) - p(n, m, \pi/2)$ is upper bounded by

$$\binom{n}{m+1} C(m, m) \int_0^{\epsilon} (1 - t^2)^{\frac{m^2-2}{2}} \left(1 - \lambda_m(t)\right)^{n-m-1} \, dt.$$

The constants $C(m, k)$ occurring in this formula describe higher moments of the volume of certain random simplices. Their definition is somewhat complicated, but we shall give it for the sake of completeness:

$$C(m, k) := \frac{(k!)^{m-k+1}}{\mathcal{O}_m^k} \text{vol}\,G_k(\mathbb{R}^m) \int_{M_k} (\text{vol}\,\Delta)^{m-k+1} \, d(S^{k-1})^{k+1},$$

where the integral is over the set M_k of all $(b_1, \ldots, b_{k+1}) \in (S^{k-1})^{k+1}$ containing the origin in their convex hull Δ. Further, $\text{vol}\,G_k(\mathbb{R}^m)$ denotes the volume of the Grassmannian consisting of the k-dimensional linear subspaces of $\mathbb{R}^m$.

By analyzing the above formulas, Bürgisser et al. (2007) proved that for a random $A \in \mathbb{R}^{n \times (m+1)}$ with independent standard normal entries ($n > m$)

$$\mathbb{E}\,(\ln \mathcal{C}(A)) \leq 2\ln(m + 1) + 3.31, \qquad (1.17)$$

which is the sharpest bound for this expectancy as of today. (Note that the number of inequalities n does not occur in the upper bound.) Previous results on this were obtained by Cheung and Cucker (2002), Cucker and Wschebor (2003), and Cheung et al. (2005).

Condition number theorems allow a probabilistic analysis of condition numbers in a systematic way by geometric tools. Let us explain this approach for the matrix condition number $\kappa(A)$ for $A \in \mathbb{R}^{m \times m}$. The first step is to replace $\kappa(A)$ by the slightly larger quantity $\kappa_F(A) := \|A\|_F \cdot \|A^{-1}\|$. Note that $\kappa(A) \leq \kappa_F(A) \leq \sqrt{m}\,\kappa(A)$. The point is that by the Eckart-Young Theorem (1.3)

$$\kappa_F(A) = \frac{\|A\|_F}{\mathrm{dist}_F(A, \Sigma)}, \tag{1.18}$$

where $\Sigma = \{B \in \mathbb{R}^{m \times m} \mid \det B = 0\}$. If the entries of A are independent standard normal, then $A/\|A\|$ is uniformly distributed on the sphere S^{m^2-1}. Since κ_F is scale-invariant, we may assume that the inputs A are chosen uniformly at random in S^{m^2-1}. We also write $\Sigma_S := \Sigma \cap S^{m^2-1}$.

The ϵ-*neighborhood* of Σ_S, for $0 < \epsilon \leq 1$, is defined as

$$T(\Sigma_S, \epsilon) := \{A \in S^{m^2-1} \mid d_S(A, \Sigma_S) < \arcsin \epsilon\},$$

where $d_S(A, \Sigma_S) := \inf\{d_S(A, B) \mid B \in \Sigma_S\}$ and $d_S(A, B)$ denotes the angular (or Riemannian) distance between A and B in S^{m^2-1}. Using $d_F(A, \Sigma) = \sin d_S(A, \Sigma_S)$ we obtain from (1.18) for $0 < \epsilon \leq 1$

$$\mathrm{Prob}\{\kappa_F(A) \geq \epsilon^{-1}\} = \frac{\mathrm{vol}\, T(\Sigma_S, \epsilon)}{\mathrm{vol}\, S^{m^2-1}}.$$

The task is therefore to compute or to estimate the volume of neighborhoods of Σ_S. This can be achieved by combining Weyl's (1939) tube formula with techniques from integral geometry, as explained in more detail in the next two sections.

It is important to realize that this approach applies to a much more general context than just the matrix condition number. In the context of one variable polynomial equation solving, one can already find the core of these ideas in Smale's early AMS bulletin article dating from 1981. This approach has been elaborated upon for the average-case probabilistic analysis of various problems by many researchers, as mentioned already in the introduction. The remainder of this survey will be devoted to show how these results on average-case analysis can be naturally refined in the sense of smoothed analysis.

Before doing so, we would like to say a word about what is known regarding the average-case analysis of the condition number μ_{norm} for polynomial equation solving introduced in Section 1.4. In Shub and Smale (1993b) it was shown that if $f \in \mathbb{P}(\mathcal{H}_d)$ is chosen uniformly at random, then we have for $0 < \epsilon \leq 1/\sqrt{n}$

$$\mathrm{Prob}\{\mu_{\mathrm{norm}}(f) \geq \epsilon^{-1}\} \leq 0.25\, n^2(n+1)N(N-1)\, d_1 \cdots d_n\, \epsilon^4, \quad (1.19)$$

where $N = \dim \mathcal{H}_d - 1$. By combining this with an improvement of (1.15), Shub and Smale (1994) derived the existence of a "nonuniform" algorithm for finding an approximate zero of $f \in \mathcal{H}_d$ in average polynomial time. The nonuniformity was due to the fact that good starting points of the homotopy were only proven to exist, but were not constructed. Beltrán and Pardo (2008) succeeded to replace the nonuniformity by randomness and described a randomized average polynomial time algorithm for this problem.

1.6 Smoothed Probabilistic Analysis

The condition numbers we have encountered so far fit within the following abstract framework. We assume our data space is a finite-dimensional real Hilbert space, say $\mathbb{R}^{p+1}$ with the standard scalar product $\langle\,,\,\rangle$. By a semi-algebraic cone $\Sigma \subseteq \mathbb{R}^{p+1}$ we understand a semi-algebraic set $\Sigma \neq \{0\}$ that is closed under multiplications with positive scalars. We interpret Σ as a set of *ill-posed inputs* and abstractly define the associated *conic condition number* $\mathcal{C}(a)$ of $a \in \mathbb{R}^{p+1} \setminus \{0\}$ as

$$\mathcal{C}(a) := \frac{\|a\|}{\mathrm{dist}(a, \Sigma)}, \quad (1.20)$$

where $\|\,\|$ and dist are the norm and distance induced by $\langle\,,\,\rangle$.

The classical matrix condition number $\kappa(A)$ is not conic since the operator norm $\|\,\|$ is not induced by an inner product. However, $\kappa(A)$ is upper bounded by $\kappa_F(A) = \|A\|_F \|A^{-1}\|$, which, due to the Eckart-Young Theorem (1.18), is conic with respect to the set $\Sigma \subseteq \mathbb{R}^{m \times m}$ of singular matrices. Likewise, by replacing the operator norm by the Frobenius norm, Renegar's condition number $C(A)$ of $A \in \mathbb{R}^{n \times (m+1)}$, cf. (1.6), can be replaced by a conic condition number, which differs from $C(A)$ by at most a factor of $\sqrt{m+1}$. Also the condition number $\mu_{\mathrm{norm}}(f)$ for polynomial equation solving can be analyzed in this general framework, as we will see soon.

Let us continue with the general discussion. Since $\mathcal{C}(\lambda a) = \mathcal{C}(a)$ for $\lambda > 0$ we restrict the input data a to the sphere S^p and set $\Sigma_S := \Sigma \cap S^p$.

Let $d_S(a, b)$ denote the angular distance between two points a and b in S^p and set $d_S(a, \Sigma_S) := \inf\{d(a, b) \mid b \in \Sigma_S\}$. We further assume that Σ is symmetric, i.e., $-\Sigma = \Sigma$, which is actually the case in the examples considered so far, except in the linear programming setting. Then it is easy to see that for $a \in S^p$ we have

$$\mathrm{dist}(a, \Sigma) = \sin d_S(a, \Sigma_S). \tag{1.21}$$

Recall that for $0 < \epsilon \leq 1$, the *ϵ-neighborhood* of Σ_S is defined as

$$T(\Sigma_S, \epsilon) := \{a \in S^p \mid d_S(a, \Sigma_S) < \arcsin \epsilon\}.$$

By the definition (1.20) we have $\mathcal{C}(a) > \epsilon^{-1}$ iff $a \in T(\Sigma_S, \epsilon)$, for $a \in S^p$. Thus an average-case analysis of $\mathcal{C}(a)$ for a chosen uniformly at random in S^p boils down to estimating the volume of $T(\Sigma_S, \epsilon)$.

The following model for a smoothed analysis of $\mathcal{C}(a)$, proposed in Bürgisser et al. (2006), naturally fits into this geometric framework. Recall that $B(\overline{a}, \sigma)$ denotes the spherical cap centered at $\overline{a}$ with angular radius $\arcsin \sigma$, for $0 \leq \sigma \leq 1$. *Uniform smoothed analysis* of $\mathcal{C}$ consists of providing good upper bounds on

$$\sup_{\overline{a} \in S^p} \mathrm{Prob}_{a \in B(\overline{a}, \sigma)}\{\mathcal{C}(a) \geq \epsilon^{-1}\},$$

where a is assumed to be chosen uniformly at random in $B(\overline{a}, \sigma)$. The geometric meaning is to provide bounds on the relative volume of the intersection of ϵ-neighborhoods of Σ_S with small spherical disks, see equation (1.2) and Figure 1.2.

The following result from Bürgisser et al. (2008) extends the previously mentioned result by Demmel (1988) from average-case to smoothed analysis. Actually, a sharper bound is proven. For more precise statements see Section 1.7.

Theorem 1.2 *Let $\mathcal{C}$ be a conic condition number with set Σ of ill-posed inputs. Assume that Σ is contained in a real algebraic hypersurface, given as the zero set of a homogeneous polynomial of degree d. Then, for all $0 < \sigma \leq 1$ and all $0 < \epsilon \leq \sigma/(p(2d+1))$ we have*

$$\sup_{\overline{a} \in S^p} \mathrm{Prob}_{a \in B(\overline{a}, \sigma)}\{\mathcal{C}(a) \geq \epsilon^{-1}\} \leq 26\, dp\, \frac{\epsilon}{\sigma},$$

$$\sup_{\overline{a} \in S^p} \mathbb{E}_{a \in B(\overline{a}, \sigma)}(\ln \mathcal{C}(a)) \leq 2\ln(dp) + 2\ln\frac{1}{\sigma} + 4.7.$$

Demmel's 1988 paper dealt with both complex and real problems. For complex problems he provided complete proofs. For real problems, Demmel's bounds rely on an unpublished (and apparently unavailable) result by Ocneanu on the volumes of tubes around real algebraic varieties. A second goal of Bürgisser et al. (2008) was to prove a result akin to Ocneanu's. We will outline this proof in Section 1.7.

The setting of conic condition numbers has a natural counterpart over the complex numbers that we want to sketch briefly. Assume that the data space is a finite-dimensional complex Hilbert space, say $\mathbb{C}^{p+1}$, with the standard Hermitian inner product $\langle \ , \ \rangle$. Fix an algebraic cone $\Sigma \subseteq \mathbb{C}^{p+1}$, i.e., a zero set of homogeneous complex polynomials, that is interpreted as a set of ill-posed inputs to some computational problem. We define the associated conic condition number $\mathcal{C}(a)$ of a nonzero $a \in \mathbb{C}^{p+1}$ as in (1.20). It should be clear that the examples of linear and polynomial equation solving have a natural formulation over $\mathbb{C}$.

Since $\mathcal{C}(a) = \mathcal{C}(\lambda a)$ for $\lambda \in \mathbb{C}^*$ it is natural to think of the inputs as elements of the complex projective space $\mathbb{P}^p := \mathbb{P}^p(\mathbb{C})$ and to define their condition number correspondingly. On the space $\mathbb{P}^p$, the Fubini-Study metric is a natural way to measure distances, angles and volumes. We do not formally define it, but just note that the induced Riemannian distance $d_R(a,b)$ between two points $a, b \in \mathbb{P}^p$ satisfies $\cos d_R(a,b) = |\langle \hat{a}, \hat{b} \rangle| / (\|\hat{a}\| \|\hat{b}\|)$, where $\hat{a}$ and $\hat{b}$ are affine representatives in $\mathbb{C}^{p+1}$ of a and b in $\mathbb{P}^p$. (Hence d_R has the meaning of an angle as d_S in the situation over $\mathbb{R}$.) Besides the Riemannian metric d_R on $\mathbb{P}^p$, one considers the so-called *projective distance* between points $a, b \in \mathbb{P}^p$ defined by

$$d_{\mathbb{P}}(a,b) = \sin d_R(a,b).$$

This is motivated by the definition of conic condition numbers. In fact, as for (1.21), one shows that the condition number of $a \in \mathbb{P}^p$ then takes the form

$$\mathcal{C}(a) = 1/d_{\mathbb{P}}(a, \Sigma)$$

where, abusing notation, Σ is interpreted now as a subset of $\mathbb{P}^p$. (In the following we will not distinguish anymore between affine representatives and their corresponding elements of $\mathbb{P}^p$.) We denote by $B_{\mathbb{P}}(a, \sigma)$ the ball of radius σ around a in $\mathbb{P}^p$ with respect to projective distance.

In what follows we assume that Σ is purely m-dimensional, that is, all of its irreducible components are of dimension m. We recall that the *degree* $\deg \Sigma$ of Σ in the sense of algebraic geometry can be defined as the number of intersection points of Σ with a linear subspace of $\mathbb{P}^p$ of dimension $p - m$ in general position.

The following general result by Bürgisser et al. (2006) gives a smoothed analysis of conic condition numbers over the complex numbers. We remark that this result, unlike Theorem 1.2, also appropriately covers the case where Σ has codimension greater than one.

Theorem 1.3 *Let* $\mathcal{C}$ *be a conic condition number with set of ill-posed inputs* $\Sigma \subset \mathbb{P}^p$ *that is purely* m-*dimensional. Then, for all* $\overline{a} \in \mathbb{P}^p$, *all* $0 < \sigma \le 1$, *and all* $0 < \epsilon \le (p - m)/(p\sqrt{2})$, *we have*

$$\mathrm{Prob}_{a \in B_{\mathbb{P}}(\overline{a}, \sigma)}\{\mathcal{C}(a) \ge \epsilon^{-1}\} \;\le\; K(p, m) \deg \Sigma \left(\frac{\epsilon}{\sigma}\right)^{2(p-m)} \left(1 + \frac{p}{p-m}\frac{\epsilon}{\sigma}\right)^{2m}$$

and

$$\mathbb{E}_{a \in B_{\mathbb{P}}(\overline{a}, \sigma)}(\ln \mathcal{C}(a)) \;\le\; \frac{\ln K(p, m) + 3 + \ln \deg \Sigma}{2(p-m)} + \ln \frac{pm}{p-m} + 2\ln \frac{1}{\sigma},$$

with the constant $K(p, m) := 2\dfrac{p^{3p}}{m^{3m}(p-m)^{3(p-m)}}$.

The proof of this result is based on ideas in Renegar (1987) and Beltrán and Pardo (2007).

1.6.1 Applications

Theorem 1.2 and Theorem 1.3 easily imply a smoothed analysis of several of the conic condition numbers we encountered earlier. The next three corollaries are from Bürgisser et al. (2006, 2008).

Corollary 1.1 *The matrix condition number* $\kappa(A)$ *for* $A \in \mathbb{R}^{m \times m}$ *satisfies for all* $0 < \sigma \le 1$

$$\sup_{\|\bar{A}\|_F = 1} \mathbb{E}_{A \in B(\bar{A}, \sigma)}(\ln \kappa(A)) \;\le\; 6\ln m + 2\ln\frac{1}{\sigma} + 4.7.$$

Proof We have $\kappa(A) \le \kappa_F(A)$, where $\kappa_F(A)$ is the conic condition number whose set Σ of ill-posed inputs is the zero set of the determinant, which is a homogeneous polynomial of degree m. Now apply Theorem 1.2. $\qquad\square$

A smoothed analysis of $\kappa(A)$ for Gaussian perturbations was previously given by Wschebor (2004) and Sankar et al. (2006) by direct methods.

We discuss now eigenvalue computations. Let $A \in \mathbb{C}^{m \times m}$ and $\lambda \in \mathbb{C}$ be a simple eigenvalue of A. Suppose that $x \in \mathbb{C}^m$ and $y \in \mathbb{C}^m$ are right and left eigenvectors associated with λ, respectively (i.e., nonzero and satisfying $Ax = \lambda x$ and $y^T A = \lambda y^T$). From the fact that λ is a simple eigenvalue, one can deduce that $\langle x, y \rangle \neq 0$, cf. Wilkinson (1972). For any sufficiently small perturbation $\delta A \in \mathbb{C}^{m \times m}$ there exists a unique eigenvalue $\lambda + \delta \lambda$ of $A + \delta A$ close to λ. We thus have

$$(A + \delta A)(x + \delta x) = (\lambda + \delta \lambda)(x + \delta x),$$

which implies up to second order terms $\delta A\, x + A\, \delta x \approx \delta \lambda\, x + \lambda\, \delta x$. By multiplying with y^T from the left we get

$$\delta \lambda = \frac{1}{\langle x, y \rangle} y^T \delta A\, x + o(\|\delta A\|).$$

Moreover, $\sup_{\|\delta A\|_F \leq 1} |y^T \delta A\, x| = \|x\|\, \|y\|$.

It therefore makes sense to define the *condition number of A for the computation of λ* as follows

$$\kappa(A, \lambda) := \frac{\|x\|\, \|y\|}{|\langle x, y \rangle|} \qquad (1.22)$$

and to set $\kappa(A, \lambda) := \infty$ if λ is a multiple eigenvalue of A. We further define the *condition number of A for eigenvalue computations* by

$$\kappa_{\mathsf{eigen}}(A) := \max_\lambda \kappa(A, \lambda),$$

where the maximum is over all the complex eigenvalues λ of A. The set of ill-posed inputs $\Sigma_{\mathsf{eigen}} := \{A \in \mathbb{C}^{m \times m} \mid \kappa_{\mathsf{eigen}}(A) = \infty\}$ consists of the matrices having multiple eigenvalues. Wilkinson (1972) proved that

$$\kappa_{\mathsf{eigen}}(A) \leq \frac{\sqrt{2}\, \|A\|_F}{\operatorname{dist}(A, \Sigma)}. \qquad (1.23)$$

Corollary 1.2 *The condition number $\kappa_{\mathsf{eigen}}(A)$ for $A \in \mathbb{C}^{m \times m}$ satisfies for all $0 < \sigma \leq 1$*

$$\sup_{\|\bar{A}\|_F = 1} \mathbb{E}_{A \in B(\bar{A}, \sigma)} (\ln \kappa_{\mathsf{eigen}}(A)) \leq 8 \ln m + 2 \ln \frac{1}{\sigma} + 5.$$

Proof According to (1.23), $2^{-1/2} \kappa_{\mathsf{eigen}}$ is bounded by the conic condition number, whose associated set Σ_{eigen} of ill-posed inputs consists of the matrices A having multiple eigenvalues. Σ_{eigen} is the zero set of the discriminant polynomial of the characteristic polynomial, which can be shown to be homogeneous of degree $m^2 - m$. Now apply Theorem 1.3.

□

We remark that it is also possible to derive a corresponding statement for the computation of real eigenvalues of real matrices. However, some care has to be taken when defining the corresponding condition number.

Our next application is concerned with the condition number $\mu_{\mathrm{norm}}(f)$ for finding an approximate solution of the multivariate polynomial equation $f(\zeta) = 0$, where $f \in \mathcal{H}_d$ (see Section 1.4).

Corollary 1.3 *For all* $\overline{f} \in \mathcal{H}_d$ *of norm one and* $0 < \sigma \leq 1$ *we have*

$$\mathbb{E}_{f \in B(\overline{f}, \sigma)}(\ln \mu_{\mathrm{norm}}(f)) \leq 3.5 \ln N + \ln \mathcal{D} + 0.5 \ln n + 2 \ln \frac{1}{\sigma} + 5.$$

where $N = \dim \mathcal{H}_d - 1$ *and* $\mathcal{D} = d_1 \cdots d_n$ *is the Bézout number.*

Shub and Smale (1993b) obtained similar estimates for the average of $\ln \mu_{\mathrm{norm}}$, see (1.19).

Proof The discriminant variety Σ consists of the systems $f \in \mathbb{P}(\mathcal{H}_d)$ having multiple zeros. It is a well-known fact that Σ is a hypersurface in $\mathbb{P}(\mathcal{H}_d)$ defined by a homogeneous polynomial of total degree at most $2n\mathcal{D}^2$, see Bürgisser et al. (2006).

Recall from (1.12) the variety Σ' of ill-posed pairs. The discriminant variety Σ is the projection of Σ' onto the first factor. By the condition number theorem (1.13) we have for all $(f, \zeta) \in V \setminus \Sigma'$

$$\mu_{\mathrm{norm}}(f, \zeta) = \frac{1}{d_{\mathbb{P}}(f, \Sigma_\zeta)},$$

where $d_{\mathbb{P}}(f, \Sigma_\zeta)$ denotes the projective distance of f to $\Sigma_\zeta := \{f' \in \mathbb{P}(\mathcal{H}_d) \mid (f', \zeta) \in \Sigma'\}$ measured in the fiber $\{f' \in \mathbb{P}(\mathcal{H}_d) \mid f'(\zeta) = 0\}$. Since $\Sigma_\zeta \subseteq \Sigma$ we have $d_{\mathbb{P}}(f, \Sigma_\zeta) \geq d_{\mathbb{P}}(f, \Sigma)$. Therefore

$$\mu_{\mathrm{norm}}(f, \zeta) \leq \frac{1}{d_{\mathbb{P}}(f, \Sigma)},$$

which implies $\mu_{\mathrm{norm}}(f) \leq 1/d_{\mathbb{P}}(f, \Sigma)$. Now apply Theorem 1.3.   □

We remark that it is also possible to derive a corresponding statement for real polynomial systems.

Let us move now to applications to condition numbers of convex optimization. When trying to directly apply Theorem 1.2 we obtain bad bounds. The reason is that the corresponding sets Σ of ill-posed inputs are semialgebraic (of codimension one). Inequalities are essential here and by enclosing Σ in algebraic hypersurfaces essential information gets lost.

Nevertheless, the ideas in the proof of Theorem 1.2 turned out to be useful for obtaining a uniform smoothed analysis of the GCC-condition number $\mathcal{C}(A)$ of the linear programming feasibility problem (1.9). (For the average-case analysis of $\mathcal{C}(A)$ see Section 1.5.) The point is that the conclusion of Theorem 1.2 is also true when Σ is the boundary of a spherical convex set.

For stating this precisely let us introduce some notation. By a *convex body* K in the sphere S^m we understand the intersection with S^m of a closed regular convex cone C in $\mathbb{R}^{m+1}$. We call $T_o(\partial K, \epsilon) := T(\partial K, \epsilon) \backslash K$ the *outer ϵ-neighborhood* of the boundary ∂K. The assertion is

$$\frac{\text{vol}\,(T_o(\partial K, \epsilon) \cap B(a, \sigma))}{\text{vol}\, B(a, \sigma)} \leq 6.5\, m\, \frac{\epsilon}{\sigma} \quad \text{if } \epsilon \leq \frac{\sigma}{2m}, \qquad (1.24)$$

and the same upper bound holds for the relative volume of the inner ϵ-neighborhood of ∂K.

The relation of this bound to Theorem 1.2 is the following. By convexity, the intersection of ∂K with a hyperequator of S^m in general position consists of at most two points. In that sense we may think of ∂K as a set of "degree" at most two. Of course this analogy has to be taken with a grain of salt. For instance, if K corresponds to a polyhedral cone C, then ∂K can be expressed as the zeroset of a polynomial equation and inequality constraints. However, the degree of this equation would be the number of facets of C, which is in general a huge number. We will outline the proof of (1.24) in Section 1.7.3.

The smoothed analysis of the GCC-condition number is performed in the following model. Fix $0 < \sigma \leq 1$ and $\bar{a}_1, \ldots, \bar{a}_n \in S^m$. Independently choose points a_i uniformly at random in the spherical caps $B(\bar{a}_i, \epsilon)$ of S^m centered at $\bar{a}_i$ with angular radius $\arcsin \sigma$. In other words, $A = (a_1, \ldots, a_n)$ is chosen uniformly at random in $B(\bar{A}, \alpha) := \prod_i B(\bar{a}_i, \epsilon)$. We recall that $\mathcal{F}_{n,m}$ denotes the set of feasible instances in $(S^m)^n$.

The following recent result is from Amelunxen and Bürgisser (2008).

Theorem 1.4 *For $n > m$ and $0 < \sigma \leq 1$ we have*

$$\sup_{\bar{A} \in (S^m)^n} \mathbb{E}_{A \in B(\bar{A}, \alpha)} \big(\ln \mathcal{C}(A) \big) = \mathcal{O}\big(\ln(\frac{nm}{\sigma}) \big).$$

For the average-case ($\sigma = 1$) we even get $\mathbb{E}\left(\ln \mathbb{C}(A)\right) = \mathcal{O}(\log m)$, as already stated in (1.17). Moreover, we have for $0 < \epsilon \le \sigma/(2m(m+1))$

$$\sup_{\bar{A} \in (S^m)^n} \text{Prob}\{A \in \mathcal{F}_{n,m},\ \mathbb{C}(A) \ge \epsilon^{-1}\} \ \le\ 6.5\, nm(m+1)\,\frac{\epsilon}{\sigma}.$$

For the infeasible case ($A \notin \mathcal{F}_{n,m}$) a slightly worse tail estimate holds.

Dunagan et al. (2003) previously gave a smoothed analysis of Renegar's condition number. The crucial ingredient of their proof is a result due to Ball (1993) about the measure of Gaussians on boundaries of convex sets in Euclidean space. Our proof of Theorem 1.4 roughly uses the same overall strategy as Dunagan et al. (2003). However, we substitute Ball's result by the volume estimate (1.24) on neighborhoods of boundaries of spherically convex sets. A relevant observation that enables us to successfully apply this estimate is the following result. Let $\mathcal{F}_{n,m}^\circ := \mathcal{F}_{n,m} \setminus \partial \mathcal{F}_{n,m}$ denote the set of strictly feasible instances.

Lemma 1.1 *Let $A = (a_1, \ldots, a_n) \in \mathcal{F}_{n,m}^\circ$ and $\mathbb{C}(A) \ge (m+1)\epsilon^{-1}$. Then there exists $i \in \{1, \ldots, n\}$ such that $a_i \in T_o(\partial K_i, \epsilon)$, where $-K_i$ is the spherical convex hull of $a_1, \ldots, a_{i-1}, a_{i+1}, \ldots, a_n$.*

The proof of the probability tail estimate in Theorem 1.4 for the feasible case is now easy. Suppose that A is chosen uniformly at random in $B(\bar{A}, \sigma)$. Lemma 1.1 yields with $t = (m+1)\epsilon^{-1}$

$$\text{Prob}\{A \in \mathcal{F}_{n,m}^\circ,\ \mathbb{C}(A) \ge t\} \le \sum_{i=1}^{n} \text{Prob}\{A \in \mathcal{F}_{n,m}^\circ,\ a_i \in T_o(\partial K_i, \epsilon)\}.$$

Note that $B(\bar{A}, \sigma) = B(\bar{A}', \sigma) \times B(\bar{a}_n, \sigma)$ where $\bar{A}' := (\bar{a}_1, \ldots, \bar{a}_{n-1})$. We bound the probability on the right-hand side for $i = n$ by an integral of probabilities conditioned on $A' := (a_1, \ldots, a_{n-1})$:

$\text{Prob}\{A' \in \mathcal{F}_{n-1,m}^\circ$ and $a_n \in T_o(\partial K_n, \epsilon)\}$

$$= \frac{1}{\text{vol}\, B(\bar{A}', \sigma)} \int_{A' \in \mathcal{F}_{n-1,m}^\circ \cap B(\bar{A}', \sigma)} \text{Prob}\{a_n \in T_o(\partial K_n, \epsilon) \mid A'\}\, dA'.$$

Fix now $A' \in \mathcal{F}_{n-1,m}^\circ$ and consider the convex set K_n in S^m. The volume bound (1.24) on the outer neighborhood of ∂K_n yields

$$\text{Prob}\{a_n \in T_o(\partial K_n, \epsilon) \mid A'\} = \frac{\text{vol}\left(T_o(\partial K_n, \epsilon) \cap B(\bar{a}_n, \sigma)\right)}{\text{vol}\, B(\bar{a}_n, \sigma)} \le 6.5\, m\, \frac{\epsilon}{\sigma}.$$

We conclude that

$$\text{Prob}\{A \in \mathcal{F}_{n,m}^\circ,\ a_n \in T_o(\partial K_n, \varphi)\} \ \le\ 6.5\, m\, \frac{\epsilon}{\sigma}.$$

The same upper bound holds for any K_i. Altogether, we obtain

$$\mathrm{Prob}\{A \in \mathcal{F}_{n,m}^\circ \text{ and } \mathcal{C}(A) \geq t\} \leq 6.5\,nm(m+1)\,\frac{1}{\sigma t},$$

which is the bound claimed in Theorem 1.4.

One of the goals of our current research with Amelunxen is to find a general result providing smoothed analysis estimates of condition numbers for convex optimization, in particular for semidefinite programming. The proof just presented heavily relies on the product structure of the cone $\mathbb{R}_+^{m+1}$ and does not generalize.

1.7 Tools from Integral Geometry

As already pointed out, uniform smoothed analysis boils down to the task of providing bounds on the relative volume of the intersection of ϵ-neighborhoods of Σ_S with small spherical disks, see Figure 1.2.

Even though the volume of neighborhoods of subsets of Euclidean spaces or spheres is a rich and thoroughly studied mathematical topic, as can be seen from the textbook by Gray (1990), further developments were needed to arrive at the results mentioned in Section 1.6. In the following we first describe some of the classic results on the volume of neighborhoods, then we discuss the principal kinematic formula and finally indicate how to combine these tools in order to prove Theorem 1.2.

1.7.1 On the volume of tubes

To warm up, assume that K is a convex compact subset of $\mathbb{R}^n$. Consider the ϵ-neighborhood K_ϵ of K consisting of the points in $\mathbb{R}^n$ having (Euclidean) distance at most ϵ from K. Steiner (1840) observed that the volume of K_ϵ is a polynomial function in ϵ: $\mathrm{vol}\,K_\epsilon = \sum_{i=0}^n c_i(K)\epsilon^i$. Clearly, $c_0(K)$ equals the volume of K, and it should be intuitively clear that $c_1(K)$ equals the $(n-1)$-dimensional volume of the boundary ∂K of K. It is an easy and instructive exercise to prove Steiner's result for convex polytopes in $\mathbb{R}^2$ and $\mathbb{R}^3$. This exercise also reveals the meaning of the coefficients $c_i(K)$. (For instance, $c_n(K)$ always equals the volume $\mathcal{O}_{n-1}/n$ of the n-dimensional unit ball.) In Minkowski's theory of convex bodies, the coefficients $c_i(K)$ are called cross-sectional measures of K (Quermassintegrale), see Bonnesen and Fenchel (1974) for more information. So the volume of the outer ϵ-neighborhood $T_o(\partial K, \epsilon)$ of ∂K satisfies $\mathrm{vol}\,T_o(\partial K, \epsilon) = \sum_{i=1}^n c_i(K)\epsilon^i$. Weyl (1939) considerably extended this observation by showing that the volume of the ϵ-neighborhood $T(M, \epsilon)$ of a compact smooth submanifold M of $\mathbb{R}^n$ is a polynomial function in ϵ, for sufficiently small values of ϵ.

For our purposes, we need to study the case where the ambient space is a sphere S^n. Weyl (1939) also analyzed this case. We will only state his result in the special case where M is a smooth oriented hypersurface of S^n. In order to do so, we first need to review a few elementary concepts from differential geometry, see for instance Thorpe (1993) or do Carmo (1992), p. 129.

We assume that a unit normal vector field ν has been chosen on M (which corresponds to the choice of an orientation of M). Let $T_x M$ denote the tangent space of M at $x \in M$. The *second fundamental form* $\mathrm{II}_M(x) \colon T_x M \times T_x M \to \mathbb{R}$ of M at x is defined as $\mathrm{II}_M(x)(u, w) :=$ $-\langle \nabla_u \nu(x), w \rangle$. Here, $\nabla_u \nu(x)$ denotes the covariant derivative of ν at x in the direction u. It can be computed by taking the derivative of $\nu \colon M \to \mathbb{R}^{n+1}$ at x in the direction u and projecting orthogonally onto $T_x S^n$. It is well known that $\mathrm{II}_M(x)$ is a symmetric bilinar form. Its eigenvalues $\kappa_1(x), \ldots, \kappa_{n-1}(x)$ are called the *principal curvatures* at x of the hypersurface M. For $1 \le i < n$ we define the *ith curvature* $K_{M,i}(x)$ of M at x as the ith elementary symmetric polynomial in $\kappa_1(x), \ldots, \kappa_{n-1}(x)$, and put $K_{M,0}(x) := 1$. In particular, $K_{M,n-1}(x) = \det L_M(x)$. We define the *integral $\mu_i(M)$ of ith curvature* and the *integral $|\mu_i|(M)$ of ith absolute curvature* of M as follows ($0 \le i \le n-1$):

$$\mu_i(M) := \int_M K_{M,i}\, dM, \quad |\mu_i|(M) := \int_M |K_{M,i}|\, dM.$$

For reasons that will soon become apparent, it is more convenient to think in terms of the *normalized integrals of (absolute) curvature* of M defined by

$$\mu_i^{\mathrm{no}}(M) := \frac{2}{\mathcal{O}_{n-i-1}\mathcal{O}_i}\, \mu_i(M), \quad |\mu_i^{\mathrm{no}}|(M) := \frac{2}{\mathcal{O}_{n-i-1}\mathcal{O}_i}\, |\mu_i|(M).$$

Note that $\mu_0^{\mathrm{no}}(M) = |\mu_0^{\mathrm{no}}|(M) = \mathcal{O}_{n-1}^{-1} \operatorname{vol} M$ is the volume of M relative to the volume of S^{n-1}.

For $0 < \epsilon \le 1$ we define the *ϵ-tube* $T^\perp(M, \epsilon)$ around M as the set of points in S^n such that there exists a great circle segment in S^n of angular length less than $\arcsin \epsilon$ that connects x with a point in M and intersects M orthogonally in that point, see Figure 1.3. Note that $T^\perp(M, \epsilon) \subseteq T(M, \epsilon)$. If M has a smooth boundary, then $T(M, \epsilon)$ is the

Fig. 1.3. ϵ-tube $T^\perp(M, \epsilon)$ and ϵ-neighborhood $T(M, \epsilon)$ around the curve M.

union of $T^\perp(M, \epsilon)$ and a "half-tube" around the boundary of ∂M. If $\partial M = \emptyset$, then $T^\perp(M, \epsilon) = T(M, \epsilon)$.

Weyl's formula now states that for sufficiently small ϵ we have

$$\operatorname{vol} T^\perp(M, \epsilon) = \sum_{\substack{0 \leq i \leq n-1 \\ i \text{ even}}} \mu_i^{\mathrm{no}}(M) \operatorname{vol} T(S^{n-i-1}, \epsilon). \tag{1.25}$$

Here S^{n-i-1} is interpreted as a subset of S^n. (There is a cancellation effect between the contributions of "outer" and "inner" neighborhoods which results in the sum being only over even indices i.) If M is an open subset of S^{n-1}, then $\mu_i^{\mathrm{no}}(M) = 0$ for $i > 0$ and (1.25) specializes to the obvious formula $\operatorname{vol} T^\perp(M, \epsilon) = \mu_0^{\mathrm{no}}(M) \operatorname{vol} T(S^{n-1}, \epsilon)$, which is asymptotically equal to $2\epsilon \operatorname{vol} M$ for $\epsilon \to 0$. For completeness, let us also mention that

$$\operatorname{vol} T(S^{n-i-1}, \epsilon) = \mathcal{O}_{n-i-1} \mathcal{O}_i \int_0^{\arcsin \epsilon} (\cos \rho)^{n-i-1} (\sin \rho)^i \, d\rho.$$

By tracing Weyl's proof, it is not hard to see that the following upper bound is valid *for all* $0 < \epsilon \leq 1$:

$$\operatorname{vol} T^\perp(M, \epsilon) \leq \sum_{i=0}^{n-1} |\mu_i^{\mathrm{no}}|(M) \operatorname{vol} T(S^{n-i-1}, \epsilon). \tag{1.26}$$

The question is now how to bound the normalized integrals $|\mu_i^{\mathrm{no}}|(M)$ of absolute curvature in specific situations. It turns out that this can be effectively done with tools from integral geometry. In a first step, we focus on $|\mu_0^{\mathrm{no}}|(M)$, that is, we need to bound the volume of M.

1.7.2 The principal kinematic formula

The orthogonal group $G := O(n+1)$ is a compact Lie group. It has an invariant Riemannian metric (induced by the Euclidean metric on the space of real $n+1$-matrices) and a corresponding invariant volume form (Haar measure). So we can talk about random elements of G chosen with respect to the uniform distribution. Note that G acts on S^n in a straightforward way.

Now suppose that $M, N \subseteq S^n$ are smooth submanifolds of dimension m and p, respectively, such that $m + p \geq n$. By transversality principles, the intersection of M with a random translate gN of N is almost surely empty or a submanifold of dimension $m + p - n$. So the volume of $M \cap gN$ in this dimension is well defined (we put $\operatorname{vol} \emptyset := 0$). A key result in integral geometry states that

$$\mathbb{E}_{g \in G} \left(\frac{\operatorname{vol}(M \cap gN)}{\mathcal{O}_{m+p-n}} \right) = \frac{\operatorname{vol} M}{\mathcal{O}_m} \cdot \frac{\operatorname{vol} N}{\mathcal{O}_p}. \qquad (1.27)$$

In fact the smoothness assumption on M and N in this formula is not important since removing lower dimensional parts does not change the volume (for instance it is sufficient to require that M, N are semialgebraic).

Santaló (1976), which is the standard reference on integral geometry, refers to (1.27) as *Poincaré's formula* (cf. §7.1 in Santaló). Apparently, Poincaré stated this result for the case of S^2 and in such form it was also known to Barbier (1860). Formula (1.27) is stated in §18.6 of Santaló's book, but a proof is only given in §15.2 for an analogous statement for Euclidean space. The book by Howard (1993) states and proves formulas like (1.27) for homogeneous spaces in great generality.

The following corollary of (1.27) allows us to reduce the estimation of volumes to counting arguments:

$$\frac{\operatorname{vol} M}{\mathcal{O}_m} = \frac{1}{2} \, \mathbb{E}_{g \in G} \left(\#(M \cap gS^{n-m}) \right). \qquad (1.28)$$

To illustrate this with a simple example, assume that M is the real zero set in S^n of a homogeneous polynomial f of degree d. Suppose that $\dim M = n - 1$. We claim that if $M \cap gS^1$ is finite, then it has at most $2d$ points. In order to see this assume w.l.o.g. $g = \operatorname{id}$ and $f(1, 0, \dots, 0) \neq 0$. Suppose that S^1 is given by $x_2 = \cdots = x_n = 0$. For each $x_0 \in \mathbb{R}$ such that $f(x_0, 1, 0, \dots, 0) = 0$ there are two points $\pm(1 + x_0^2)^{-1/2}(x_0, 1, 0, \dots, 0)$ in $M \cap S^1$ and these are all the points in $M \cap S^1$. Equation (1.28) then implies that

$$\frac{\operatorname{vol} M}{\mathcal{O}_{n-1}} = \frac{1}{2} \, \mathbb{E}_{g \in G} \left(\#(M \cap gS^1) \right) \leq d \cdot \operatorname{Prob}_{g \in G} \{ M \cap gS^1 \neq \emptyset \} \leq d.$$

More generally, we obtain for $a \in S^n$ and $0 < \sigma \leq 1$

$$\frac{\text{vol}\,(M \cap B(a,\sigma))}{\mathcal{O}_{n-1}} \leq d \cdot \text{Prob}_{g \in G}\{M \cap B(a,\sigma) \cap gS^1 \neq \emptyset\}$$

$$\leq d \cdot \text{Prob}_{g \in G}\{B(ga,\sigma) \cap S^1 \neq \emptyset\} = d\,\frac{\text{vol}\,T(S^1,\sigma)}{\mathcal{O}_n}.$$

We remark that a statement analogous to (1.27) holds for complex projective spaces $\mathbb{P}^n(\mathbb{C})$. The same argument as before then shows that for a complex m-dimensional algebraic subvariety M of $\mathbb{P}^n(\mathbb{C})$ we have $\text{vol}\,M = \deg M \cdot \text{vol}\,(\mathbb{P}^m(\mathbb{C}))$.

Formula (1.27) has a natural extension involving the normalized integrals $\mu_i^{\text{no}}(M)$ of curvature. This is called the principal kinematic formula of integral geometry and is considered the most important result of integral geometry. We will only need the following special case extending (1.28). As before let M be an oriented smooth hypersurface M in S^n. Then we have for $0 \leq i \leq n - 1$

$$\mu_i^{\text{no}}(M) = \mathbb{E}_{g \in G}\left(\mu_i^{\text{no}}(M \cap gS^{i+1})\right). \qquad (1.29)$$

Note that for almost all g, $M \cap gS^{i+1}$ is either empty or a smooth hypersurface of the sphere gS^{i+1} (with a canonical orientation inherited from M). With this interpretation $\mu_i^{\text{no}}(M \cap gS^{i+1})$ is well defined (setting $\mu_i^{\text{no}}(\emptyset) := 0$).

The general principal kinematic formula for spheres is so beautiful that we cannot resist stating it here. Also, this could be useful for future applications, when the set of ill-posed inputs has higher codimension.

Let $M \subseteq S^n$ be a smooth submanifold of dimension m. For $x \in M$ let $S_x := S(T_x M^\perp)$ denote the sphere of unit normal vectors v in $T_x S^n$ that are perpendicular to $T_x M$. Let us denote by $K_{M,i}(x,v)$ the ith elementary symmetric polynomial in the eigenvalues of the second fundamental form of the embedding $M \hookrightarrow S^n$ at x in direction v, see do Carmo (1992), p. 128. We now define the *normalized integral* $\mu_i^{\text{no}}(M)$ of *ith curvature* of M as ($0 \leq i \leq m$):

$$\mu_i^{\text{no}}(M) := \frac{1}{\mathcal{O}_{m-i}\mathcal{O}_{n-m+i-1}} \int_{x \in M} \int_{v \in S_x} K_{M,i}(x,v)\,dS_x(v)\,dM(x).$$

This value is easily seen to vanish if i is odd (consider $v \mapsto -v$). It follows from Weyl (1939) that $\mu_i^{\text{no}}(M)$ is a relative isometric invariant of M in the sense that $\mu_i^{\text{no}}(M)$ can be written as an integral over M of a function whose value at $x \in M$ only depends on the difference of the values at x of the curvature tensor of M and the curvature tensor of S^n restricted to M.

Weyl's formula, extending (1.25), states that for sufficiently small c we have

$$T^{\perp}(M, \epsilon) \;=\; \sum_{\substack{0 \le i \le m \\ i \text{ even}}} \mu_i^{\text{no}}(M) \operatorname{vol} T(S^{m-i}, \epsilon). \tag{1.30}$$

The principal kinematic formula for spheres is best stated in terms of the *curvature polynomial* $\mu^{\text{no}}(M; X)$ of M defined as

$$\mu^{\text{no}}(M; X) \;:=\; \sum_{i=0}^{m} \mu_i^{\text{no}}(M) \, X^i,$$

where X denotes a formal variable. The degree of $\mu^{\text{no}}(M; X)$ is at most the dimension m of M. For example we have $\mu^{\text{no}}(S^m; X) = 1$.

The *principal kinematic formula* says that for smooth submanifolds M and N of S^n having dimension m and p, respectively, such that $m + p \ge n$, we have

$$\mathbb{E}_{g \in G}\big(\mu^{\text{no}}(M \cap gN; X)\big) \;\equiv\; \mu^{\text{no}}(M; X) \cdot \mu^{\text{no}}(N; X) \bmod X^{m+p-n+1}.$$

Here, the expectation on the left-hand side is defined coefficientwise, while on the right-hand side we have a *polynomial multiplication modulo* $X^{m+p-n+1}$. This makes perfect sense as $m + p - n$ is the expected dimension of $M \cap gN$. We note that the principal kinematic formula contains (1.29) as a special case (for even i).

It is not at all easy to locate the principal kinematic formula for spheres in the above explicit form in the literature. Santaló in his book attributes the principal kinematic formula in the plane to Blaschke, and in Euclidean spaces to Chern (1966) and Federer (1959). An elementary and unconventional introduction to geometric probability and the kinematic formula for Euclidean spaces can be found in Klain and Rota (1997). The normalization of integrals of curvatures leading to the simple formula of reduced polynomial multiplication was discovered by Nijenhuis (1974), again for Euclidean space. Santaló derives the principal kinematic formula for the special case of intersections of domains in spheres, but he does not care about the scaling coefficients. In fact, the principal kinematic formulas for submanifolds of spheres and Euclidean spaces take exactly the same form. An indication of this at first glance astonishing fact can be found, somewhat hidden, in Santaló's book on page 320. The situation was clarified by Howard (1993), who gave a unified treatment of kinematic formulas in homogeneous spaces. But

Howard does not care about the scaling constants either. For the purpose of explicitly bounding the volumes of tubes, a good understanding of the scaling factors is relevant. To the best of our knowledge, the general principal kinematic formula for spheres in the above form was first explicitly stated in Bürgisser (2007).

1.7.3 Bounding integrals of absolute curvature

The basic idea is best explained with the example of a hypersurface in $\mathbb{R}^n$. The approach is inspired by Spivak (1979), p. 409ff. Suppose that f is a polynomial of degree d with compact zero set $Z \subseteq \mathbb{R}^n$ such that the gradient of f does not vanish on Z. Consider the Gauss map $\nu\colon Z \to S^{n-1}$, $x \mapsto \operatorname{grad} f(x)/\|\operatorname{grad} f(x)\|$. The Jacobian determinant of ν at $x \in Z$ yields the Gaussian curvature: $K_{Z,n-1}(x) = \det D\nu(x)$. Let $\varphi(y) := \#\nu^{-1}(y)$ denote the size of the fiber of $y \in S^{n-1}$ and put $\varphi(\pm y) := \varphi(y) + \varphi(-y)$. The transformation formula implies

$$\int_Z |\det D\nu|\, dZ = \int_{S^{n-1}} \varphi(y)\, dS^{n-1} = \int_{S^{n-1}} \varphi(-y)\, dS^{n-1}.$$

Hence we obtain $\int_Z |\det D\nu|\, dZ = \frac{1}{2} \int_{S^{n-1}} \varphi(\pm y)\, dS^{n-1}$.

A point $x \in \mathbb{R}^n$ satisfies $\nu(x) = (\pm 1, 0, \ldots, 0)$ iff

$$f(x) = 0, \partial_2 f(x) = 0, \ldots, \partial_n f(x) = 0.$$

If y is a regular point of ν, then all real solutions of this system of equations are nondegenerate, hence they are isolated in $\mathbb{C}^n$. We conclude $\varphi(\pm y) \le d(d-1)^{n-1}$ from Bézout's theorem, which is a standard result from algebraic geometry, see Mumford (1976). This estimate holds for any regular value $y \in S^{n-1}$. We therefore obtain that

$$\int_Z |K_{Z,n-1}|\, dZ \le \frac{\mathcal{O}_{n-1}}{2} d(d-1)^{n-1}.$$

Note that this bound is sharp for $d = 2$ and $Z = S^{n-1}$.

The previous reasoning can be extended to hypersurfaces in S^n as follows. Suppose now that $f \in \mathbb{R}[X_0, \ldots, X_n]$ is homogeneous of degree d with zero set $M \subseteq S^n$ such that the derivative of the restriction of f to S^n does not vanish on M. Then M is a compact smooth hypersurface of S^n oriented by the gradient of f. We claim that

$$|\mu_{n-1}^{\mathrm{no}}|(M) \le d(d-1)^{n-1}. \tag{1.31}$$

Before showing this bound, let us illustrate it with a simple example.

The zero set of $f = \sum_{i=1}^{n} X_i^2 - \epsilon^2 X_0^2$ consists of two small circles in S^n centered at $(\pm 1, 0, \ldots, 0)$ with a radius going to zero as $\epsilon \to 0$. For each of the circles C_ϵ, the total Gaussian curvature $\mu_{n-1}(C_\epsilon) = |\mu_{n-1}|(C_\epsilon)$ converges to $\mathcal{O}_{n-1}$ as $\epsilon \to 0$. Hence $|\mu_{n-1}^{no}|(C_\epsilon) \to 1$. This shows that (1.31) is a sharp bound for $d = 2$.

In order to prove (1.31), consider the Gauss map $\nu \colon M \to S^n$ defined as before. For simplicity, we assume that the image N of ν is a smooth hypersurface of S^n (this can be achieved by removing lower dimensional parts). Again put $\varphi(y) := \#\nu^{-1}(y)$ for $y \in N$. The transformation formula gives

$$|\mu_{n-1}|(M) = \int_M |\det D\nu| \, dM = \int_N \varphi \, dN = \sum_{\ell \in \mathbb{N}} \ell \operatorname{vol} F_\ell,$$

where $F_\ell := \{ y \in N \mid \varphi(y) = \ell \}$. Poincaré's formula (1.28) implies

$$\operatorname{vol} F_\ell = \frac{\mathcal{O}_{n-1}}{2} \, \mathbb{E}_{g \in G} (\#(F_\ell \cap gS^1)).$$

Therefore,

$$\sum_{\ell \in \mathbb{N}} \ell \operatorname{vol} F_\ell = \frac{\mathcal{O}_{n-1}}{2} \, \mathbb{E} \left(\sum_{\ell \in \mathbb{N}} \ell \, \#(F_\ell \cap gS^1) \right) = \frac{\mathcal{O}_{n-1}}{2} \, \mathbb{E} \left(\#\nu^{-1}(gS^1) \right).$$

Now gS^1 intersects N transversally for almost all $g \in G$. To simplify notation suppose this is the case for $g = \mathrm{id}$. A point $x \in \mathbb{R}^{n+1}$ lies in $\nu^{-1}(S^1)$ iff it satisfies the following system of equations

$$\sum_{i=0}^{n} x_i^2 - 1 = 0, \quad f(x) = 0, \quad \partial_2 f(x) = \cdots = \partial_n f(x) = 0.$$

By Bézout's theorem, the number of solutions to this system of equations is bounded by $2 \, d \, (d-1)^{n-1}$. Altogether, $\#\nu^{-1}(gS^1) \leq 2 \, d \, (d-1)^{n-1}$ for almost all g and the assertion (1.31) follows.

Similarly, one shows that if K is a convex body in S^n with smooth boundary ∂K, then $|\mu_{n-1}^{no}|(\partial K) \leq 1$, which is an optimal bound. The argument is as before, replacing Bézout's theorem by the fact that if $\partial K \cap gS^1$ is finite, then it consists of at most two points by convexity. (Compared to (1.31) we save here a factor 2 since K does not contain diametral points.)

Now let the hypersurface M of S^n be given as before as the zero set of the homogeneous polynomial f of degree d. Let $a \in S^n$ and $0 < \sigma \leq 1$. We can bound the ith integral of absolute curvature $|\mu_i^{no}|(M \cap B(a, \sigma))$

in terms of the degree d and the dimension parameters n, i as follows:

$$|\mu_i^{no}|(M \cap B(a, \sigma)) \leq 2\, d\, (d-1)^i\, \frac{\text{vol}\, T(S^{i+1}, \sigma)}{\mathcal{O}_n}. \qquad (1.32)$$

In order to show this, put $U := M \cap B(a, \sigma)$ and let U_+ be the set of points of U where $K_{M,i}$ is positive and U_- be the set of points of U where $K_{M,i}$ is negative. Then $|\mu_i|(U) = |\mu_i(U_+)| + |\mu_i(U_-)|$.

Let $g \in G$ be such that M intersects gS^{i+1} transversally. We apply the bound (1.31) to the hypersurface $M \cap gS^{i+1}$ of the sphere gS^{i+1}, which yields $|\mu_i^{no}|(M \cap gS^{i+1}) \leq d(d-1)^i$. By monotonicity we obtain

$$|\mu_i^{no}(U_+ \cap gS^{i+1})| \leq |\mu_i^{no}|(U_+ \cap gS^{i+1}) \leq |\mu_i^{no}|(M \cap gS^{i+1}) \leq d(d-1)^i.$$

The kinematic formula (1.29) applied to U_+ implies that

$$\begin{aligned}
|\mu_i^{no}(U_+)| &\leq \mathbb{E}_{g \in G}\left(|\mu_i^{no}(U_+ \cap gS^{i+1})|\right) \\
&\leq d(d-1)^i\, \text{Prob}_{g \in G}\{U_+ \cap gS^{i+1} \neq \emptyset\} \\
&\leq d(d-1)^i\, \text{Prob}_{g \in G}\{B(a, \sigma) \cap gS^{i+1} \neq \emptyset\} \\
&= d(d-1)^i\, \frac{\text{vol}\, T(S^{i+1}, \sigma)}{\mathcal{O}_n}.
\end{aligned}$$

The same upper bound holds for $|\mu_i(U_-)|$ and hence the assertion (1.32) follows.

A similar reasoning shows $|\mu_i^{no}|(\partial K \cap B(a, \sigma)) \leq \mathcal{O}_n^{-1}\text{vol}\, T(S^{i+1}, \sigma)$ for a convex body K in S^n with smooth boundary ∂K.

We outline now the proof of Theorem 1.2. By plugging in the estimate (1.32) into the upper bound on tube volumes (1.26), we obtain

$$\frac{\text{vol}\, T^\perp\big(M \cap B(a, \sigma), \epsilon\big)}{\text{vol}\, B(a, \sigma)} \leq 2d \sum_{i=0}^{n-1} d^i\, \frac{\text{vol}\, T(S^{i+1}, \sigma)}{\text{vol}\, B(a, \sigma)}\, \frac{\text{vol}\, T(S^{n-i-1}, \epsilon)}{\mathcal{O}_n}$$

and after some estimations one can arrive at the estimate

$$\frac{\text{vol}\, T^\perp\big(M \cap B(a, \sigma), \epsilon\big)}{\text{vol}\, B(a, \sigma)} \leq 4 \sum_{k=0}^{n-1} \binom{n}{k} \left(\frac{d\epsilon}{\sigma}\right)^k + \frac{2n\mathcal{O}_n}{\mathcal{O}_{n-1}} \left(\frac{d\epsilon}{\sigma}\right)^n.$$

From this inequality, estimates of the volume of the set $T(M, \epsilon) \cap B(a, \sigma)$ can be deduced by noting that the latter set is contained in the ϵ-tube around $M \cap B(\pm a, \sigma + \epsilon)$. Another problem is that M is assumed to be smooth, but the real algebraic hypersurface M' in the statement of Theorem 1.2 may have singularities. Fortunately, this can be easily dealt with by a perturbation argument. By some further estimations, one finally arrives at

$$\frac{\text{vol}\,(T(M',\epsilon) \cap B(a,\sigma))}{\text{vol}\,B(a,\sigma)} \leq 26\,dn\,\frac{\epsilon}{\sigma}$$

for $\epsilon \leq \sigma/(n(2d+1))$, as claimed in Theorem 1.2. For details we refer to the original paper by Bürgisser et al. (2008). We note that the bound (1.24) on the volume of ϵ-neighborhoods of ∂K follows in a similar way.

In order to deduce from the above the bound on the expectation stated in Theorem 1.2, we use the general observation that a tail bound of the form

$$\text{Prob}\{X \geq t\} \leq Kt^{-\alpha} \quad \text{for all } t \geq t_0 > 0$$

for a nonnegative absolutely continuous random variable X such that $K, \alpha > 0$ implies

$$\mathbb{E}\,(\ln X) \leq \ln t_0 + \frac{1}{\alpha}\,(\ln K + 1)\,.$$

We finally remark that the proof of Theorem 1.3, dealing with the situation over $\mathbb{C}$, is more direct and avoids curvatures. However, it is not possible to extend those arguments to the situation over $\mathbb{R}$.

Acknowledgements The surveys by Smale (1997) and Cucker (2002) were very helpful in writing this article. I thank Dennis Amelunxen for useful comments. This work was supported by DFG grant BU1371/2-1.

References

D. Amelunxen and P. Bürgisser (2008), 'Uniform smoothed analysis of a condition number of linear programming', arXiv:0803.0925.

L. Amodei and J.-P. Dedieu (2008), *Analyse Numérique Matricielle*, Mathématiques Appliquées pour le Master/SMAI. Dunod, to appear.

K. Ball (1993), 'The reverse isoperimetric problem for Gaussian measure', *Discrete Comput. Geom.* **10**(4), 411–420.

E. Barbier (1860), 'Note sur le problème de l'aguille et le jeu du joint couvert', *J. Math. Pures et Appl.* **5**(2), 273–286.

A. Belloni, R. M. Freund and S. Vempala (2007), 'An efficient re-scaled perceptron algorithm for conic systems', in *Proc. of 20th Conf. on Computational Learning Theory*, San Diego, 2007.

C. Beltrán and L. M. Pardo (2007), 'Estimates on the distribution of the condition number of singular matrices', *Found. Comput. Math.* **7**(1), 87–134.

C. Beltrán and L. M. Pardo (2008), 'On Smale's 17th problem: a probabilistic positive solution', *Found. Comput. Math.* **8**(1), 1–43.

C. Beltrán and M. Shub, 'Complexity of Bézout's theorem VII: Distance estimates in the condition metric', *Found. Comput. Math.*, to appear.

L. Blum (1990), 'Lectures on a theory of computation and complexity over
the reals (or an arbitrary ring)', in *Lectures in the Sciences of Complexity
II*, E. Jen (ed.), 1–47, Addison-Wesley.

L. Blum, F. Cucker, M. Shub and S. Smale (1998), *Complexity and Real
Computation*, Springer.

L. Blum and M. Shub (1986), 'Evaluating rational functions: infinite precision
is finite cost and tractable on average', *SIAM J. Comput.* **15**(2), 384–398.

L. Blum, M. Shub and S. Smale (1989), 'On a theory of computation and
complexity over the real numbers: NP-completeness, recursive functions
and universal machines', *Bull. Amer. Math. Soc. (N.S.)* **21**(1), 1–46.

T. Bonnesen and W. Fenchel (1974), *Theorie der Konvexen Körper*, Springer-
Verlag, Berlin.

K.-H. Borgwardt (1982), 'The average number of pivot steps required by the
simplex-method is polynomial', *Z. Oper. Res. Ser.* **26**(5), 157–177.

S. Boyd and L. Vandenberghe (2004), *Convex Optimization*, Cambridge University Press.

P. Bürgisser (2007), 'Average Euler characteristic of random real algebraic
varieties', *C. R. Math. Acad. Sci. Paris* **345**(9), 507–512.

P. Bürgisser, F. Cucker, and M. Lotz (2007), 'Coverage processes on spheres
and condition numbers for linear programming', arXiv:0712.2816.

P. Bürgisser, F. Cucker and M. Lotz (2006), 'General formulas for the
smoothed analysis of condition numbers', *C. R. Acad. Sci. Paris, Ser. I*
343, 145–150.

P. Bürgisser, F. Cucker and M. Lotz (2006), 'Smoothed analysis of complex
conic condition numbers', *J. Math. Pures et Appl.* **86**, 293–309.

P. Bürgisser, F. Cucker and M. Lotz (2008), 'The probability that a slightly
perturbed numerical analysis problem is difficult', *Math. Comp.* **77**, 1559–
1583.

M. P. do Carmo (1992), *Riemannian Geometry*, Birkhäuser.

S.-S. Chern (1966), 'On the kinematic formula in integral geometry',
J. Math. Mech. **16**, 101–118.

D. Cheung and F. Cucker (2001), 'A new condition number for linear programming', *Math. Program.* **91**(1, Ser. A), 163–174.

D. Cheung and F. Cucker (2002), 'Probabilistic analysis of condition numbers
for linear programming', *J. Optim. Theory Appl.* **114**(1), 55–67.

D. Cheung, F. Cucker, and R. Hauser (2005), 'Tail decay and moment estimates of a condition number for random linear conic systems', *SIAM
J. Optim.* **15**(4), 1237–1261.

D. Cheung, F. Cucker and J. Peña (2008), 'A condition number for multifold
conic systems', *SIAM J. Optim.* **19**(1), 261–280.

F. Cucker (2002), 'Real computations with fake numbers', *J. Complexity*
18(1), 104–134.

F. Cucker and J. Peña (2002), 'A primal-dual algorithm for solving polyhedral
conic systems with a finite-precision machine', *SIAM J. Optim.* **12**(2),
522–554.

F. Cucker and M. Wschebor (2003), 'On the expected condition number of
linear programming problems', *Numer. Math.* **94**(3), 419–478.

J. Demmel (1987), 'On condition numbers and the distance to the nearest
ill-posed problem', *Numer. Math.* **51**, 251–289.

J. Demmel (1988), 'The probability that a numerical analysis problem is difficult', *Math. Comp.* **50**, 449–480.

J. Dunagan, D. A. Spielman, and S.-H. Teng (2003), 'Smoothed Analysis of Renegar's Condition Number for Linear Programming', arXiv:cs.DS/0302011v2.

J. Dunagan and S. Vempala (2004), 'A simple polynomial-time rescaling algorithm for solving linear programs', in *Proceedings of the 36th Annual ACM Symposium on Theory of Computing*, 315–320, New York, ACM.

C. Eckart and G. Young (1936), 'The approximation of one matrix by another of lower rank', *Psychometrika* **1**, 211–218.

A. Edelman (1988), 'Eigenvalues and condition numbers of random matrices', *SIAM J. Matrix Anal. Appl.* **9**, 543–556.

A. Edelman (1992), 'On the distribution of a scaled condition number', *Math. Comp.* **58**, 185–190.

H. Federer (1959), 'Curvature measures', *Trans. Amer. Math. Soc.* **93**, 418–491.

R. M. Freund and J. R. Vera (1999), 'Some characterizations and properties of the "distance to ill-posedness" and the condition measure of a conic linear system', *Math. Program.* **86**(2, Ser. A), 225–260.

J.-L. Goffin (1980), 'The relaxation method for solving systems of linear inequalities', *Math. Oper. Res.* **5**(3), 388–414.

A. Gray (1990), *Tubes*, Addison-Wesley, Redwood City, CA.

M. Grötschel, L. Lovász and A. Schrijver (1988), *Geometric Algorithms and Combinatorial Optimization*, volume 2 of *Algorithms and Combinatorics: Study and Research Texts*, Springer-Verlag, Berlin.

M. R. Hestenes and E. Stiefel (1952), 'Methods of conjugate gradients for solving linear systems', *J. Research Nat. Bur. Standards* **49**, 409–436.

N. J. Higham (1996), *Accuracy and Stability of Numerical Algorithms*, SIAM, Philadelphia, PA.

R. Howard (1993), 'The kinematic formula in Riemannian homogeneous spaces', *Mem. Amer. Math. Soc.* **106**.

S. Janson (1986), 'Random coverings in several dimensions', *Acta Math.* **156**, 83–118.

L. G. Khachyian (1979), 'A polynomial algorithm in linear programming', *Dokl. Akad. Nauk SSSR* **244**, 1093–1096.

D. A. Klain and G.-C. Rota (1997), *Introduction to Geometric Probability*, Cambridge University Press, Cambridge.

E. Kostlan (1988), 'Complexity theory of numerical linear algebra', *J. Comput. Appl. Math.* **22**, 219–230.

D. Mumford (1976), *Algebraic Geometry I: Complex Projective Varieties*, Springer.

Y. Nesterov and A. Nemirovskii (1994), *Interior-Point Polynomial Algorithms in Convex Programming*, SIAM Studies in Applied Mathematics, **13**, SIAM, Philadelphia, PA.

A. Nijenhuis (1974), 'On Chern's kinematic formula in integral geometry', *J. Differential Geometry* **9**, 475–482.

J. Renegar (1987), 'On the efficiency of Newton's method in approximating all zeros of systems of complex polynomials', *Math. of Oper. Research* **12**, 121–148.

J. Renegar (1995a), 'Incorporating condition measures into the complexity theory of linear programming', *SIAM J. Optim.* **5**(3), 506–524.

J. Renegar (1995b), 'Linear programming, complexity theory and elementary functional analysis', *Math. Programming* **70**(3, Ser. A), 279–351.

F. Rosenblatt (1962), *Principles of Neurodynamics. Perceptrons and the Theory of Brain Mechanisms*, Spartan Books, Washington, D.C.

A. Sankar, D. A. Spielman, and S.-H. Teng (2006), 'Smoothed analysis of the condition numbers and growth factors of matrices', *SIAM J. Matrix Anal. Appl.* **28**(2), 446–476.

L. A. Santaló (1976), *Integral Geometry and Geometric Probability*, Addison-Wesley, Reading, Mass.

M. Shub, 'Complexity of Bézout's theorem VI: Geodesics in the condition (number) metric', *Found. Comput. Math.*, to appear.

M. Shub and S. Smale (1993a), 'Complexity of Bézout's theorem. I. Geometric aspects', *J. Amer. Math. Soc.* **6**(2), 459–501.

M. Shub and S. Smale (1993b), 'Complexity of Bézout's theorem II: volumes and probabilities', in *Computational Algebraic Geometry*, F. Eyssette and A. Galligo (eds.), **109**, *Progress in Mathematics*, 267–285, Birkhäuser.

M. Shub and S. Smale (1994), 'Complexity of Bézout's theorem V: polynomial time', *Theoretical Computer Science* **133**, 141–164.

M. Shub and S. Smale (1996), 'Complexity of Bézout's theorem IV: probability of success; extensions', *SIAM J. Numer. Anal.* **33**, 128–148.

S. Smale (1981), 'The fundamental theorem of algebra and complexity theory', *Bull. Amer. Math. Soc.* **4**, 1–36.

S. Smale (1983), 'On the average number of steps of the simplex method of linear programming', *Math. Programming* **27**(3), 241–262.

S. Smale (1997), 'Complexity theory and numerical analysis', *Acta Numerica* **6**, 523–551.

S. Smale (2000), 'Mathematical problems for the next century', In *Mathematics: Frontiers and Perspectives*, 271–294, AMS, Providence, RI.

H. Solomon (1978), *Geometric Probability*, SIAM, Philadelphia, PA.

D. A. Spielman and S.-H. Teng (2001), 'Smoothed analysis of algorithms: why the simplex algorithm usually takes polynomial time', in *Proceedings of the Thirty-Third Annual ACM Symposium on Theory of Computing*, 296–305, New York, ACM.

D. A. Spielman and S.-H. Teng (2002), 'Smoothed analysis of algorithms', in *Proceedings of the International Congress of Mathematicians*, volume I, 597–606.

D. A. Spielman and S.-H. Teng (2003), 'Smoothed analysis of termination of linear programming algorithms', *Math. Programm. Series B* **97**, 375–404.

D. A. Spielman and S.-H. Teng (2004), 'Smoothed analysis: Why the simplex algorithm usually takes polynomial time', *Journal of the ACM* **51**(3), 385–463.

D. A. Spielman and S.-H. Teng (2006), 'Smoothed analysis of algorithms and heuristics', in *Foundations of Computational Mathematics, Santander 2005*, L. M. Pardo, A. Pinkus, E. Süli, M. J. Todd (eds.), Cambridge University Press, 274–342.

M. Spivak (1979), *A Comprehensive Introduction to Differential Geometry. Vol. III*, Publish or Perish Inc.

J. Steiner (1840), 'Über parallele Flächen', *Monatsbericht der Akademie der Wissenschaften zu Berlin*, 114–118. Also Werke vol. 2 (1882), 171–176.

T. Tao and V. Vu (2007), 'The condition number of a randomly perturbed matrix', *Proceedings 39th Annual ACM Symposium on Theory of Com-*

puting, 248–255.

J. A. Thorpe (1994), *Elementary Topics in Differential Geometry,* Springer-Verlag, New York.

A. M. Turing (1948), 'Rounding-off errors in matrix processes', *Quart. J. Mech. Appl. Math.* **1**, 287–308.

S. A. Vavasis and Y. Ye (1995), 'Condition numbers for polyhedra with real number data', *Oper. Res. Lett.* **17**(5), 209–214.

J. von Neumann and H. H. Goldstine (1947), 'Numerical inverting of matrices of high order', *Bull. Amer. Math. Soc.* **53**, 1021–1099.

J. G. Wendel (1962), 'A problem in geometric probability', *Math. Scand.* **11**, 109–111.

H. Weyl (1939), 'On the volume of tubes', *Amer. J. Math.* **61**(2), 461–472.

J. H. Wilkinson (1963), *Rounding Errors in Algebraic Processes,* Prentice-Hall, Englewood Cliffs, N.J.

J. H. Wilkinson (1972), 'Note on matrices with a very ill-conditioned eigenproblem', *Numer. Math.* **19**, 176–178.

H. Woźniakowski (1977), 'Numerical stability for solving nonlinear equations', *Numer. Math.* **27**(4), 373–390.

M. Wschebor (2004), 'Smoothed analysis of $\kappa(a)$', *J. of Complexity* **20**, 97–107.

2

A World of Binomials

Alicia Dickenstein

Departamento de Matemática, FCEN
Universidad de Buenos Aires
Ciudad Universitaria, Pab. I
C1428EGA Buenos Aires, Argentina
e-mail: alidick@dm.uba.ar

Abstract

Binomial systems are basic blocks in the study of general polynomial systems, as well as in the structure of hypergeometric differential equations and the dynamics of mass action kinetic systems. In this article we highlight the basic facts about binomials, on binomials and complexity of polynomial system solving, and on the duals of binomial varieties, i.e., discriminants of sparse polynomial equations.

2.1 Introduction

A binomial in n variables is a polynomial with two terms $ax^\alpha + bx^\beta$, where $x = (x_1, \ldots, x_n)$, $\alpha \neq \beta \in \mathbb{N}^n$, x^α means as usual $x_1^{\alpha_1} \cdots x_n^{\alpha_n}$, and a, b lie in a coefficient ring k, usually the field of real or complex numbers. A binomial ideal I in $k[x_1, \ldots, x_n]$ is an ideal generated by binomials, that is to say the set of all polynomial "consequences" of a given set of binomials $b_1, \ldots, b_r$

$$I = \{\sum_{i=1}^{r} h_i b_i \; : \; h_i \in k[x_1, \ldots, x_n], \; i = 1, \ldots, r\},$$

and so all the solutions of the system $b_1(x) = \cdots = b_r(x) = 0$ are also zeros of all the polynomials in the ideal I.

The family of binomial ideals is at the same time rich, a good source of concrete examples, and constitutes a step below general ideals. In fact, we could isolate two other such families of ideals: the linear ideals (i.e., the world of linear algebra) and the monomial ideals (essentially tied to combinatorics and divisibility questions). The study of binomial ideals conjugates a blend of combinatorics and linear algebra. We refer the reader to the paper Eisenbud, Sturmfels (1996) for a thorough study.

If we allow trinomials, it is well known that we can represent any polynomial system iterating the following process: let $f = m_1 + m_2 + $

$m_3 + m_4$ be any quatrinomial in any number of variables. Then, we can add two variables z_1, z_2 to name partial sums of the monomials in f so that

$$f = 0 \Leftrightarrow m_1 + m_2 - z_1 = m_3 + m_4 - z_2 = z_1 + z_2 = 0.$$

Also, if we allow the combination of binomials and linear forms, we again recover *all* polynomial systems, as we describe in Corollary 2.1 and Proposition 2.4.

Binomial ideals are basic ingredients in the theory of toric varieties Sturmfels (1996), in the study of semigroup algebras Delorme (1976), Miller, Sturmfels (2005), in the homogenized version of hypergeometric systems of partial differential equations Gelfand, Kapranov, Zelevinsky (1989), Gelfand, Graev, Retakh (1992), Gelfand, Kapranov, Zelevinsky (1990), Adolphson (1994), Saito, Sturmfels, Takayama (2000), Dickenstein, Matusevich, Sadykov (2005), Dickenstein, Matusevich, Miller (2006) and in the modeling of mass action kinetics dynamics Feinberg (1979), Feinberg (1989), Horn (1974), Horn, Jackson (1972), Sontag (2001), Gatermann (2001), Gatermann, Wolfrum (2005), Gunawardena (2003), Craciun, Dickenstein, Shiu, Sturmfels (2007), Conradi, Flockerzi, Raisch, Stellin (2007), Adleman, Gopalkrishnan, Huang, Moisset, Reishus (2008), in the homotopy continuation methods for polynomial system solving Huber, Sturmfels (1995), Verschelde, Verlinden, Cools (1994), and they also provide worst case complexity bounds Mayr, Meyer (1982). Binomial ideals arise in other domains of applied mathematics such as integer programming Thomas (1995), Conti, Traverso (1991), and statistics Diaconis, Sturmfels (1998), Pachter, Sturmfels (2005), Geiger, Meek, Sturmfels (2006), Drton, Sullivant, Sturmfels (2009).

The aim of this article is to highlight some basic facts about binomial systems and to hint at the ubiquity and richness of their occurrence. We also include a brief discussion of sparse discriminants, see Gelfand, Kapranov, Zelevinsky (1994), (also called toric or A-discriminants), i.e., the equations of the duals of binomial (toric) varieties. They describe those parameters in a sparse family of polynomials where singularities occur and they also describe the singularities of hypergeometric differential systems, see Gelfand, Kapranov, Zelevinsky (1989), Adolphson (1994).

2.2 Basics about binomials

In this section we summarize five basic results about the structure of the zeros of binomial ideals over the torus $T^n = (k^*)^n$, that is, of zeros in k^n with all non-zero coordinates. These features are the core of all applications of binomial ideals.

Clearly, any system of the form

$$y^{\alpha_j} - y^{\beta_j} = 0, \quad j = 1, \dots, r, \tag{2.1}$$

admits the solution $(1, \dots, 1) \in T^n$. The next proposition shows that this is the general shape of binomial systems which have a common root in the torus.

We consider a general system of binomial equations

$$a_j \, x^{\alpha_j} + b_j \, x^{\beta_j} = 0, \quad j = 1, \dots, r, \tag{2.2}$$

with $a_j, b_j \in k$, $(a_j, b_j) \neq (0, 0)$ for all j.

Proposition 2.1 (First basic fact about binomials) *A binomial system as in* (2.2) *has a common solution* **c** *in the torus* T^n *if and only if, up to the diagonal action of the torus* T^n *on* k^n *by coordinatewise multiplication, the system can be written with coefficients* $1, -1$. *More explicitly, given a common zero* $c \in T^n$, *in new coordinates* $y_i = \frac{x_i}{c_i}$, $i = 1, \dots, n$, *the system looks (up to multiplication by non-zero constants) like*

$$y^{\alpha_j} - y^{\beta_j} = 0, \quad j = 1, \dots, r.$$

Proof The proof is straightforward. We can assume that $a_j, b_j \in k^*$. Then

$$a_j x^{\alpha_j} + b_j x^{\beta_j} = a_j c^{\alpha_j} y^{\alpha_j} + b_j c^{\beta_j} y^{\beta_j} = \kappa(y^{\alpha_j} - y^{\beta_j}),$$

where $\kappa := a_j c^{\alpha_j} = -b_j c^{\beta_j}$.   □

So, the coefficients of binomial systems with toric roots are not relevant (as long as they are non-zero), and all the action is in the exponents. The next question is then when a binomial system has a solution with all non-zero coordinates. To make the statements more transparent, we assume that k is an algebraically closed field, e.g., $k = \mathbb{C}$, but one can consider the case $k = \mathbb{R}$ under additional hypotheses (see Remark 2.1 below).

Proposition 2.2 (Second basic fact about binomials) *A binomial system as in (2.2) has a solution $c \in T^n$ if and only if $a_j, b_j \neq 0$ for all j, and for each linear relation*

$$\sum_{j=1}^{r} \lambda_j (\alpha_j - \beta_j) = 0, \quad \lambda_j \in \mathbb{Z}, \tag{2.3}$$

it holds that

$$\prod_{j=1}^{r} \left(\frac{-b_j}{a_j} \right)^{\lambda_j} = 1. \tag{2.4}$$

Moreover, it is enough to check these conditions for a basis of the space of relations.

Proof Given a common root $c \in T^n$, from the equalities $a_j x^{\alpha_j} + b_j x^{\beta_j} = 0, j = 1, \ldots, r$, we see that both $a_j, b_j \in k^*$ and $c^{\alpha_j - \beta_j} = -\frac{b_j}{a_j}$ for all j. It is clear that any linear relation (2.3) among the exponents gets translated into a multiplicative relation among the quantities $-\frac{b_j}{a_j}$.

Reciprocally, assume that $a_j, b_j \neq 0$ for all j and that for each linear relation (2.3), the corresponding equality (2.4) holds. Assume $\{\alpha_1 - \beta_1, \ldots, \alpha_s - \beta_s\}$, $s \leq r$, is a basis of the linear space generated by all the vectors $\alpha_j - \beta_j$. We claim that it is enough to see that the equations $a_j x^{\alpha_j} + b_j x_j^{\beta} = 0, j = 1, \ldots, s$, have a common solution $c \in T^n$. If this happens, take any $\ell = s+1, \ldots, r$ and write $\alpha_\ell - \beta_\ell = \sum_{j=1}^{s} \lambda_j (\alpha_j - \beta_j)$, that is: $(\alpha_\ell - \beta_\ell) + \sum_{j=1}^{s} (-\lambda_j)(\alpha_j - \beta_j) = 0$. Then, we have that

$$c^{\alpha_\ell - \beta_\ell} = \prod_{j=1}^{s} \left(c^{\alpha_j - \beta_j} \right)^{\lambda_j} = \prod_{j=1}^{s} \left(-\frac{b_j}{a_j} \right)^{\lambda_j} = -\frac{b_\ell}{a_\ell}.$$

Now, consider the matrix $A \in \mathbb{Z}^{s \times n}$ with rows $\alpha_1 - \beta_1, \ldots \alpha_s - \beta_s$ and let $S \in \mathbb{Z}^{s \times n}$ be its Smith Normal form. Then, S is a diagonal matrix with $s_{i,i} | s_{i+1,i+1}$ and all $s_{i,i} \neq 0$ (given that A is a matrix of full rank s by our assumption of linear independence), and there exist left and right unimodular multipliers $U \in \mathbb{Z}^{s \times s}$, $V \in \mathbb{Z}^{n \times n}$ such that $A = USV$. As the matrices U, V are invertible over $\mathbb{Z}$, they define invertible monomial changes of coordinates in T^s and T^n, respectively. It is clear that for any choice of a vector κ of non-zero constants $\kappa = (\kappa_1, \ldots, \kappa_s)$, we can solve $(c')^S = \kappa$ (in obvious vector notation) by taking $s_{i,i}$-th roots. It is enough to take $\kappa = (-\frac{b_1}{a_1}, \ldots, -\frac{b_s}{a_s})^{U^{-1}}$ and $c = (c')^{V^{-1}}$ to get the desired solution. $\square$

Remark 2.1 The computation of the Smith Normal form $A = USV$ of an integer matrix A can be carried out using the polynomial algorithm presented in Kannan, Bachem (1979), modified to work with rectangular matrices in the way the authors suggest. The product of the diagonal entries of S equals the greatest common divisor g of all maximal minors of A. So, if $g = 1$ we do not need to require in Proposition 2.2 that k is an algebraically closed field. It is also straightforward from the proof that the statement of the proposition is valid for $k = \mathbb{R}$ if $a_j b_j < 0$ for all j (so that $-\frac{b_j}{a_j} \in \mathbb{R}_{>0}$). Similar arguments apply to the following results in this section.

In contrast with the simplicity of finding all roots in the torus of a binomial system, we now show that *any* sparse polynomial system on the torus T^n is equivalent either to a system of *binomials and linear forms* or to a system of *binomials in linear forms*.

We consider a sparse polynomial system of r polynomials in n variables with exponents $m_1, \ldots, m_N \in N^n$ and coefficients a_j^i in k:

$$f_i = \sum_{j=1}^{N} a_j^i x^{m_j} = 0, \quad i = 1, \ldots r. \tag{2.5}$$

In fact, the same arguments in the rest of this section hold for integer exponents in $\mathbb{Z}^n$, and it is convenient not to restrict our exponents to the first orthant. The roots in the torus are unchanged if we divide by the monomial m_1, and so one can assume that $m_1 = 0$ if necessary.

Given an integer vector $\lambda \in \mathbb{Z}^n$, we denote by $\lambda^+, \lambda^- \in \mathbb{Z}^n$ the vectors of positive and negative entries of λ, i.e.,

$$\lambda_i^+ = \max\{\lambda_i, 0\}, \quad i = 1, \ldots, n, \quad \lambda^+ - \lambda^- = \lambda.$$

We rephrase the first basic fact in terms of elimination of variables.

Proposition 2.3 (Third basic fact about binomials) *Given $y = (y_1, \ldots, y_N) \in T^N$, there exists $x \in T^n$ such that $y = (x^{m_1}, \ldots, x^{m_N})$ if and only if for any λ in the integer kernel I of the $n \times N$-integer matrix with columns $m_1, \ldots, m_N$ it holds that*

$$y^\lambda = 1, \qquad or \quad y^{\lambda^+} - y^{\lambda^-} = 0. \tag{2.6}$$

Moreover, it is enough to check this condition for all λ in a basis of I.

Proof We can write the equality $y = (x^{m_1}, \ldots, x^{m_N})$ as the following systems of binomial equations

$$x^{m_1} - y_1 = \cdots = x^{m_N} - y_N.$$

Then, it has the form (2.2) with $r = N$, $a_j = 1$, $b_j = -y_j$, $\alpha_j = m_j$, and $\beta_j = 0$ for all j. The result is just a particular case of Proposition 2.2 $\square$

Corollary 2.1 (Fourth basic fact about binomials) *System* (2.5) *is equivalent over* T^n *(with variables* $(x_1, \ldots, x_n)$*) to the following* system *of linear forms and binomials over* T^N *(with variables* $(y_1, \ldots, y_N)$*):*

$$\ell_j = \sum_{i=1}^{N} a_j^i y_j = 0, \quad i = 1, \ldots r, \quad y^{\lambda_+} - y^{\lambda_-} = 0, \lambda \in I, \qquad (2.7)$$

where I *denotes, as before, the integer kernel of the* $n \times N$*-integer matrix* M *with columns* $m_1, \ldots, m_N$.

Remark 2.2 Corollary 2.1 (over $\mathbb{R}^n$) appears as Proposition 1 of Section·4 in Blum, Shub, Smale (1989) for dense polynomials $f_1, \ldots, f_r$ of degree d, that is $N = \binom{n+d}{d}$ and $m_1, \ldots, m_N$ are all the integer points in the d-th dilate of the unit simplex in $\mathbb{R}^n$. In this case, one can take all binomials $y^{\lambda_+} - y^{\lambda_-}$ of degree 2 (corresponding to the identities $x^{m_i + m_j} = x^{m_i} x^{m_j}$). This property is used by Blum, Shub and Smale in the Main Theorem in Section 6 of that paper, where they give the analogue of Cook's Theorem for $\mathbb{R}$ by showing that the 4-feasibility problem over $\mathbb{R}$ is NP-complete over $\mathbb{R}$. But in the sparse case, binomials of higher degree are unavoidable, so we need to add new variables y_j in order to reduce their degree to 2.

We now introduce *Gale Duality* for polynomial systems. This mechanism is present, for instance, in the proof of the toric residue mirror conjecture in Szenes, Vergne (2004) and it has been exploited to get bounds for the number of positive real solutions (or solutions in the real torus $(\mathbb{R}^*)^n$) in terms of the number N of monomials in Bihan, Sottile (2007), Bihan, Sottile (2008). The study of affine solutions (i.e., solutions which can have zero coordinates) poses greater technical difficulties.

In a nutshell, Gale duality amounts to the following: the solutions of system (2.7) can be thought of as either the solutions of the system given by the restriction of the linear forms ℓ_j in (2.7) to the zero set of the binomial equations, i.e., to the original system (2.5) as asserted in Corollary 2.1, or as the solutions of restriction of the binomials in (2.7) to the linear space with equations $\ell_1 = \cdots = \ell_r = 0$, i.e., the system of binomials in linear forms described in (2.9) below.

To state this equivalence explicitly, we call K the kernel of the $n \times N$ matrix (a_j^i) of the coefficients of the input system (2.5) and we form a matrix V whose *columns* give a basis of K. Denote by $b_1, \ldots, b_s$ the *row* vectors of V. Then, any N-tuple $y \in K$ is of the form

$$y = (\langle b_1, t \rangle, \ldots, \langle b_s, t \rangle), \qquad (2.8)$$

where $t = (t_1, \ldots, t_{\dim K})$ and $\langle b_i, t \rangle = \sum_j b_{ij} t_j$. Recall that I denotes the integer kernel (or nullspace) of the matrix with columns m_j, that is $\lambda \in \mathbb{Z}^n$ lies in I if and only if $\sum_j \lambda_j m_j = 0$.

Proposition 2.4 (Fifth basic result about binomials: Gale duality) *There exist $x \in T^n$ satisfying system (2.5) if and only if there exist $t \in k^N$ in the complement of the hyperplane arrangement defined by $\{\langle b_i, t \rangle = 0\}$ for $i = 1, \ldots, s$, satisfying the following system:*

$$g_\lambda(t) = \prod_{\lambda_i > 0} \langle b_i, t \rangle^{\lambda_i^+} - \prod_{\lambda_i < 0} \langle b_i, t \rangle^{\lambda_i^-} = 0, \quad \lambda \in I, \qquad (2.9)$$

or, equivalently, if (2.9) is satisfied for the vectors λ in the columns of V.

Proof By Corollary 2.1, we know that there exists a solution $x \in T^n$ of (2.5) if and only if there exists a solution $y \in T^N$ of (2.7). Observe that such a vector y lies in $K \cap T^N$. These vectors are precisely parametrized as in (2.8) by vectors $t \in k^N$ in the complement of the hyperplane arrangement defined by $\{\langle b_i, t \rangle = 0\}$ for $i = 1, \ldots, s$, and the result follows. $\qquad \square$

Observe that the torus T^n is equal to the complement in k^n of the hyperplane arrangement defined by the coordinate hyperplanes $\{x_i = 0\}, i = 1, \ldots, n$. As we have already remarked, if zero coordinates of the solutions are allowed, the difficulty of the problem increases, both at the theoretical and computational level. We illustrate this in the next section with the simplest possible case. We refer to Sturmfels (1997) for the combinatorial issues that come into play to choose a finite system of generators in Proposition 2.2 if one allows zero coordinates. Also, we refer to Dickenstein, Sturmfels (2002), Ojeda, Piedra (2000), and Dickenstein, Matusevich, Miller (2008) for the primary decomposition of general binomial ideals.

2.3 Counting solutions to binomial systems

In this section we consider a square system of n binomial equations in n variables, that is, a sparse system of the form

$$a_j\, x^{\alpha_j} + b_j\, x^{\beta_j} = 0, \quad j = 1, \ldots, n, \tag{2.10}$$

with a_j, b_j in an algebraically closed field k, $\alpha_j, \beta_j \in \mathbb{N}^n$. We highlight some results from Cattani, Dickenstein (2007), which show the algebraic and combinatorial difficulty of finding all affine roots and not just all toric roots, and we refer the reader to that paper for further details. Given a vector $v \in k^n$, its support is the set of indices $\mathrm{supp}(v) := elm\{i \in \{1, \ldots, n\} : v_i \neq 0\}$. Thus, the torus T^n can be described as those vectors with empty support and if we consider the torus action on k^n by coordinatewise multiplication

$$(t_1, \ldots, t_n) \cdot (x_1, \ldots, x_n) := (t_1 x_1, \ldots, t_n x_n), \quad t \in T^n, x \in k^n,$$

two vectors $v, v' \in k^n$ lie in the same torus orbit if and only if $\mathrm{supp}(v) = \mathrm{supp}(v')$.

Call $M \in \mathbb{Z}^{n \times n}$ the matrix with rows $\alpha_1 - \beta_1, \ldots, \alpha_n - \beta_n$ and set $\delta := |\det(M)|$. When $\delta \neq 0$, the number of solutions in the torus T^n equals $\delta > 0$, independently of the value of the coefficients, as can be extracted from the proof of Proposition 2.2. This is a very special case of the well known bound of Bernstein-Kouchnirenko-Khovanskii, see Gelfand, Kapranov, Zelevinsky (1994).

When $\delta = 0$, it is possible to decide in polynomial time (in the size of the sparse input) whether for generic coefficients the system has no solutions in the torus. This is shown in Cattani, Dickenstein (2007), as an application of the basic results in Section 2.2. Moreover, it is proved in Theorem 2.12 of that paper that it is possible to decide in polynomial time if system (2.10) has a finite number of solutions in k^n. Note that in principle, the number of possible supports of a solution vector is 2^n.

We now show, with similar arguments, that it is possible to determine in polynomial time whether the zero set of the system in affine space k^n is empty or not (assuming the coefficients belong to a computable field). I thank J. M. Rojas for posing this question. More generally, it is possible to check in polynomial time if the zero set V in affine space k^n of any binomial system (2.2), not necessarily square, is empty or not.

Proposition 2.5 *The following algorithm decides whether $V = \emptyset$:*
Input: $a_j x^{\alpha_j} + b_j x^{\beta_j}, j = 1, \ldots, r; a_j, b_j \in k$, *not both 0.*

Output: *NO (there are no common zeros) or YES (there is at least a common zero).*

Algorithm:

STEP 1 Identify the subset I of all indices i for which the x_i coordinate of any solution to the input system of binomials is necessarily non-zero. This is done inductively as follows: Set $I_0 = \emptyset$ and, for $\ell \geq 1$, let $I_\ell = I_\ell^{(1)} \cup I_\ell^{(2)}$, where

$$I_\ell^{(1)} := \bigcup \{\operatorname{supp}(\alpha_j) \, : \, a_j, b_j \neq 0 \ \ and \ \ \operatorname{supp}(\beta_j) \subset I_{\ell-1}\},$$

$$I_\ell^{(2)} := \bigcup \{\operatorname{supp}(\beta_j) \, : \, a_j, b_j \neq 0 \ \ and \ \ \operatorname{supp}(\alpha_j) \subset I_{\ell-1}\}.$$

Then $I = \bigcup_\ell I_\ell$.

STEP 2 Identify the equations containing only the variables in I. Assume that these are the first r' equations. If one of these equations is a monomial (i.e., $a_j = 0$ or $b_j = 0$), output NO. Otherwise, go to the next step.

STEP 3 Compute a basis $\lambda^{(1)}, \ldots, \lambda^{(s)}$ of the kernel of the $|I| \times r'$ matrix with jth column given by the Ith coordinates of $(\alpha_j - \beta_j)$ for $j = 1, \ldots, r'$ (that is, take the ith coordinates with i in I; note that the other coordinates of these vectors are 0).

STEP 4 Verify if

$$\prod_{j=1,\ldots,r'} (-b_j/a_j)^{\lambda_j^{(i)}} = 1$$

for all $i = 1, \ldots, s$. If the answer is no, answer NO. If the answer is yes, answer YES: it is possible to construct a solution setting $x_i = 0$ for all i not in I, by Proposition 2.2.

We come back to a square system of binomials (2.10) with a finite number of zeros. Techniques from commmutative algebra allow us to reduce any such system to a *normal form*, via the process known as parametric reduction. Theorem 3.15 in Cattani, Dickenstein (2007) gives a precise root counting (with multiplicity) and shows that for "generic" exponents the number of solutions can be computed in polynomial time. Similar techniques allow to count the number of roots without multiplicity. As we mentioned before, it is possible to determine over an algebraically closed field the number of solutions in the torus by a determinant computation (these roots are always simple, see Eisenbud, Sturmfels (1996)). Also, the number of isolated affine solutions is bounded by the Bezout bound of the product of the degrees of the input polynomials. However, if we allow M to be any integer matrix (even if $\det(M) \neq 0$) counting the number of solutions to a square binomial system with or without multiplicity is #P-complete by Cattani, Dickenstein (Theorem 4.3, 2007) by relating the problem of counting zeros of binomial systems to the known #P-complete problem of counting the independent sets in a bipartite graph, Provan, Ball (1983). We recall that the notion of #P-completeness is the enumerative analog of NP-completeness, see Valiant (1979).

Indeed, given any bipartite graph $G = (V, E)$, $V = V_1 \coprod V_2$, with $E \cap (V_1 \times V_1) = \emptyset$, $E \cap (V_2 \times V_2) = \emptyset$, $V = \{1, \ldots, n\}$, consider the following n binomials in n variables

$$p_i = x_i - x_i^2, \quad i \in V_1; \qquad p_j = x_j - \left(\prod_{(i,j) \in E} x_i \right) x_j^2, \quad j \in V_2.$$

The number of zeros of this system is finite. Moreover, $V(p_1, \ldots, p_n) \subset \{0, 1\}^n$ and its cardinal equals the number of independent sets of G, i.e., subsets S of V such that $(i, j) \notin E$ for all $i, j \in S$. All roots are simple and determined by their supports, so that the main complexity is based on deciding which are the possible zero and non-zero coordinates of the solutions. In some sense, the richness of binomial systems lies in their structure and is "orthogonal" to numerical analysis, i.e., to the behaviour of the system in terms of the coefficients.

Associating polynomials to graphs, in particular in relation with the search for independent (also called stable) sets is not new, see for instance the interesting paper Lovász (1994).

2.4 Toric varieties and discriminants

Given finite sets $A_1, \ldots, A_n \subset \mathbb{Z}^n$ and sparse polynomials $f_1, \ldots, f_n$ with these supports

$$f_i(\mathbf{a^i}, x) = \sum_{\alpha \in A_i} \mathbf{a}_\alpha^i x^\alpha,$$

Gelfand, Kapranov, Zelevinsky (1994) proved that there exists (for "general" configurations) an irreducible integer polynomial Δ_A (defined up to sign) in the vector of coefficients $\mathbf{a} = (\mathbf{a^1}, \ldots, \mathbf{a^n}) \in \mathbb{C}^N$ of the polynomials, which vanishes whenever there exists a zero $x \in T^n$ of $f_1(\mathbf{a^1}, x), \ldots, f_n(\mathbf{a^n}, x)$ which is not simple (that is, where the Jacobian determinant $J_f(\mathbf{a}, x)$ also vanishes). So, $\Delta_A = 0$ describes the closure

of the variety of *ill-posed systems*, and the size of the distance of a coefficient vector to it is basic for numerical continuation and numerical
stability, see Demmel (1987), Shub, Smale (1993), Malajovich, Rojas
(2004).

The homogeneous polynomial Δ_A is called the *mixed discriminant*
associated to the support sets $A_1, \ldots, A_n$. Computing Δ_A is an elimination problem. Denote by Z the incidence variety of tuples $(x, t, \mathbf{a})$,
$x, t \in T^n$ such that

$$f_1(\mathbf{a^1}, x) = \cdots = f_n(\mathbf{a^n}, x) = 0, \qquad (2.11)$$

and moreover

$$\sum_i \frac{\partial}{\partial x_j}(f_i(\mathbf{a^i}, x))t_i = 0, \quad j = 1, \ldots, n. \qquad (2.12)$$

Note that the existence of a non-trivial solution to equations (2.12) is
equivalent to the vanishing of the Jacobian determinant $J_f(\mathbf{a}, x)$.

The discriminant variety $\{\Delta_A(\mathbf{a}) = 0\}$ is the closure of the first projection π_1 in the following correspondence:

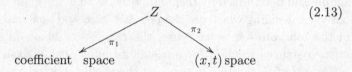

$$(2.13)$$

But the projection π_2 is much easier to understand, since its fibers
are linear spaces in the variables $\mathbf{a}$. This allows us to find a *rational parametrization* of the discriminant variety, as expressed in Proposition 2.6 below.

When $n = 1$ we recover the classical notion of discriminant of a univariate polynomial of fixed degree. Another important special case corresponds to the case when $A_i = d_i \Delta_n \cap \mathbb{Z}^n$ are the lattice points of a
dilate of the standard n-simplex Δ_n, i.e., f_i is just a generic polynomial
of degree d_i.

Example 2.1 (Wilkinson polynomial) Consider the Wilkinson polynomial

$$W_{20} = \prod_{i=1}^{20}(x + i) = \sum_{j=0}^{20} c_j x^j,$$

which is well known for its numerical instability, see Wilkinson (1994).
For instance, it clearly has 20 real roots, but the polynomial $W_{20}(x) +$

$10^{-9}x^{19}$ – obtained by adding an apparently small perturbation in the coefficient of x^{19} – has only 12 real roots and 4 pairs of complex roots, which do not "seem" to have small imaginary part. For instance, one of these pairs is approximately equal to $-16.57173899 \pm 0.8833156071i$. On the other hand, if we subtract $10^{-9}x^{19}$ from W_{20} we get a polynomial with 14 real zeros.

We think that this unstable behaviour could be explained by the fact that the vector of coefficients $\mathbf{w} = (20!, \ldots, 210, 1)$ of W_{20} is very close not only to the variety $\Delta_A = 0$ of ill-posed polynomials, but also very close to a *singular point* of the discriminant variety $\Delta_A = 0$, where $A = \{0, 1, \ldots, 20\}$. This variety is a hypersurface in an affine space of dimension 21, and its singularities have codimension one, i.e., they define a 19-dimensional variety. We have experimented with the following 2-dimensional family of polynomials of degree 20

$$W(a, b, x) := W_{20}(x) + ax^{19} + bx^{18}.$$

The corresponding discriminant $D(a, b)$ (which is a specialization of Δ_A) defines a singular curve traced inside the discriminant locus. The singularities of this curve $D(a, b) = 0$ are close to the point $a = b = 0$, i.e., to the vector of coefficients $\mathbf{w}$ of the Wilkinson polynomial.

Figure 2.1 features sample points of $D(a, b) = 0$ inside a small box around the origin, which is the point lying in the intersection of the two coordinate arrows. The marks near the ends of these arrows indicate distance 10^{-9} from $(0, 0)$ and we see several branches of $D(a, b) = 0$ very close to the origin: 4 to the right (which are crossed when moving a from 0 to 10^{-9}, causing a drop of 8 in the number of real roots), another 3 close to the left (which cause a drop of 6 in the number of real roots when we decrease a), plus two other branches still more close to the left. These drawings were done by Bernard Mourrain using the subdivision solver of Mathemagix, see Mathemagix (2008), by means of the package *subdivix* developed by Elias Tsigaridas and Bernard Mourrain, and visualized with the software Axel developed by Julien Wintz, see Axel Modeler (2008). As the figure suggests, once we increment the parameter a from 0 to $2 \cdot 10^{-10}$, we already get a polynomial with only $16 = 20 - 2 \cdot 2$ real roots (as we cross 2 of the branches). Also, it suggests that there are intersections of these branches close to $(0, 0)$, thus giving singularities of $D(a, b) = 0$ near the origin.

Considering the distance not just to the variety of ill posed problems $\Delta_A = 0$ but also to its singular locus would correspond in the case of conditioning of square $m \times m$ matrices in linear algebra, to consider not

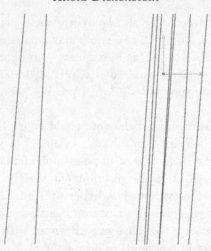

Fig. 2.1. Branches of $D(a, b) = 0$ close to the origin

only the smallest and greatest singular values (or the distance to matrices of rank at most $m-1$) but also the behaviour of the intermediate ones (or the distance to matrices of different ranks strictly smaller than $m - 1$).

Mixed discriminants are a particular case of general sparse discriminant (also known as A-discriminants, Gelfand, Kapranov, Zelevinsky (1994)) and define the dual variety of the toric (binomial) variety associated to the given supports. Given an integer matrix $A \in \mathbb{Z}^{d \times N}$, we can associate to it a monomial rational map

$$\varphi_A(t) = (t^{a_1}, \ldots, t^{a_N}), \quad t \in T^d,$$

where $a_1, \ldots, a_N \in \mathbb{Z}^d$ denote the column vectors of A. We can compose this map with the projection $p : \mathbb{C}^N \backslash \{0\} \to \mathbb{P}^{N-1}(\mathbb{C})$ and consider the variety X_A defined as the closure of the image of $p \circ \varphi_A$. This is the projective *toric* variety associated to A. This variety is cut out by binomials, as a generalization of Proposition 2.3. Note however that the number of equations needed to describe the closure of the image, that is, the number of generators of the *toric ideal* I_{X_A} of polynomials vanishing on X_A:

$$\langle x^u - x^v \ : \ \text{for all} \ u, v \in \mathbb{N}^n \ \text{such that} \ Au = Av \ \text{and} \ \sum_i u_i = \sum_i v_i \rangle,$$

can be much larger than the dimension of $\ker_{\mathbb{Z}}(A)$, as we remarked at the end of Section 2.2. Moreover, any irreducible projective variety cut out by binomials is of this form, Sturmfels (1996).

The associated *dual variety* X_A^* of X_A consists of the closure of those hyperplanes in the dual projective space which intersect X_A non-transversally at a regular point. Independently of the dimension of X_A, the expected codimension of X_A^* is one, i.e., X_A^* is in general a hypersurface. When this is the case, it has – up to sign – a unique defining equation with integer coefficients and content one: this equation is the sparse or A-discriminant Δ_A associated to A.

In more down to earth terms, consider the family of generic polynomials with exponents in A:

$$f_A(\mathbf{a}, x) = \sum_{\alpha \in A} a_\alpha x^\alpha.$$

Then, X_A^* is the closure of those vectors of coefficients $\mathbf{a}$ for which the hypersurface $\{x \in T^n : f_A(\mathbf{a}, x) = 0\}$ is not smooth, that is, for which there exists $x \in T^n$ with

$$f_A(\mathbf{a}, x) = \frac{\partial}{\partial x_1}(f_A(\mathbf{a}, x)) = \cdots = \frac{\partial}{\partial x_d}(f_A(\mathbf{a}, x)) = 0.$$

The exact conditions under which X_A^* is a hypersurface, that is, under which the A-discriminant is non-trivial (by convention, it is set equal to 1 when $\operatorname{codim}(X_A^*) > 1$) are quite delicate and completely detailed only in certain cases (see Dickenstein, Feichtner, Sturmfels (2007), Bourel, Dickenstein, Rittatore (2008) and references therein). These conditions are certainly satisfied by any "general" configuration.

The precise description of the singular locus of the discriminant variety is only known in the univariate case, see Chipalkatti, D'Andrea (2007), in the case of hyperdeterminants, see Weyman, Zelevinsky (1996), and for particular discriminants that occur in configuration spaces and classical hypergeometric differential equations.

The mixed discriminant associated with the supports $A_1, \ldots, A_n \subset \mathbb{Z}^n$ corresponds – under some general assumptions – to the A-discriminant of the Cayley configuration $A \subset \mathbb{Z}^{2n \times N}$ defined by

$$A = \mathcal{C}(A_1, \ldots, A_n) = e_1 \times A_1 \cup \cdots \cup e_n \times A_n.$$

As usual, $\{e_1, \ldots, e_n\}$ denotes the canonical basis in $\mathbb{Z}^n$. More generally, given s finite subsets $A_1, \ldots, A_s$ of $\mathbb{Z}^n$, let $f_1(\mathbf{a}, x), \ldots, f_s(\mathbf{a}, x)$ be generic polynomials with these supports. The associated Cayley configuration

$$A = \mathcal{C}(A_1, \ldots, A_s) = e_1 \times A_1 \cup \cdots \cup e_s \times A_s$$

lies in $\mathbb{Z}^{(s+n) \times N}$. When $1 \leq s \leq n$, the corresponding A-discriminant vanishes at those vectors of coefficients $\mathbf{a}$ for which the variety

$$\{x \in T^n \; : \; f_1(\mathbf{a}, x) = \cdots = f_s(\mathbf{a}, x) = 0\}$$

has a point where the rank of the Jacobian matrix of $f_1, \ldots, f_s$ is smaller than s. When $s = n + 1$ and the family of supports is *essential*, Sturmfels (1994), the discriminant of a Cayley configuration $\mathcal{C}(A_1, \ldots, A_{n+1})$ is precisely the $(A_1, \ldots, A_{n+1})$-resultant of $f_1, \ldots, f_{n+1}$. That is, it vanishes whenever the overdetermined system of sparse polynomials $f_1(\mathbf{a}, x) = \cdots = f_{n+1}(\mathbf{a}, x) = 0$ has a toric root. Sparse resultants are thus special cases of sparse discriminants. When $s > n + 1$, the discriminant variety X_A^* is not a hypersurface and then Δ_A is trivial.

As we remarked above, it is easy to understand the projection π_2 in (2.13). In other words, we can easily describe for each fixed $x \in T^n$ all the vectors of coefficients $\mathbf{a}$ which satisfy (2.11) and (2.12) by solving a linear system. This gives the following result, which is a restatement of Proposition 4.1 in Dickenstein, Feichtner, Sturmfels (2007) (see also Kapranov (1991, Theorem 2.1)).

Proposition 2.6 (Horn-Kapranov parametrization) *Assume $A \in \mathbb{Z}^{d \times N}$ and call as before $a_1, \ldots, a_N$ its column vectors. The map β_A which sends a point (u, t) with $u \in \mathrm{Ker}_{\mathbb{C}}(A)$ and $t \in T^d$, to the componentwise product $(u_1 t^{a_1}, \ldots, u_N t^{a_N})$, rationally parametrizes the discriminant variety X_A^*, i.e., X_A^* coincides with the closure of the image of β_A. Moreover, we can parametrize the linear space $\mathrm{Ker}_{\mathbb{C}}(A)$ and compose β_A with this map.*

For any fixed value of $t_0 \in T^n$ and any u in the kernel of A, the vectors of coefficients $(u_1 t_0^{a_1}, \ldots, u_N t_0^{a_N})$ correspond to all the systems which have a singular intersection at the point $(t_0)^{-1} = ((t_0)_1^{-1}, \ldots, (t_0)_d^{-1})$.

Example 2.2 Consider the matrix

$$A := \begin{pmatrix} 1\,1\,1\,0\,0\,0 \\ 0\,0\,0\,1\,1\,1 \\ 6\,0\,0\,0\,3\,1 \\ 0\,3\,1\,6\,0\,0 \end{pmatrix}.$$

A is the Cayley matrix associated with the 2 planar configurations $A_1 = \{(6, 0), (0, 3), (0, 1)\}$, and $A_2 = \{(0, 6), (3, 0), (1, 0)\}$. The A-discriminant

$\Delta_A(y_1, \ldots, y_6)$ is the *mixed discriminant* of the family of polynomials

$$\begin{cases} h_1(y;t,s) := y_1 t^6 + y_2 s^3 + y_3 s^1 \\ h_2(y;t,s) := y_4 s^6 + y_5 t^3 + y_6 l^1. \end{cases}$$

$\Delta_A(y) = 0$ whenever there exists a common zero $(s,t) \in (\mathbf{k}^*)^2$ which is not simple. The Horn-Kapranov parametrization of $X_A^* = (\Delta_A(y) = 0)$ is given by $(\lambda_1, \lambda_2 \in \mathbb{C}, \, t_1, t_2, t, s \in \mathbb{C}^*)$:

$$\begin{aligned} y_1 &= 2\lambda_1 t_1 t^6, & y_2 &= (\lambda_1 - 6\lambda_2) t_1 s^3, \\ y_3 &= (-3\lambda_1 + 6\lambda_2) t_1 s, & y_4 &= 2\lambda_2 t_2 s^6, \\ y_5 &= (-6\lambda_1 + \lambda_2) t_2 t^3, & y_6 &= (6\lambda_1 - 3\lambda_2) t_2 t. \end{aligned}$$

Setting $t_1 = t_2 = t = s = 1$, we get a parametrization of the kernel of A. The specialization $\Delta_A(1, a, -1, 1, b, -1)$ of the polynomial $\Delta_A(y)$ to the family of bivariate polynomials $h_1(a,b;t,s) = t^6 + as^3 - s$, $h_2(a,b;t,s) = s^6 + bt^3 - t$ with two parameters (a,b), equals

$$82754024941868680778822139064668229594467072\, a^{47} b^{33} +$$

$$245197110938870165270584115747165124724334688\, a^{46} b^{39} -$$

$$245197110938870165270584115747165124724334688\, b^{46} a^{39} +$$

$$23662740309026457547478521970718496800134567046336 0\, a^{28} b^7 +$$

$$17631004810327637966335552676449435712814331054687500\, a^4 b^{11} +$$

53 additional monomial terms of comparable size. It is a polynomial of degree 90 with 58 monomials and big integer coefficients.

As the previous example shows, A-discriminants are in general complicated integer polynomials with huge coefficients. They carry fascinating combinatorial information, which was basically described in Gelfand, Kapranov, Zelevinsky (1994). In principle, one can compute Δ_A by standard methods in elimination, but in practice we reach the limit of current computations very easily. Discriminants of univariate polynomials, as well as discriminants of codimension two configurations (via Gale duality) can be computed with Sylvester determinants, see Dickenstein, Sturmfels (2002). Note that expressing the discriminant as the determinant of a matrix gives a good way of computing its value for any choice of numerical constants, by computing a numerical determinant. That it, one can avoid the computation of the costly symbolic determinant. Discriminants can also be computed as the main factor of the determinant of a corresponding Cayley-Koszul complex, see Gelfand, Kapranov,

Zelevinsky (1994). Again, it is computationally complicated to "disentangle" this information. The A-discriminant is in general the main factor in the greatest common divisor of the minors of a rectangular matrix.

A main question in the area is to represent in general the vanishing of the A-discriminant as the rank drop of a matrix and to use this information to estimate the distance of a vector of coefficients to the discriminant locus and to its singular part. Instead, we can try to get a first *combinatorial approximation*, which can nonetheless give us information about discrete invariants such as dimension and degree (and asymptotics), by computing its *Newton polytope* $N(\Delta_A)$ or its *tropicalization* $\tau(X_A^*)$.

Recall that the Newton polytope $N(f)$ of a non-zero polynomial f is the polytope with integer vertices defined as the convex hull of the exponents occurring in f with non-zero coefficients. Equivalent information is encoded in the tropicalization of f (or of the hypersurface $\{f = 0\}$), which is a pure dimensional polyhedral fan dual to $N(f)$ minus its vertices, that is, the support of the tropicalization is the codimension one skeleton of the normal fan of $N(f)$. We now give the general definitions.

Given a weight $w \in \mathbb{R}^n$ and a non-zero polynomial $f = \sum_{c \in C} \gamma_c x^c$, $\gamma_c \neq 0$, $C \subset \mathbb{Z}^n$ (so that $N(f)$ is the convex hull of C), we define the initial polynomial of f with respect to w as the subsum of monomials in f which lie in the face of $N(f)$ where $\langle w, \cdot \rangle$ is minimized:

$$\text{in}_w f = \sum_{w \cdot c \, \min} \gamma_c x^c.$$

Given an algebraic variety Y in n-dimensional space, we denote by I_Y the ideal of all the polynomials vanishing on Y. Its initial ideal with respect to w is defined as the ideal generated by the initial polynomials in I_Y:

$$\text{in}_w(I_Y) = \langle \text{in}_w f : f \neq 0 \in I_Y \rangle.$$

The *tropicalization* $\tau(Y)$ of a variety Y is (as a set)

$$\tau(Y) = \{ w \in \mathbb{R}^n : \text{in}_w(I_Y) \text{ does not contain a monomial} \}.$$

The tropicalization $\tau(Y)$ is a pure dimensional polyhedral fan of the same dimension as Y, see Bieri, Groves (1984), and it can also be defined via valuations as in Einsiedler, Kapranov, Lind (2006). For the computation of the tropicalization of general varieties we refer to Jensen (2008). In order to recover more sensible information (for instance, to recover the exact Newton polytope in the hypersurface case), we need to decorate this set with intersection theoretic information, that is with a multiplicity attached to each of the cones in the polyhedral fan $\tau(Y)$, as developed in Sturmfels, Tevelev (2008).

The main theorems in Dickenstein, Feichtner, Sturmfels (2007) allow us to compute the tropicalization of the discriminantal varieties. Theorem 1.1 in that paper asserts that the tropicalization of the A-discriminant is the Minkowski sum of the tropicalization $\mathcal{B}(A)$ of the kernel of A (called the *co-Bergman fan of A*) and the (classical) row space of the $d \times N$-matrix A. This is the tropical version of the Horn-Kapranov parametrization in Proposition 2.6. The tropical linear space $\mathcal{B}(A)$ can be computed as follows. Consider the geometric lattice $\mathcal{L}(A)$ whose elements are the sets of zero-entries of the vectors in the kernel, ordered by inclusion, and denote by $\mathcal{C}(A)$ the set of proper maximal chains in $\mathcal{L}(A)$. Represent these chains as $(N-d-1)$-element subsets of $\{0,1\}^N$. The tropicalization of the kernel of A equals

$$\mathcal{B}(A) := \tau(\text{kernel}(A)) = \bigcup_{\sigma \in \mathcal{C}(A)} \mathbb{R}_{\geq 0}\, \sigma \,.$$

This polyhedral complex only depends on the pattern of zero and non-zero minors of A, see Sturmfels (2002), and it is easy to describe explicitly when the matrix A is uniform, i.e., when all its maximal minors are non-zero. This is not the case for Cayley matrices when $s > 1$, but one can adapt the hypothesis of uniformity to get a general description when all minors which could be non-zero are in fact non-zero.

Theorem 1.2 in Dickenstein, Feichtner, Sturmfels (2007) allows to recover the extreme monomials of the A-discriminant, from which we can read its degree. The exponent of x_i in the initial monomial $\text{in}_w(\Delta_A)$ equals the number of intersection points of the halfray

$$w + \mathbb{R}_{>0}e_i$$

with the tropical discriminant $\tau(X_A^*)$, counting multiplicities:

$$\deg_{x_i}\big(\text{in}_w(\Delta_A)\big) = \sum_{\sigma \in \mathcal{B}(\ker A)_{i,w}} \big|\, \det\big(A^T, \sigma_1, \ldots, \sigma_{N-d-1}, e_i\big)\,\big| \,.$$

where $\mathcal{B}(\ker A)_{i,w}$ is the subset of $\mathcal{C}(A)$ consisting of all chains such that the row space of the matrix A has non-empty intersection with the cone $\mathbb{R}_{>0}\{\sigma_1, \ldots, \sigma_{N-d-1}, -e_i, -w\}$.

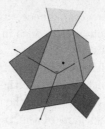

Fig. 2.2. Rays from w intersecting the tropical discriminant

This result provides an algorithm, which we implemented, to compute the degree of the A-discriminant (while the previous result can be turned into an algorithm to compute its dimension). In the smooth case, there are intrinsic formulas for the degree of the dual variety by Katz, Kleiman and Holme, see Gelfand, Kapranov, Zelevinsky (1994), and recently Matsui, Takeuchi (2008) addressed the singular case.

We illustrate these theorems in the simplest meaningful case.

Example 2.3 (The discriminant of a cubic polynomial in 1 variable) Consider the 2×4 integer matrix

$$A := \begin{pmatrix} 1 & 1 & 1 & 1 \\ 0 & 1 & 2 & 3 \end{pmatrix}.$$

Its associated toric variety X_A is known as the *twisted cubic*. A generic polynomial with exponents in A, in variables (s, t), is of the form:

$$f_A(x; s, t) = s\,(x_1 t^0 + x_2 t^1 + x_3 t^2 + x_4 t^3).$$

The associated A-discriminant Δ_A coincides with the discriminant of a generic polynomial $f_A(x; 1, t)$ of degree 3 in one variable t, which is in this case easy to compute:

$$\Delta_A = 27\,x_1^2 x_4^2 + 4\,x_1 x_3^3 + 4\,x_2^3 x_4 - x_2^2 x_3^2 - 18\,x_1 x_2 x_3 x_4.$$

The Newton polytope $N(\Delta_A)$ is the convex hull of the exponent vectors $\alpha_1 = (2, 0, 0, 2), \alpha_2 = (1, 0, 3, 0), \alpha_3 = (0, 3, 0, 1),\ \alpha_4 = (0, 2, 2, 0), \alpha_5 = (1, 1, 1, 1)$, in $\mathbb{R}^4$. Note however that the discriminant has two homogeneities read in the rows of the matrix A: the linear functions

$$\langle (1, 1, 1, 1), \alpha_i \rangle, \quad \langle (0, 1, 2, 3), \alpha_i \rangle$$

take the same values (4 and 6, respectively) for any $i = 1, \ldots, 5$. So,

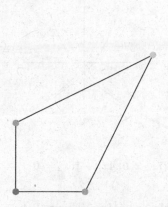
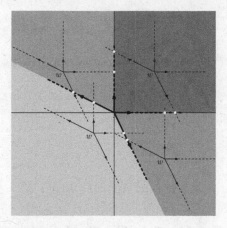

Fig. 2.3. Newton polygon, tropicalization and extreme monomials of the discriminant of a degree 3 polynomial: $x_2^3 x_4$, $x_2^2 x_3^2$, $x_1 x_3^3$, $x_1^2 x_4^2$, enumerated in counterclockwise order starting at the left upper vertex

$N(\Delta_A)$ is a polygon lying in a two dimensional plane in $\mathbb{R}^4$. It is straightforward to check that $\alpha_1, \ldots, \alpha_4$ are vertices and α_5 is an interior point.

We want to explain how the previous theorems allow us to predict the monomials in Δ_A, even if we could not compute them (as happens in the general case). We can check, for instance, that the vector $(-1, -1, -1, 0) = (0, 0, 0, 1) + (-1, -1, -1, -1)$ lies in the tropicalization $\tau(\{\Delta_A = 0\}) = \tau(\Delta_A)$ because $\mathrm{in}_{(-1,-1,-1,0)}(\Delta_A) = 4 x_1 x_3^3 - x_2^2 x_3^2$ is not a monomial. Theorem 1.1 allows to compute the tropicalization

$$\tau(\Delta_A) = \left(\cup_{i=1}^4 \mathbb{R}_{\geq 0} e_i \right) + \langle (1, 1, 1, 1), (0, 1, 2, 3) \rangle,$$

(and all the multiplicities are equal to 1). To visualize it in Figure 2.3, we mod out by the row space of A to obtain a two-dimensional representation. We choose a basis $\{(1, -2, 1, 0), (0, 1, -2, 1)\}$ of the kernel of A and we project $\pi : \tau(\Delta_A) \to \mathbb{R}^2$, so that $b_1 := \pi(e_1) = (1, 0), b_2 := \pi(e_2) = (-2, 1), b_3 := \pi(e_3) = (1, -2), b_4 := \pi(e_4) = (0, 1)$, and the image of the tropicalization is the union of the four positive rays generated by these vectors. To compute a monomial in Δ_A we pick a point $w \notin \pi(\tau(\Delta_A))$, for instance a point in the interior of the positive cone $\mathcal{C}$ generated by b_2 and b_3. We now "place" the projection $\pi(\tau(\Delta_A))$ at w and we see which rays emanating from there intersect $\pi(\tau(\Delta_A))$. The intersections are given by the point $(w + \mathbb{R}_{\geq 0} b_1) \cap \mathbb{R}_{\geq 0} b_3$, with $|\det(b_1, b_3)| = 2$ and by the point $(w + \mathbb{R}_{\geq 0} b_4) \cap \mathbb{R}_{\geq 0} b_2$, with $|\det(b_2, b_4)| = 2$. Therefore, the vertex of $N(\Delta_A)$ dual to $\mathcal{C}$ is the point $(2, 0, 0, 2)$.

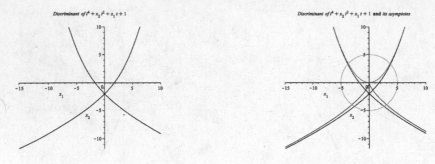

Fig. 2.4. Left: $\Delta_4 = 0$,   Right: $\Delta_4 = 0$, $4x_2^3 + 27x_1^2 = 0$, $4x_1^2 - 16x_2 = 0$

Example 2.4 (The discriminant of a quartic equation) Figure 2.4 shows, again in the simplest meaningful example, how the initial polynomials with respect to the weights in the tropicalization provide good asymptotic approximations. In order to be able to make a planar drawing, we consider the specialized family of polynomials $t^4 + x_2 t^2 + x_1 t + 1$, depending on two parameters (x_1, x_2). The corresponding discriminant equals

$$\Delta_4 = 4x_2^3 x_1^2 + 27x_1^4 - 16x_2^4 + 128x_2^2 - 144x_2 x_1^2 - 256.$$

The weights $w_1 = (-3, -2)$ and $w_2 = (-1, 2)$ lie in the tropicalization $\tau(\Delta_4)$ and the corresponding initial polynomials equal $in_{w_1}(\Delta_4) = x_1^2(4x_2^3 + 27x_1^2)$, $in_{w_2}(\Delta_4) = x_2^3(4x_1^2 - 16x_2)$. For (x_1, x_2) outside the ball of radius 5, the zero sets of these initial polynomials give good approximations of the discriminant locus $\Delta_4 = 0$. This qualitative behaviour is nicely explained in the general case in more abstract terms in Khovanskii (1988), Tevelev (2007). It would be interesting to get accurate quantitative formulations.

Another application of mixed discriminants is the determination of real roots. As we mentioned in Example 2.1, the number of real roots of real polynomials is constant when the coefficients vary on each connected component of the complement of the zeros of the real mixed discriminant, given that for the number of real roots to increase, two complex roots should merge.

Example 2.5 (Example 2.2 revisited) The two parameter family of

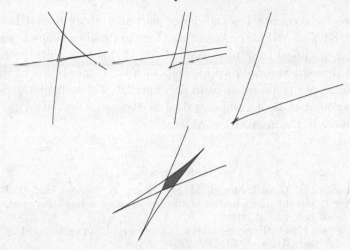

Fig. 2.5. Values of (a, b) in the dark region give 5 real roots

real bivariate trinomials

$$h_1(a, b; t, s) = t^6 + a\,s^3 - s$$
$$h_2(a, b; t, s) = s^6 + b\,t^3 - t$$

gives a simple family of bivariate trinomials which have the maximum possible number 5 (Li, Rojas, Wang (2003)) of real roots for $a = b = \frac{44}{31}$ (Dickenstein, Rojas, Rusek, Shih (2007)). In fact, the area of the set of points $(a, b) \in \mathbb{R}^2$ such that the system has 5 positive real simple roots is smaller than 5.701×10^{-7}. An implicit plot of $\Delta_A(1, a, -a, 1, b, -1) = 0$ has poor quality, but instead we can draw it efficiently using the dehomogenized version of the Horn-Kapranov parametrization in Proposition 2.6. Figure 2.5 shows a sequence of 4 plots, drawn on a logarithmic scale and successively magnified up to a factor of about 1700, of the real part of the discriminant variety $\Delta_A(1, a, -1, 1, b, -1) = 0$. The recent work Moroz, Rouillier (2008) explores generalizations of this result. Moreover, we get in Dickenstein, Rojas, Rusek, Shih (2007) an explicit good upper bound on the number of chambers of the complement of the real points in a dehomogenized A-discriminant (not just a mixed discriminant) for general configurations A of codimension two, i.e., $n + 3$ general lattice points in $\mathbb{Z}^n$. This number is smaller than $\frac{26}{5}(n + 4)^6$, which is independent of the coordinates of A.

Acknowledgements I acknowledge partial support from UBACYT, CONICET and ANPCyT, Argentina. I am grateful to Felipe Cucker for his bibliographical help, to Enrique Tobis for his help with the figures, and to Bernard Mourrain for his help with the computation of the singular locus of the discriminant in Example 2.1. I also thank the Society for the Foundations of Computational Mathematics for the invitation to participate in the meeting FoCM'08.

References

L. Adleman, M. Gopalkrishnan, M.-D. Huang, P. Moisset and D. Reishus (2008), 'On the mathematics of the law of mass action', Preprint. Available at: arXiv:0810.1108.

A. Adolphson (1994),'Hypergeometric functions and rings generated by monomials', *Duke Math. J.* **73**, 269–290.

Axel Modeler (2008), `http://axel.inria.fr`.

R. Bieri and J. Groves (1984), 'The geometry of the set of characters induced by valuations', *J. Reine Angewandte Math.* **347**, 168-195.

F. Bihan and F. Sottile (2007), 'New fewnomial upper bounds from Gale dual polynomial systems', *Mosc. Math. J.* **7**, 387–407.

F. Bihan and F. Sottile (2008), 'Gale duality for complete intersections', *Ann. Inst. Fourier* **58**, 877–891.

L. Blum, M. Shub and S. Smale (1989), 'On a theory of computation and complexity over the real numbers: NP-completeness, recursive functions and universal machines', *Bull. Amer. Math. Soc.* **21**, 1–46.

M. Bourel, A. Dickenstein and A. Rittatore (2008), 'Self-dual projective toric varieties', preprint. Available at: arXiv:0805.3259.

E. Cattani and A. Dickenstein (2007), 'Counting solutions to binomial complete intersections', *J. Complexity* **23**, 82–107.

J. Chipalkatti and C. D'Andrea, with an appendix by A. Abdesselam,(2007), 'On the Jacobian ideal of the binary discriminant', *Collect. Math.* **58**, 155–180.

P. Conti and C. Traverso (1991), 'Buchberger algorithm and linear programming', in: Applied Algebra, Algebraic Algorithms, and Error Protecting Codes, *Lect. Notes in Comp. Sci.* **539**, 130–139, Springer-Verlag.

C. Conradi, D. Flockerzi, J. Raisch and J. Stellin (2007), 'Subnetwork analysis reveals dynamic features of complex (bio)chemical networks', *PNAS* **104**, 19175–19180.

G. Craciun, A. Dickenstein, A. Shiu and B. Sturmfels (2007), 'Toric dynamical systems', to appear: *J. Symbolic Comp.*.

C. Delorme (1976), 'Sous-monoïdes d'intersection complète de N', *Ann. Sci. Ecole Norm. Sup.* **4**, 145–154.

J. Demmel (1987), 'On condition numbers and the distance to the nearest ill-posed problem', *Numer. Math.* **51**, 251–289.

P. Diaconis and B. Sturmfels (1998), 'Algebraic algorithms for generating from conditional distributions', *Ann. Stat.* **26** , 363–397.

A. Dickenstein, E. M. Feichtner and B. Sturmfels (2007), 'Tropical discriminants, *J. Amer. Math. Soc.* **20**, 1111–1133.

A. Dickenstein, L. F Matusevich and E. Miller (2006), 'Binomial D-modules', Preprint. Available at: arXiv.org:math/0610353.

A. Dickenstein, L. F Matusevich and E. Miller (2008), 'Combinatorics of binomial primary decomposition', to appear: *Math. Z..*

A. Dickenstein, L. F. Matusevich and T. Sadykov (2005), 'Bivariate hypergeometric D-modules', *Adv. Math.* **196**, 78–123.

A. Dickenstein, J. M. Rojas, K. Rusek and J. Shih (2007), 'Extremal real algebraic geometry and A-discriminants', *Mosc. Math. J.* **7**, Special Issue on the occasion of Askold Khovanskii's 60th birthday.

A. Dickenstein and B. Sturmfels (2002), 'Elimination theory in codimension two', *J. Symbolic Comp.* **34**, 119–135.

M. Drton, S. Sullivant and B. Sturmfels (2009), *Oberwolfach Lectures on Algebraic Statistics*, Oberwolfach Seminars, Birkhäuser.

D. Eisenbud and B. Sturmfels (1996), 'Binomial ideals', *Duke Math. J.* **84**, 1–45.

M. Einsiedler, M. Kapranov and D. Lind (2006), 'Non-archimedean amoebas and tropical varieties', *J. Reine Angewandte Math.* **601**, 139–157.

M. Feinberg (1979), *Lectures on Chemical Reaction Networks*. Notes of lectures given at the Mathematics Research Center of the University of Wisconsin in 1979. Available at:
`www.che.eng.ohio-state.edu/~FEINBERG/LecturesOnReactionNetworks`

M. Feinberg (1989), 'Necessary and sufficient conditions for detailed balancing in mass action systems of arbitrary complexity', *Chem. Eng. Sci.* **44**, 1819–1827.

K. Gatermann (2001), 'Counting stable solutions of sparse polynomial systems in chemistry', *Contemp. Math.* **286**, Symbolic Computation: Solving Equations in Algebra, Geometry and Engineering, (Editors E. Green et al.), 53–69.

K. Gatermann and M. Wolfrum (2005), 'Bernstein's second theorem and Viro's method for sparse polynomial systems in chemistry', *Adv. Appl. Math.* **34**, 252–294.

D. Geiger, C. Meek and B. Sturmfels (2006), 'On the toric algebra of graphical models', *Ann. Stat.* **34**, 1463–1492.

I. M. Gelfand, M. I. Graev and V. S. Retakh (1992), 'General hypergeometric systems of equations and series of hypergeometric type', *Russian Math. Surveys* **47**, 1–88.

I. M. Gelfand, M. M. Kapranov and A. V. Zelevinsky (1989), 'Hypergeometric functions and toric varieties', *Funktsional. Anal. i Prilozhen.* **23**, 12–26.

I. M. Gelfand, M. M. Kapranov and A. V. Zelevinsky (1990), 'Generalized Euler integrals and A-hypergeometric function', *Adv. Math.* **84**, 255–271.

I. M. Gelfand, M. M. Kapranov and A. V. Zelevinsky (1994), *Discriminants, Resultants and Multidimensional Determinants*, Birkhäuser.

J. Gunawardena, 'Chemical reaction network theory for in-silico biologists', Technical Report, 2003. Available at:
`http://vcp.med.harvard.edu/papers/crnt.pdf`.

F. Horn and R. Jackson (1972), 'General mass action kinetics', *Arch. Rat. Mech. Anal.* **47**, 81–116.

F. Horn (1974), 'The dynamics of open reaction systems. Mathematical aspects of chemical and biochemical problems and quantum chemistry', *SIAM-AMS Proceedings*, **VIII**, 125–137.

B. Huber and B. Sturmfels (1995), 'A polyhedral method for solving sparse polynomial systems', *Math. Comp.* **64**, 1541–1555

A. N. Jensen (2008), 'Computing Gröbner fans and tropical varieties in Gfan', in: *Software for Algebraic Geometry*, IMA Volumes in Mathematics and Applications, Vol. 148, M. Stillman, J. Verschelde and N. Takayama (Eds.), Springer, New York, 33–46.

R. Kannan and A. Bachem (1979), 'Polynomial algorithms for computing the Smith and Hermite normal forms of an integer matrix', *SIAM J. Comp.* **8**, 499–507.

M. Kapranov (1991), 'A characterization of A-discriminantal hypersurfaces in terms of the logarithmic Gauss map', *Math. Ann.* **290**, 277–285.

A. G. Khovanskii (1988), 'Algebra and mixed volumes', Addendum 3 in the book: Yu. D. Burago and V. A. Zalgaller, *Geometric Inequalities*, Grundlehren, 285. Springer-Verlag, Berlin.

T. Y. Li, J. M. Rojas and X. Wang (2003), 'Counting real connected components of trinomial curves intersections and m-nomial hypersurfaces', *Discrete Comp. Geom.* **30**, 379–414.

L. Lovász (1994), 'Stable sets and polynomials', Graphs and Combinatorics (Qawra, 1990), *Discrete Math.* **124**, 137–153.

G. Malajovich and J. M. Rojas (2004), 'High probability analysis of the condition number of sparse polynomial systems', *Theor. Comp. Sci.* **315**, 525–555.

Mathemagix (2008) 'http://www.mathemagix.org/'.

E. W. Mayr and A. R. Meyer (1982), 'The complexity of the word problems for commutative semigroups and polynomial ideals', *Adv. Math.* **46**, 305–329.

E. Miller and B. Sturmfels (2005), *Combinatorial Commutative Algebra*, Graduate Texts in Mathematics, vol. 227, Springer-Verlag, New York.

G. Moroz and F. Rouillier (2008), Extended abstract, ADG 2008, Shangai.

I. Ojeda and R. Piedra (2000), 'Cellular binomial ideals. Primary decomposition of binomial ideals', *J. Symbolic Comp.* **30**, 383–400.

J. S. Provan and M. O. Ball (1983), 'The complexity of counting cuts and of computing the probability that a graph is connected', *SIAM J. Comput.* **12**, 777–788.

L. Pachter and B. Sturmfels, editors, (2005), *Algebraic Statistics for Computational Biology*, Cambridge University Press, Cambridge.

Y. Rabinovich, A. Sinclair and A. Wigderson (1992), 'Quadratic dynamical systems', *Proc. 33rd Annual Symposium on Foundations of Computer Science (FOCS)*, 304–313.

M. Saito, B. Sturmfels and N. Takayama (2000), *Gröbner Deformations of Hypergeometric Differential Equations*, Springer-Verlag, Berlin.

M. Shub and S. Smale (1993), 'Complexity of Bézout's theorem. I. Geometric aspects', *J. Amer. Math. Soc.* **6**, 459–501.

E. Sontag (2001), 'Structure and stability of certain chemical networks and applications to the kinetic proofreading model of T-cell receptor signal transduction', *IEEE Trans. Automat. Control* **46**, 1028–1047.

B. Sturmfels (1994), 'On the Newton polytope of the resultant', *J. Alg. Comb.* **3**, 207–236.

B. Sturmfels (1996), *Gröbner Bases and Convex Polytopes*, Volume 8, University Lecture Series, AMS, Providence, RI.

B. Sturmfels (1997), 'Equations defining toric varieties', Algebraic Geometry (Santa Cruz 1995), *Proc. Sympos. Pure Math.* **62**, Part 2, AMS, Providence, RI, 437–449.

B. Sturmfels (2002), *Solving Systems of Polynomial Equations*, CBMS Regional Conference Series in Mathematics 97, AMS, Providence, RI.

B. Sturmfels and E. Tevelev (2008), 'Elimination Theory for Tropical Varieties', *Math. Research Letters* **15**, 543–562.

A. Szenes and M. Vergne (2004), 'Toric reduction and a conjecture of Batyrev and Materov', *Invent. Math.* **158**, 453–495.

Y. Matsui and K. Takeuchi (2008), 'A geometric degree formula for A-discriminants and Euler obstructions of toric varieties', preprint. Available at: arXiv:0807.3163.

E. Tevelev (2007), 'Compactifications of subvarieties of tori', *Amer. J. Math* **129**, 1087–1104

R. Thomas (1995),'A geometric Buchberger algorithm for integer programming', *Math. Oper. Res.* **20**, 864–884.

L. G. Valiant (1979), 'The complexity of enumeration and reliability problems', *SIAM J. Comput.* **8**, 410–421.

J. Verschelde, P. Verlinden, and R. Cools (1994), 'Homotopies exploiting Newton polytopes for solving sparse polynomial systems', *SIAM J. Numer. Anal.* **31**, 915–930.

J. Weyman and A. Zelevinsky (1996), 'Singularities of hyperdeterminants', *Ann. Inst. Fourier, Grenoble* **46**, 591–644

J. H. Wilkinson (1994), *Rounding Errors in Algebraic Processes*, Dover.

3

Linear and Nonlinear Subdivision Schemes in Geometric Modeling

Nira Dyn

School of Mathematical Sciences
Tel Aviv University
Tel Aviv, Israel
e-mail: niradyn@post.tau.ac.il

Abstract

Subdivision schemes are efficient computational methods for the design, representation and approximation of 2D and 3D curves, and of surfaces of arbitrary topology in 3D. Such schemes generate curves/surfaces from discrete data by repeated refinements. While these methods are simple to implement, their analysis is rather complicated.

The first part of the paper presents the "classical" case of linear, stationary subdivision schemes refining control points. It reviews univariate schemes generating curves and their analysis. Several well-known schemes are discussed.

The second part of the paper presents three types of nonlinear subdivision schemes, which depend on the geometry of the data, and which are extensions of univariate linear schemes. The first two types are schemes refining control points and generating curves. The last is a scheme refining curves in a geometry-dependent way, and generating surfaces.

3.1 Introduction

Subdivision schemes are efficient computational tools for the generation of functions/curves/surfaces from discrete data by repeated refinements. They are used in geometric modeling for the design, representation and approximation of curves and of surfaces of arbitrary topology. A linear stationary scheme uses the same linear refinement rules at each location and at each refinement level. The refinement rules depend on a finite number of *mask coefficients*. Therefore, such schemes are easy to implement, but their analysis is rather complicated.

The first subdivision schemes were devised by G. de Rahm (1956)

68

for the generation of functions with a first derivative everywhere and a second derivative nowhere. In fact, in most cases, the limits generated by subdivision schemes are convolutions of a smooth function with a fractal one.

This paper consists of two parts. The first part is a review of "classical" subdivision schemes, namely linear, stationary schemes refining control points. The second part considers several constructions of nonlinear schemes, taking into account the geometry of the refined objects. The nonlinear schemes are all extensions of "good" linear univariate schemes.

In the first part we review only linear univariate subdivision schemes (schemes generating curves from initial points). Although the main application of subdivision schemes is in generating surfaces, we limit the presentation here to these schemes. The main reasons for this choice are two. The theory of univariate schemes is much more complete and easier to present, and it contains most of the aspects of the theory of schemes generating surfaces. The understanding of this theory is a first essential step towards the understanding of the theory of schemes generating surfaces. Also, the schemes that are extended in the second part to nonlinear schemes are univariate.

Section 3.2 is devoted to the presentation of stationary linear univariate subdivision schemes. Important examples such as the B-spline schemes and the interpolatory 4-point scheme are discussed in detail. Among these important examples are the schemes that are extended in the second part to nonlinear schemes. A sketch of tools used for analysis of convergence and smoothness of linear univariate schemes is given in §3.2.2. The relation between subdivision schemes and the construction of wavelets is briefly discussed in §3.2.3.

The second part of the paper consists of three sections. In Section 3.3 linear schemes are extended to refine manifold-valued data. This is done in two steps. First, the refinement rules of any convergent linear scheme are presented (in several possible ways) in terms of repeated binary averages. This is demonstrated by several examples in §3.3.1. Then in §3.3.2 the manifold-valued subdivision schemes are constructed, either by replacing every linear binary average in the linear refinement rules by a geodesic average, yielding a geodesic analogous scheme, or by replacing every linear binary average in the linear refinement rules by its projection to the manifold, yielding a projection analogue scheme. The manifold-valued subdivision schemes constructed in this way are analyzed in §3.3.3 via their *proximity* to the linear schemes from which they are derived.

In Section 3.4 two data-dependent extensions of the linear interpolatory 4-point scheme are discussed. In both extensions the refinement is adapted to the local geometry of the four points used. These two geometric (data-dependent) 4-point schemes are effective in case of an initial control polygon with edges of significantly different length. For such initial control polygons the limits generated by the linear 4-point scheme have unwanted features (artifacts), while the geometric 4-point schemes tend to generate artifact free limits.

The last section deals with repeated refinements of curves for the generation of a surface. Here the scheme that is extended is the quadratic B-spline scheme (called the Chaikin algorithm). It is used to refine curves by first constructing a *geometric correspondence* between pairs of curves, and then applying the refinement to corresponding points. Since this is work in progress only partial results are presented.

3.2 Linear Subdivision Schemes for the Generation of Curves

In this section we discuss stationary, linear schemes, generating curves by repeated refinements of points. Such a subdivision scheme is defined by a finite set of real coefficients called a *mask*

$$\mathbf{a} = \{a_i \in \mathbb{R}, \ i \in \sigma(\mathbf{a}) \subset \mathbb{Z}\},$$

where $\sigma(\mathbf{a})$ denotes the finite support of the mask. The scheme with the mask $\mathbf{a}$ is denoted by $S_{\mathbf{a}}$.

The refinement rules of $S_{\mathbf{a}}$ have the form

$$(S_{\mathbf{a}}\mathcal{P})_\alpha = \sum_{\beta \in \mathbb{Z}} a_{\alpha - 2\beta} P_\beta, \quad \alpha \in \mathbb{Z}, \tag{3.1}$$

where $\mathcal{P}$ denotes the polygonal line through the points $\{P_i\}_{i \in \mathbb{Z}} \subset \mathbb{R}^d$, with $d \geq 1$.

Note that there are two different refinement rules in (3.1), one corresponding to odd α and one to even α involving the odd, respectively even, coefficients of the mask.

Remark 3.1 In this section we consider schemes operating on points defined on $\mathbb{Z}$, although, in geometric applications the schemes operate on finite sets of points. Due to the finite support of the mask, our considerations apply directly to closed curves and also to 'open' ones, except in a finite zone near the boundary.

The initial points to be refined are called *control points*, and the corresponding polygonal line through them is called a *control polygon*. These terms are used also for the points and the corresponding polygonal line at each refinement level. The subdivision scheme is a sequence of repeated refinements of control polygons,

$$\{\mathcal{P}^{k+1} = S_\mathbf{a}\mathcal{P}^k, \ k \geq 0\}, \tag{3.2}$$

with $\mathcal{P}^0$ the initial control polygon. The refinements in (3.2) are stationary, since at each refinement level the same scheme $S_\mathbf{a}$ is used. Material on nonstationary subdivision schemes can be found in the review paper Dyn & Levin (2002).

A subdivision scheme is termed *uniformly convergent* if the sequence of control polygons $\{\mathcal{P}^k\}$ converges uniformly on compact sets of $\mathbb{R}^d$. This is the notion of convergence relevant to geometric applications.

Each control polygon in (3.2) has a parametric representation as the vector in $\mathbb{R}^d$ of the piecewise linear interpolant to the data $\{(i2^{-k}, P_i^k),$ $i \in \mathbb{Z}\}$, and the uniform convergence is analyzed in the setting of continuous functions (see, e.g., Dyn (1992)).

For $\mathcal{P}^0 \subset \mathbb{R}^d$ with $d = 1$, the scheme converges to a univariate function, while in the case $d \geq 2$ the scheme converges to a curve in $\mathbb{R}^d$.

The convergence of a scheme $S_\mathbf{a}$ implies the existence of a *basic limit function* $\phi_\mathbf{a}$, which is the limit obtained from the initial data, $\delta_i^0 = 0$ everywhere on $\mathbb{Z}$ except $\delta_0^0 = 1$.

It follows from the linearity and uniformity of (3.1) that the limit $S_\mathbf{a}^\infty \mathcal{P}^0$, obtained from any initial control polygon $\mathcal{P}^0$ passing through the initial control points $\{P_\alpha^0 \in \mathbb{R}^d, \ \alpha \in \mathbb{Z}\}$, can be written in terms of integer translates of $\phi_\mathbf{a}$, as

$$(S_\mathbf{a}^\infty \mathcal{P}^0)(t) = \sum_{\alpha \in \mathbb{Z}} P_\alpha^0 \phi_\mathbf{a}(t - \alpha), \quad t \in \mathbb{R}. \tag{3.3}$$

Equation (3.3) is a parametric representation of a curve in $\mathbb{R}^d$ for $d \geq 2$.

By the stationarity (3.2) of the subdivision scheme, $S_\mathbf{a}^\infty \delta^1 = \phi_\mathbf{a}(2t)$ with $\delta_i^1 = 0$ for all $i \in \mathbb{Z}/2$ except $\delta_0^1 = 1$. Thus we get from (1.3) and the observation

$$S_\mathbf{a}^\infty \delta^0 = S_\mathbf{a}^\infty (S_\mathbf{a} \delta^0) = S_\mathbf{a}^\infty \Big(\sum_{\alpha \in \sigma(\mathbf{a})} a_\alpha \delta_{\cdot - \alpha}^1 \Big),$$

that $\phi_{\mathbf{a}}$ satisfies the *refinement equation* (*two-scale relation*)

$$\phi_{\mathbf{a}}(t) = \sum_{\alpha \in \mathbb{Z}} a_\alpha \phi_{\mathbf{a}}(2t - \alpha). \tag{3.4}$$

It is easy to obtain from (3.1) or from (3.4) that the support of $\phi_{\mathbf{a}}$ equals the convex hull of $\sigma(\mathbf{a})$ (see, e.g., Cavaretta, Dahmen & Micchelli (1991)).

Remark 3.2 The discussion above leads to the conclusion that a convergent subdivision scheme gives rise to a unique continuous basic limit function, satisfying a refinement equation with the mask of the scheme. The converse is not generally true. Yet a continuous function satisfying a refinement equation is a basic limit function of a convergent subdivision scheme if its integer shifts are linearly independent: see Cavaretta, Dahmen & Micchelli (1991).

3.2.1 The main types of schemes

The first subdivision schemes in geometric modelling were proposed for easy and quick rendering of B-spline curves. A B-spline curve has the form

$$C(t) = \sum_i P_i B_m(t - i) \tag{3.5}$$

with $\{P_i\} \subset \mathbb{R}^d$ the control points, and B_m a B-spline of degree m with integer knots, namely $B_m|_{[i,i+1]}$ is a polynomial of degree m for each $i \in \mathbb{Z}$, $B_m \in C^{m-1}(\mathbb{R})$, and B_m has compact support $[0, m+1]$.

Equation (3.5) is a parametric representation of a B-spline curve. The B-spline curves (3.5) are a powerful design tool, since their shape is similar to the shape of the control polygon $\mathcal{P}$ corresponding to the control points in (3.5): see, e.g., Prautzsch, Boehm & Paluszny (2002).

Such curves can be well approximated by the control polygons generated by the repeated refinements (3.2), using the mask $\mathbf{a}^{[m]}$ with coefficients

$$a_i^{[m]} = 2^{-m} \binom{m+1}{i}, \quad i = 0, \ldots, m+1. \tag{3.6}$$

The repeated refinements of a B-spline scheme of degree m are thus

$$P_i^{\ell+1} = \sum_j a_{i-2j}^{[m]} P_j^\ell, \quad i \in \mathbb{Z}, \quad \ell = 0, 1, 2, \ldots . \tag{3.7}$$

In (3.7) we use the convention $a_i^{[m]} = 0$, $i \notin \{0, 1, \ldots, m+1\}$.

By the convergence analysis, presented in §3.2.2, it is easy to check that the subdivision scheme with the mask (3.6) is convergent. It converges to B-spline curves of degree m, since B_m is the basic limit function of $S_{\mathbf{a}^{[m]}}$.

This can be easily concluded, in view of Remark 3.2, from the definition of B-splines, by observing that B_m satisfies a refinement equation (3.4) with the mask $\mathbf{a}^{[m]}$ (see, e.g., Dyn (1992)), and that the integer translates of B_m are linearly independent.

Thus the control polygon $\mathcal{P}^k = S_{\mathbf{a}}^k \mathcal{P}$ approximates $C(t)$ for k large enough, and $C(t)$ can be easily rendered by rendering its approximation $\mathcal{P}^k$. In practice a large enough k is about 4 as can be seen in Figure 3.1.

The first scheme of this type was devised in Chaikin (1974) for a fast geometric rendering of quadratic B-spline curves. It has the refinement rules

$$P_{2i}^{k+1} = \frac{3}{4}P_{i-1}^k + \frac{1}{4}P_i^k, \quad P_{2i+1}^{k+1} = \frac{1}{4}P_{i-1}^k + \frac{3}{4}P_i^k. \quad (3.8)$$

Figure 3.1 illustrates three refinement steps with this scheme, applied to a closed initial control polygon. Chaikin scheme is extended to nonlinear schemes in §3.3 and in §3.5.

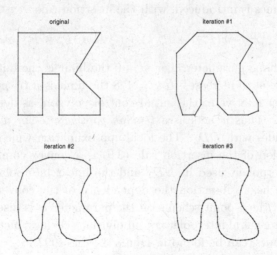

original   iteration #1

iteration #2   iteration #3

Fig. 3.1. Refinements of a polygon with Chaikin scheme

The schemes for general B-spline curves were introduced and investigated in Cohen, Lyche & Riesenfeld (1980). All other subdivision schemes can be regarded as generalizations of the B-spline schemes.

The B-spline schemes generate curves with a shape similar to the shape of the initial control polygons, but do not pass through the initial control points. Schemes with limit curves that interpolate the initial control points were introduced in the late 1980's. These schemes are called *interpolatory*, and have the even refinement rule $P_{2i}^{k+1} = P_i^k$. Thus the control points at refinement level k are contained in those at refinement level $k + 1$. The odd refinement rule, called in case of interpolatory schemes the *insertion rule*, is designed by a local approximation based on nearby control points.

The first schemes of this type were introduced by Deslauriers & Dubuc (1988) and by Dyn, Gregory & Levin (1987). In the first paper the insertion rule for the point P_{2i+1}^{k+1} is obtained by first interpolating the data $\{((i + j), P_{i+j}^k), \ j = -N + 1, \ldots, N\}$ by a vector polynomial of degree $2N - 1$, and then sampling it at the point $i + \frac{1}{2}$. This is done for a fixed positive N. Regarding N as a parameter, this construction yields a one-parameter family of convergent interpolatory subdivision schemes, with masks of increasing support, and with basic limit functions of increasing smoothness (see also the book Daubechies (1992)). We denote the scheme in this family corresponding to N by DD_N.

In the second paper above, a one-parameter family of 4-point interpolatory schemes is introduced, with the insertion rule

$$P_{2i+1}^{k+1} = -w(P_{i-1}^k + P_{i+2}^k) + (\frac{1}{2} + w)(P_i^k + P_{i+1}^k). \qquad (3.9)$$

Here w is a shape parameter. For $w = 0$ the limit is the initial control polygon, while as w increases to $w = 1/8$ the limit is a C^1 curve which becomes looser relative to the initial control polygon, as demonstrated in Figure 3.3. Thus w acts as a *tension parameter*. For $w = \frac{1}{16}$ this scheme coincides with DD_2. The local approximation which is used in the construction of the insertion rule (3.9) is a convex combination of the cubic interpolant used in DD_2 and the linear interpolant used in DD_1. In the next subsection the dependence of the convergence and smoothness of the 4-point scheme on the parameter w is discussed.

More about general interpolatory subdivision schemes, including multivariate schemes, can be found in Dyn & Levin (1992).

While in the case of the B-spline schemes the limit was known and convergence was guaranteed, in the case of the interpolatory schemes, analysis tools had to be developed. The analysis in Deslauriers & Dubuc (1988) and Daubecheis (1992) and in references therein is mainly in the Fourier domain, while in Dyn, Gregory & Levin (1987) it is done in the geometric domain, based on the symbol of the scheme. Hints on this method of analysis, which was further developed in Dyn, Gregory & Levin (1991), are given in the next subsection.

3.2.2 Analysis of convergence and smoothness

Given the coefficients of the mask of a scheme, one would like to be able to determine if the scheme is convergent, and what is the smoothness of the resulting basic limit function (which is the generic smoothness of the limits generated by the scheme in view of (3.3)). Such analysis tools are also essential for the design of new schemes. We sketch here the method of convergence and smoothness analysis in Dyn, Gregory & Levin (1991): see also Dyn (1992).

An important tool in the analysis is the symbol of a scheme $S_\mathbf{a}$ with the mask $\mathbf{a} = \{a_\alpha : \alpha \in \sigma(\mathbf{a})\}$,

$$a(z) = \sum_{\alpha \in \sigma(\mathbf{a})} a_\alpha z^\alpha, \tag{3.10}$$

introduced in Cavaretta, Dahmen & Micchelli (1991). In the following we use also the notation $S_{a(z)}$ for $S_\mathbf{a}$.

A first step towards the convergence analysis is the derivation of the necessary condition for uniform convergence,

$$\sum_{\beta \in \mathbb{Z}} a_{\alpha - 2\beta} = 1, \quad \alpha = 0 \text{ or } 1 \,(\mathrm{mod}\ 2), \tag{3.11}$$

The condition in (3.11) is derived easily from the refinement step

$$(S_\mathbf{a}^{k+1} \mathcal{P})_\alpha = \sum_{\beta \in \mathbb{Z}} a_{\alpha - 2\beta} (S_\mathbf{a}^k \mathcal{P})_\beta, \quad \alpha \in \mathbb{Z},$$

for k large enough so that for all $\ell \geq k$, $\|S_\mathbf{a}^\ell \mathcal{P} - S_\mathbf{a}^\infty \mathcal{P}\|_\infty$ is small enough.

The necessary condition (3.11) implies that we have to consider symbols satisfying

$$a(1) = 2, \quad a(-1) = 0. \tag{3.12}$$

Condition (3.12) is equivalent to

$$a(z) = (1 + z)q(z) \quad \text{with} \quad q(1) = 1. \tag{3.13}$$

The scheme with symbol $q(z)$, $S_\mathbf{q}$, satisfies $S_\mathbf{q} \Delta = \Delta S_\mathbf{a}$, where Δ is the difference operator

$$\Delta \mathcal{P} = \{(\Delta \mathcal{P})_i = P_i - P_{i-1}, \quad i \in \mathbb{Z}\}. \tag{3.14}$$

A necessary and sufficient condition for the convergence of S_a is the contractivity of the scheme S_q, namely S_a is convergent if and only if $S_q^\infty \mathcal{P} \equiv 0$ for any $\mathcal{P}$. The contractivity of S_q is equivalent to the existence of a positive integer L, such that $\|S_q^L\|_\infty < 1$. This condition can be checked for a given L by algebraic operations on the symbol $q(z)$.

From practical geometrical reasons, only small values of L have to be considered. A small value of L guarantees "visual convergence" of $\{\mathcal{P}^k\}$ to $S_a^\infty \mathcal{P}^0$, already for small k, as the distances between consecutive control points contract to zero fast. A good scheme corresponds to $L = 1$ as the B-spline schemes, or to $L = 2$ as the 4-point scheme.

The smoothness analysis relies on the result that if the symbol of a scheme has a factorization

$$a(z) = \left(\frac{1+z}{2}\right)^\nu b(z)\,,\tag{3.15}$$

such that the scheme S_b is convergent, then S_a is convergent and its limit functions are related to those of S_b by

$$D^\nu (S_a^\infty \mathcal{P}) = S_b^\infty \Delta^\nu \mathcal{P},\tag{3.16}$$

with D the differentiation operator.

Thus, each factor $(1 + z)/2$ multiplying a symbol of a convergent scheme adds one order of smoothness. This factor is termed the *smoothing factor*.

The relation between (3.15) and (3.16) is a particular instance of the "algebra of symbols" (see, e.g., Dyn & Levin (1995)). If $a(z), b(z)$ are two symbols of converging schemes, then S_c with the symbol $c(z) = \frac{1}{2} a(z) b(z)$ is convergent, and

$$\phi_c = \phi_a * \phi_b\,.\tag{3.17}$$

Example (*B*-spline schemes.)   In view of (3.6), the symbol of the scheme generating B-spline curves of degree m is

$$a(z) = (1 + z)^{m+1}/2^m\,.\tag{3.18}$$

All B-spline schemes with $m \geq 1$ converge, since $q(z) = \frac{(1+z)^m}{2^m}$ and $\|S_{q(z)}\|_\infty = \frac{1}{2}$. The known smoothness of the limit functions generated by the m-th degree B-spline scheme can be concluded easily, using the tools of analysis presented in this subsection. The factor $b(z) = \frac{(1+z)^2}{2}$ corresponds to the scheme S_b generating the initial control polygon as the limit curve, which is continuous, while the factors $\left(\frac{1+z}{2}\right)^{m-1}$ add smoothness, so that $S_{a^{[m]}}^\infty \mathcal{P}^0 \in C^{m-1}$ for $m \geq 1$.

Example (The 4-point scheme.) The analysis sketched above is the tool by which the following results were obtained for the interpolatory 4-point scheme with the insertion rule (3.9). The symbol of the scheme can be written as

$$a_w(z) = \frac{1}{2z}(z+1)^2 \left[1 - 2wz^{-2}(1-z)^2(z^2+1)\right]. \qquad (3.19)$$

The range of w for which $S_{a_w(z)}$ is convergent is the range for which $S_{q_w(z)}$ with symbol $q_w(z) = a_w(z)/(1+z)$ is contractive. The condition $\|S_{q_w(z)}\|_\infty < 1$ holds in the range $-3/8 < w < (-1+\sqrt{13})/8$, while the condition $\|S^2_{q_w(z)}\|_\infty < 1$ holds in the range $-3/8 < w < (-1+\sqrt{17})/8$. Thus $S_{a_w(z)}$ is convergent in the range $|w| < 3/8$. In fact it was shown by M. J. D. Powell that the convergence range is $|w| \leq \frac{1}{2}$.

To find a range of w where $S_{a_w(z)}$ generates C^1 limits, the contractivity of $S_{c_w(z)}$ with $c_w(z) = 2a_w(z)/(1+z)^2$ has to be investigated. It is easy to check that $\|S_{c_w(z)}\|_\infty \geq 1$, but that $\|S^2_{c_w(z)}\|_\infty < 1$ for $0 < w < (\sqrt{5}-1)/8$. Only a year ago the maximal positive w for which the limit is C^1 was obtained in Hechler, Mößner & Reif (2008).

The limit of $S_{a_w(z)}$ is not C^2 even for $w = 1/16$, although for $w = 1/16$ the symbol is divisible by $(1+z)^3$. It is shown in Daubechies & Lagarias (1992), by other methods, that the basic limit function for $w = 1/16$, restricted to its support, has a second derivative only at the non-dyadic points there.

The conditions for smoothness given above are only sufficient. Yet, there is a large class of convergent schemes for which the factorization in (3.15) is necessary for generating C^ν limit functions. This class contains the B-spline schemes and the interpolatory schemes. See, e.g., Dyn & Levin (2002).

3.2.3 Subdivision schemes and the construction of wavelets

Any convergent subdivision scheme $S_\mathbf{a}$ defines a sequence of nested spaces in terms of its basic limit function $\phi_\mathbf{a}$. For every $k \in \mathbb{Z}$ define the space

$$V_k = \text{span}\{\phi_\mathbf{a}(2^k(\cdot - i)), \quad i \in 2^{-k}\mathbb{Z}\}$$
$$= \text{span}\{\phi_\mathbf{a}(2^k \cdot -i), \quad i \in \mathbb{Z}\}. \qquad (3.20)$$

Then in view of the refinement equation (3.4) satisfied by $\phi_\mathbf{a}$, these spaces are nested, namely

$$\cdots \subset V_{-1} \subset V_0 \subset V_1 \subset V_2 \subset \cdots \qquad (3.21)$$

Such a bi-infinite sequence of spaces is the framework in which wavelets are constructed. It is called a multiresolution analysis when the integer translates of $\phi_{\mathbf{a}}$ constitute a Riesz basis of V_0 (see Daubechies (1992)). In the wavelets literature one starts from a solution of (3.4), termed a *scaling function*, which is not necessarily the limit of a converging subdivision scheme, as indicated by Remark 3.2.

The choice of the mask coefficients for the construction of wavelets depends on the properties required from the wavelets. For example the orthonormal wavelets of Daubechies are generated by masks (*filters*) that are related to the masks of the DD_N schemes. The symbols of these masks, denoted by $\tilde{a}_N(z)$, satisfy $|\tilde{a}_N(z)|^2 = a_N(z)$, with $a_N(z)$ the symbol of the scheme DD_N. This relation between the symbols can be expressed in terms of the corresponding scaling functions as

$$\phi_{\tilde{\mathbf{a}}_{\mathbf{N}}} * \phi_{\tilde{\mathbf{a}}_{\mathbf{N}}}(-\cdot) = \phi_{\mathbf{a}_{\mathbf{N}}} \; ,$$

showing that the integer translates of $\phi_{\tilde{\mathbf{a}}_{\mathbf{N}}}$ constitute an orthonormal system due to the interpolatory nature of $\phi_{\mathbf{a}_{\mathbf{N}}}$, which vanishes at all integers except at zero where it is 1.

The next sections bring several constructions of nonlinear schemes, based on linear schemes. In §3.3.1 more information on linear schemes, needed for the construction of schemes on manifolds, is presented.

3.3 Curve Subdivision Schemes on Manifolds

To design subdivision schemes for curves on a manifold, we require that the control points generated at each refinement level are on the manifold, and that the limit of the sequence of corresponding control polygons is also on the manifold. Such schemes are nonlinear.

The first approach to this problem is presented in Rahman, Drori, Stodden, Donoho & Schröder (2005). It is based on adapting a linear univariate subdivision scheme $S_{\mathbf{a}}$. Given control points $\{P_i\}$ on the manifold, let $\mathcal{P} = \{P_i\}$ denote the corresponding control polygon, and let T denote the adaptation of $S_{\mathbf{a}}$ to the manifold. Then the point $(T\mathcal{P})_i$ is defined by first executing the linear refinement step of $S_{\mathbf{a}}$ on the projections of the points $\{P_j, \; a_{i-2j} \neq 0\}$ to a tangent plane at a chosen point P_i^*, and then projecting the obtained point to the manifold. This can be written as

$$(T\mathcal{P})_i = \psi_{P_i^*}^{-1}\Big(\sum_{j \in \mathbb{Z}} a_{i-2j} \psi_{P_i^*}(P_j)\Big), \tag{3.22}$$

where P_i^* is some chosen "center" of the points $\{P_j\,,\ a_{i-2j} \neq 0\}$, and $\psi_{P_i^*}$ is the projection from the manifold to the tangent plan at P_i^*. Recently it was shown in Xie & Yu (2008a) and Xie & Yu (2008b), that with a proper choice of the "center" point many properties of the linear scheme, such as convergence, smoothness and approximation order, are shared by the nonlinear scheme derived from it.

Here we discuss two other constructions of subdivision schemes on manifolds from converging linear schemes. These constructions are based on the observation that the refinement rules of any convergent linear scheme can be calculated by repeated binary averages: see Wallner & Dyn (2005).

3.3.1 Linear schemes in terms of repeated binary averages

A linear scheme for curves, S, is defined by two refinement rules of the form,

$$(S\mathcal{P})_j = \sum_i a_{j-2i}P_i \,, \quad j = 0 \text{ or } 1 \ (\mathrm{mod}\ 2)\,, \tag{3.23}$$

where $\mathcal{P} = \{P_i\}$. As discussed in §3.2.2, any convergent linear scheme is affine invariant, namely $\sum_i a_{j-2i} = 1$. It is shown in Wallner & Dyn (2005), that for a convergent linear scheme each of the refinement rules in (3.23) is expressible, in a non-unique way, by repeated binary averages. A reasonable choice is a symmetric representation relative to the topological relations in the control polygon.

For example, with the notation $Av_\alpha(P,Q) = (1-\alpha)P + \alpha Q\,,\ \alpha \in \mathbb{R},\ P,Q \in \mathbb{R}^d$, the insertion rule of the interpolatory 4-point scheme (3.9) can be rewritten as

$$P_{2j+1}^k = Av_{\frac{1}{2}}\Big(Av_{(-2w)}(P_j, P_{j-1})\,,\ Av_{(-2w)}(P_{j+1}, P_{j+2})\Big)\,,$$

or as

$$P_{2j+1}^k = Av_{(-2w)}\Big(Av_{\frac{1}{2}}(P_j, P_{j+1})\,,\ Av_{\frac{1}{2}}(P_{j-1}, P_{j+2})\Big)\,.$$

Refinement rules represented in this way are termed hereafter *refinement rules in terms of repeated binary averages*.

Among the linear schemes there is a class of *factorizable schemes* for which the symbol $a(z) = \sum_i a_i z^i$ can be written as a product of linear real factors. For such a scheme, the control polygon obtained by one refinement step of the form (3.1) can be computed by several simple global steps, uniquely determined by the factors of the symbol.

To be more specific, let us consider a symbol of the form

$$a(z) = z^{-\ell}(1+z)\frac{1+x_1 z}{1+x_1} \cdots \frac{1+x_m z}{1+x_m} . \tag{3.24}$$

Note that this symbol corresponds to an affine invariant scheme since $a(1) = 2$, and $a(-1) = 0$. In fact, the form of the symbol in (3.24) is general for converging factorizable schemes.

Let $\mathcal{P}^k$ denote the control polygon at refinement level k, corresponding to the control points $\{P_i^k\}$. Each simple step in the execution of the refinement $S_a \mathcal{P}^k$ corresponds to one factor in (3.24). The first step in calculating the control points at level $k+1$ corresponds to the factor $1 + z$, and consists of elementary refinement:

$$P_{2i}^{k+1,0} = P_{2i+1}^{k+1,0} = P_i^k . \tag{3.25}$$

This step is followed by m averaging steps corresponding to the factors $\frac{1+x_j z}{1+x_j}$, $j = 1, \ldots, m$. The averaging step dictated by the factor $\frac{1+x_j z}{1+x_j}$ is,

$$P_i^{k+1,j} = \frac{1}{1+x_j}(P_i^{k+1,j-1} + x_j P_{i-1}^{k+1,j-1}), \quad i \in \mathbb{Z}. \tag{3.26}$$

The control points at level $k+1$ are $P_i^{k+1} = P_{i+\ell}^{k+1,m}$, $i \in \mathbb{Z}$.

The execution of the refinement step by several simple global steps is equivalent to the observation that

$$S_{a(z)}\mathcal{P}^k = \tau^\ell R_{\frac{1+x_1 z}{1+x_1}} \cdots R_{\frac{1+x_m z}{1+x_m}} S_{(1+z)}\mathcal{P}^k , \tag{3.27}$$

with $(R_{b+cz}\mathcal{P})_i = bP_i + cP_{i-1}$ and $(\tau\mathcal{P})_i = P_{i+1}$. The equality (3.27) follows from the representation of (3.1) as the formal equality

$$\sum_i (S_a \mathcal{P})_i z^i = a(z) \sum_j P_j z^{2j} , \tag{3.28}$$

with $\mathcal{P} = \{P_i\}$. The formal equality in (3.28) is in the sense of equality between the coefficients of the same power of z on both sides of (3.28).

We term the execution of the refinement step by several simple global steps, based on the factorization of the symbol to linear factors, *global refinement procedure by repeated averaging*.

An important family of factorizable schemes is that of the B-spline schemes, with symbols given by (3.18). Note that the factors in (3.18) corresponding to repeated averaging are all of the form $\frac{1+z}{2}$. Thus the global refinement procedure by repeated averaging is equivalent to the algorithm of Lane & Riesenfeld (1980). Also, it follows from the fact that all factors in (3.18) except for the factor $1 + z$ are smoothing factors, that any B-spline scheme is optimal, in the sense that it has a mask of minimal support among all schemes with the same smoothness.

The interpolatory 4-point scheme given by the insertion rule (3.9) with $w = \frac{1}{16}$ (which is the DD_2 scheme) is also factorizable. Its symbol has the form (see, e.g., Dyn (2005))

$$a(z) = (1 + z)^4 (3 + \sqrt{3}z)(3 - \sqrt{3}z)/48. \tag{3.29}$$

3.3.2 Construction of subdivision schemes on manifolds

The two constructions of nonlinear schemes on manifolds in Wallner & Dyn (2005) start from a convergent linear scheme, S, given either by local refinement rules in terms of repeated binary averages, or by a global refinement procedure in terms of repeated binary averages.

The first construction of a subdivision T on a manifold M, *analogous to S*, replaces every binary average in the representation of S, by a corresponding geodesic average on M. Thus $Av_\alpha(P, Q)$ is replaced by $gAv_\alpha(P, Q)$, where $gAv_\alpha(P, Q) = c(\alpha\tau)$, with $c(t)$ the geodesic curve on M from P to Q, satisfying $c(0) = P$ and $c(\tau) = Q$. The resulting subdivision scheme is termed a geodesic subdivision scheme.

The second construction uses a smooth projection mapping onto M, and replaces every binary average by its projection onto M. The resulting nonlinear scheme is termed a projection subdivision scheme. In the case of a surface in $\mathbb{R}^3$, a possible choice of the projection mapping is the orthogonal projection onto the surface.

Note that for a factorizable scheme the analogous manifold schemes obtained from its representation in terms of the global refinement procedure by repeated averaging, depend on the order of the linear factors corresponding to binary averages in (3.24). Yet for the B-spline schemes there is one geodesic analogous scheme, and one projection analogous scheme obtained from this representation since all the factors in (3.18), except the factor $(1 + z)$, are identical.

Example In this example the linear scheme is the Chaikin algorithm of (3.8)

$$P_{2j}^{k+1} = Av_{\frac{1}{4}}(P_{j-1}^k, P_j^k), \quad P_{2j+1}^{k+1} = Av_{\frac{3}{4}}(P_{j-1}^k, P_j^k), \tag{3.30}$$

with the symbol

$$a(z) = (1 + z)^3 / 4. \tag{3.31}$$

The different adaptations to the manifold case are:

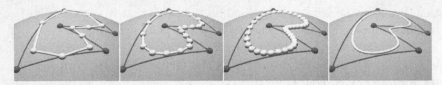

Fig. 3.2. Geodesic B-Spline subdivision of degree three. From left to right: Tp, T^2p, T^3p, $T^\infty p$.

(i) Chaikin geodesic scheme, derived from (3.30):

$$P_{2j}^{k+1} = gAv_{\frac{1}{4}}(P_{j-1}^k, P_j^k)\,, \quad P_{2j+1}^{k+1} = gAv_{\frac{3}{4}}(P_{j-1}^k, P_j^k)\,.$$

(ii) Chaikin geodesic scheme, derived from (3.31):

$$P_{2i}^{k+1,0} = P_{2i+1}^{k+1,0} = P_i^k\,, \quad P_i^{k+1,j} = gAv_{\frac{1}{2}}(P_i^{k+1,j-1}, P_{i-1}^{k+1,j-1}),$$

for $j = 1, 2$.

(iii) Chaikin projection scheme derived from (3.30):

$$P_{2j}^{k+1} = G(Av_{\frac{1}{4}}(P_{j-1}^k, P_j^k))\,, \quad P_{2j+1}^{k+1} = G(Av_{\frac{3}{4}}(P_{j-1}^k, P_j^k))\,.$$

(iv) Chaikin projection scheme derived from (3.31):

$$P_{2i}^{k+1,0} = P_{2i+1}^{k+1,0} = P_i^k\,, \quad P_i^{k+1,j} = G(Av_{\frac{1}{2}}(P_i^{k+1,j-1}, P_{i-1}^{k+1,j-1}),$$

for $j = 1, 2$.

In the above G is a specific smooth projection mapping to the manifold M. Figure 3.2 displays a curve on a sphere, created from a finite number of initial control points on the sphere, by a geodesic analogous scheme to a third degree B-spline scheme.

3.3.3 Analysis of convergence and smoothness by proximity

The analysis of convergence and smoothness of the geodesic and the projection schemes we present is based on their proximity to the linear scheme from which they are derived, and on the smoothness properties of this linear scheme. We limit the discussion to C^1 and C^2 smoothness. To formulate the proximity conditions we introduce some notation.

For a control polygon $\mathcal{P} = \{P_i\}$, we define

$$\Delta\mathcal{P} = \{P_i - P_{i-1}\}, \ \Delta^\ell\mathcal{P} = \Delta(\Delta^{\ell-1}\mathcal{P}), \ d_\ell(\mathcal{P}) = \max_i \|(\Delta^\ell\mathcal{P})_i\|. \quad (3.32)$$

The difference between two control polygons $\mathcal{P} = \{P_i\}$, $\mathcal{Q} = \{Q_i\}$, is $\mathcal{P} - \mathcal{Q} = \{P_i - Q_i\}$.

With this notation the two proximity relations of interest to us are the following:

Definition 3.1

(i) Two schemes S and T are in 0-proximity if

$$d_0(S\mathcal{P} - T\mathcal{P}) \leq Cd_1(\mathcal{P})^2,$$

for all control polygons $\mathcal{P}$ with $d_1(\mathcal{P})$ small enough.

(ii) Two schemes S and T are in 1-*proximity* if

$$d_1(S\mathcal{P} - T\mathcal{P}) \leq C[d_1(\mathcal{P})d_2(\mathcal{P}) - d_1(\mathcal{P})^3],$$

for all control polygons $\mathcal{P}$ with $d_1(\mathcal{P})$ small enough.

In these definitions C is a generic constant.

From the 0-proximity condition we can deduce the convergence of T from the convergence of the linear scheme S. Furthermore, under mild conditions on S, we can also deduce that if S generates C^1 limit curves then T generates C^1 limit curves whenever it converges.

In Wallner (2006), results on C^2 smoothness of the limit curves generated by T are obtained, based on 0-proximity and 1-proximity of T and S, and some mild conditions on S, in addition to S being C^2.

The 0-proximity and the 1-proximity conditions hold for the manifold schemes of subsection 3.3.2, when S is a convergent linear subdivision scheme and when M is a smooth manifold. Moreover, for M a compact manifold or a surface with bounded normal curvatures, the two proximity conditions hold uniformly for all $\mathcal{P}$ such that $d_1(\mathcal{P}) < \delta$, with a global δ.

Examples The B-spline schemes with symbol (3.18) for $m \geq 3$, generate C^2 curves and satisfy the mild conditions necessary for deducing that the limit curves of their manifold analogous schemes are also C^2. On the other hand the linear 4-point scheme generates limit curves which are only C^1, a property that is shared by its manifold analogous schemes.

Further analysis of manifold-valued subdivision schemes can be found in a series of papers by Wallner and his collaborators, see, e.g., Wallner, Nava Yazdani & Grohs (2007), Grohs & Wallner (2008), and in the works Xie & Yu (2007), Xie & Yu (2008a), Xie & Yu (2008b).

3.4 Geometric 4-Point Interpolatory Subdivision Schemes

The refinement step of a linear scheme $(S_{\mathbf{a}}\mathcal{P})_j = \sum_i a_{j-2i}P_i$, $j \in \mathbb{Z}$, is applied separately to each component of the control points. Therefore these schemes are insensitive to the geometry of the control polygons. For control polygons with edges of similar length, this insensitivity is not problematic. Yet, the limit curves generated by linear schemes, in case the initial control polygon has edges of significantly different length, have artifacts, namely geometric features which do not exist in the initial control polygon. This can be seen in the upper right figure of Figure 3.3 and in the second column of Figure 3.4. Data dependent schemes can cure this problem.

Here we present two geometric versions of the 4-point scheme, which are data dependent. The first is based on adapting the tension parameter to the geometry of the 4-points involved in the insertion rule, while the second is based on a geometric parametrization of the control polygon at each refinement level.

3.4.1 Adaptive tension parameter

In this section we present a nonlinear version of the linear 4-point interpolatory scheme introduced in Marinov, Dyn & Levin (2005), which adapts the tension parameter to the geometry of the control points.

It is well known that the linear 4-point scheme with the refinement rules

$$P_{2j}^{k+1} = P_j^k, \quad P_{2j+1}^{k+1} = -w(P_{j-1}^k + P_{j+2}^k) + (\frac{1}{2} + w)(P_j^k + P_{j+1}^k) , \quad (3.33)$$

where w is a fixed tension parameter, has the following attributes:

- It generates "good" curves when applied to control polygons with edges of comparable length.
- It generates curves which become smoother (have greater Hölder exponent of the first derivative), the closer the tension parameter is to $1/16$.
- Only for very small values of the tension parameter, it generates a curve which preserves the shape of an initial control polygon with edges of significantly different length. (Recall that the control polygon itself corresponds to the generated curve with zero tension parameter.)

We first write the refinement rules in (3.33) in terms of the edges

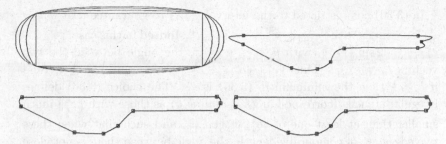

Fig. 3.3. Curves generated by the linear 4-point scheme: (Upper left) the effect of different tension parameters, (upper right) artifacts in the curve generated with $w = \frac{1}{16}$, (lower left) artifact-free but visually non-smooth curve generated with $w = 0.01$. Artifact-free and visually smooth curve generated in a nonlinear way with adaptive tension parameters (lower right).

$\{e_j^k = P_{j+1}^k - P_j^k\}$ of the control polygon, and relate the inserted point P_{2j+1}^{k+1} to the edge e_j^k. The insertion rule can be written in the form,

$$P_{e_j^k} = M_{e_j^k} + w_{e_j^k}(e_{j-1}^k - e_{j+1}^k) \qquad (3.34)$$

with $M_{e_j^k}$ the midpoint of e_j^k and $w_{e_j^k}$ the adaptive tension parameter. Defining $d_{e_j^k} = w_{e_j^k}(e_{j-1}^k - e_{j+1}^k)$ as the displacement from $M_{e_j^k}$, we control its size by choosing $w_{e_j^k}$ according to a geometrical criterion.

In Marinov, Dyn & Levin (2005) there are various geometrical criteria, all of them guaranteeing that the inserted control point $P_{e_j^k}$ is different from the boundary points of the edge e_j^k, and that the length of each of the two edges replacing e_j^k is bounded by the length of e_j^k. This is achieved if $w_{e_j^k}$ is chosen so that

$$\|d_{e_j^k}\| \le \frac{1}{2}\|e_j^k\| . \qquad (3.35)$$

All the criteria restrict the value of the tension parameter $w_{e_j^k}$ to the interval $(0, \frac{1}{16}]$, such that a tension close to $1/16$ is assigned to *regular stencils* namely, stencils of four points with three edges of almost equal length, while the *less regular* the stencil is, the closer to zero is the tension parameter assigned to it.

A *natural* choice of an adaptive tension parameter obeying (3.35) is

$$w_{e_j^k} = \min\left\{\frac{1}{16}, c\,\frac{\|e_j^k\|}{\|e_{j-1}^k - e_{j+1}^k\|}\right\}, \text{ with a fixed } c \in \left[\frac{1}{8}, \frac{1}{2}\right). \qquad (3.36)$$

In (3.36) c is restricted to the interval $[\frac{1}{8}, \frac{1}{2})$ to guarantee that $w_{e_j^k} = \frac{1}{16}$ for stencils with $\|e_{j-1}^k\| = \|e_j^k\| = \|e_{j+1}^k\|$. Indeed in this case, $\|e_{j-1}^k - e_{j+1}^k\| = 2\sin\frac{\theta}{2}\|e_j^k\|$, with θ, $0 \le \theta \le \pi$, the angle between the two vectors e_{j-1}^k, e_{j+1}^k. Thus $\|e_j^k\|/\|e_{j-1}^k - e_{j+1}^k\| = (2\sin\frac{\theta}{2})^{-1} \ge \frac{1}{2}$, and if $c \ge \frac{1}{8}$ then the minimum in (3.36) is $\frac{1}{16}$. The choice (3.36) defines irregular stencils (corresponding to small $w_{e_j^k}$) as those with $\|e_j^k\|$ much smaller than at least one of $\|e_{j-1}^k\|, \|e_{j+1}^k\|$, and such that when these two edges are of comparable length, the angle between them is not close to zero.

The convergence of this geometric 4-point scheme, and the continuity of the limits generated, follow from a result in Levin (1999). There it is proved that the 4-point scheme with variable tension parameter is convergent, and that the limits generated are continuous, whenever the tension parameters are restricted to the interval $[0, \tilde{w}]$, with $\tilde{w} < \frac{1}{8}$.

But we cannot apply the result in Levin (1999) on C^1 limits of the 4-point scheme with variable tension parameter to the geometric 4-point scheme defined by (3.34) and (3.36), since the tension parameters used during this subdivision process are not bounded away from zero.

Nevertheless, many simulations indicate that the curves generated by this scheme are C^1 (see Marinov, Dyn & Levin (2005)).

3.4.2 Geometric parametrization of the control polygons

In this subsection we present a geometric 4-point scheme, which is introduced and investigated in Dyn, Floater & Hormann (2008). The idea for the geometric insertion rule of the point P_{2i+1}^k comes from the insertion rule of the DD_2 scheme (see §3.2.1).

The insertion rule of the DD_2 scheme is obtained by sampling the vector cubic polynomial, interpolating the data $\{((i + j), P_{i+j}^k), \ j = -1, 0, 1, 2\}$ at the point $i + \frac{1}{2}$. From this point of view, the linear scheme corresponds to a uniform parametrization of the control polygon at each refinement level. This approach fails when the initial control polygon has edges of significantly different length. Yet the use of the centripetal parametrization, instead of the uniform parametrization, leads to a geometric 4-point scheme with artifact-free limit curves, as can be seen in Figure 3.4.

The centripetal parametrization, which is known to be effective for interpolation of control points by a cubic spline curve (see Floater (2008)),

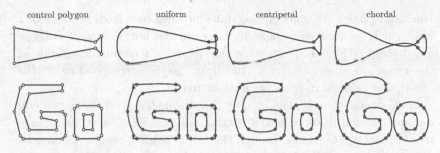

control polygon   uniform   centripetal   chordal

Fig. 3.4. Comparisons between 4-point schemes based on different parametrizations.

has the form $\mathbf{t_{cen}}(\mathcal{P}) = \{t_i\}$, with

$$t_0 = 0, \quad t_i = t_{i-1} + \|P_i - P_{i-1}\|_2^{\frac{1}{2}}, \qquad (3.37)$$

where $\| \cdot \|_2$ is the Euclidean norm, and $\mathcal{P} = \{P_i\}$.

Let $\mathcal{P}^k$ be the control polygon at refinement level k, and let $\{t_i^k\} = \mathbf{t_{cen}}(\mathcal{P}^k)$. The refinement rules for the geometric 4-point scheme, based on the centripetal parametrization are:

$$P_{2i}^{k+1} = P_i^k, \quad P_{2i+1}^{k+1} = \pi_{k,i}\Big(\frac{1}{2}(t_i^k + t_{i+1}^k)\Big),$$

with $\pi_{k,i}$ the vector of cubic polynomials, satisfying the interpolation conditions

$$\pi_{k,i}(t_{i+j}^k) = P_{i+j}^k, \quad j = -1, 0, 1, 2.$$

Note that this construction can be done with any parametrization. In fact in Dyn, Floater & Hormann (2008) the chordal parametrization $(t_{i+1} - t_i = \|P_{i+1} - P_i\|_2)$ is also investigated, but found to be inferior to the centripetal parametrization (see Figure 3.4).

In contrast to the analysis of the schemes on manifolds, the method of analysis of the geometric 4-point schemes based on the chordal and centripetal parametrizations is rather ad-hoc. It is shown in Dyn, Floater & Hormann (2008) that the centripetal and chordal schemes are well defined, in the sense that any inserted point is different from the end points of the edge to which it corresponds, and that both schemes are convergent to continuous limit curves. Although numerical simulations indicate that both schemes generate C^1 curves, as does the linear 4-point scheme, there is no proof of such a property for the geometric schemes.

Another type of information on the limit curves, which is relevant to the absence or presence of artifacts, is available in Dyn, Floater &

Hormann (2008). Bounds on the Hausdorff distance from sections of a limit curve to their corresponding edges in the initial control polygon are derived. These bounds give a partial qualitative understanding of the empirical observation that the limit curves corresponding to the centripetal parametrization are artifact free.

Let C denote a curve generated by the scheme based on the centripetal parametrization from an initial control polygon $\mathcal{P}^0$. Since the scheme is interpolatory, C passes through the initial control points. Denote by $C|_{e_i^0}$ the section of C starting at P_i^0 and ending at P_{i+1}^0. Then

$$haus(C|_{e_i^0}, e_i^0) \le \frac{5}{7} \|e_i^0\|_2.$$

Thus the section of the curve cannot be too far from a short edge. On the other hand the corresponding bound in the linear case has the form

$$haus(C|_{e_i^0}, e_i^0) \le \frac{3}{13} \max\{\|e_j^0\|_2 , \quad |j - i| \le 2\},$$

and a section of the curve can be rather far from its corresponding short edge, if this edge has a long neighboring edge. In the case of the chordal parametrization the bound is even worse

$$haus(C|_{e_i^0}, e_i^0) \le \frac{11}{5} \max\{\|e_j^0\|_2 , \quad |j - i| \le 2\}.$$

Comparisons of the performance of the three 4-point schemes, discussed in this section, are given in Figure 3.4.

3.5 Geometric Refinement of Curves

The scheme discussed in this subsection is designed and investigated in Dyn, Elber & Itai (2008). It is a nonlinear extension of the quadratic B-spline scheme S_a corresponding to the symbol given by (3.18) with $m = 2$. S_a is a linear scheme generating a curve from a set of initial control points, yet its extension presented here generates surfaces, as the refined objects are not control points but control curves. This new nonlinear scheme repeatedly refines a set of control curves, taking into account the geometry of the curves, so as to generate a limit surface which is related to the geometry of the initial control curves.

In fact, a surface can be generated from an initial set of curves $\{C_i\}_{i=0}^n$ using S_a in a linear way. The initial curves have to be parametrized in some reasonable way to yield the set $\{C_i(s), \ s \in [0, 1]\}_{i=0}^n$, and then S_a is applied to each control polygon of the form $\mathcal{P}_s = \{C_i(s)\}_{i=0}^n$

corresponding to a fixed s in $[0,1]$. This is equivalent to refining the curves with $S_\mathbf{a}$ to obtain the refined set of control curves $S_\mathbf{a}\mathcal{P}_s$, $s \in [0,1]$. The limit surface is obtained by repeated refinements of the control curves, and has the form

$$(S_\mathbf{a}^\infty \mathcal{P}_s)(t), \quad s \in [0,1], \; t \in \mathbb{R}.$$

The quality of the generated surface depends on the quality of the parametrization of the initial curves, as a reasonable parametrization for each set of refined curves at each refinement level:

$$\{S_\mathbf{a}^k \mathcal{P}_s, \; s \in [0,1]\}, \quad k = 0,1,2,\ldots$$

In the nonlinear scheme the control curves at each refinement level are parametrized, taking into account their geometry, and $S_\mathbf{a}$ is applied to all control polygons generated by points on the curves corresponding to the same parameter value.

The parametrization of the curves at refinement level k, $\{C_i^k\}_{i=0}^{n_k}$ is in terms of a vector of *correspondences* $\tau = (\tau_0, \ldots, \tau_{n_k-1})$ between pairs of consecutive curves. A correspondence between two curves is a one-to-one and onto continuous map from the points of one curve to the points of the other. Thus τ determines a parametrization of the curves in terms of the points of C_0^k, in the sense that all the points $P_0, \ldots, P_{n_k}$, with $P_0 \in C_0^k$ and $P_{i+1} = \tau_i(P_i) \in C_{i+1}$, correspond to the same parameter value.

The convergence of the nonlinear scheme is proved for initial curves contained in a compact set in $\mathbb{R}^3$. This condition is also satisfied by all control curves generated by the nonlinear scheme, due to the refinement rules of the curves. The correspondence used is a geometrical correspondence τ^* defined by,

$$\tau_i^* = \arg \min_{\tau \in T^k(C_i^k, C_{i+1}^k)} \max\{\|\tau(P) - P\|_2, \quad P \in C_i^k\},$$

where $T^k(C_i^k, C_{i+1}^k)$ is a set of allowed correspondences. This set depends on all the curves at refinement level k in a rather mild way. We omit here the technical details. It is shown in Dyn, Elber, & Itai (2008) that if the initial curves are *admissible for subdivision*, namely the sets $T^0(C_i^0, C_{i+1}^0)$ for $i = 0, \ldots, n_0 - 1$ are nonempty, then the sets $T^k(C_i^k, C_{i+1}^k)$ for $i = 0, \ldots, n_k - 1$ are also nonempty for all $k > 0$.

With the above notation, the refinement step at refinement level k can be written as:

- for $i = 0, \ldots, n_k - 1$,

(i) compute τ_i^*

(ii) for each $P \in C_i^k$, define

$$Q_i(P) = \frac{3}{4}P + \frac{1}{4}\tau_i^*(P), \quad R_i(P) = \frac{1}{4}P + \frac{3}{4}\tau_i^*(P)$$

(iii) define two refined curves

$$C_{2i}^{k+1} = \{Q_i(P), \quad P \in C_i\}, \quad C_{2i+1}^{k+1} = \{R_i(P), \quad P \in C_i\}$$

- $n_{k+1} = 2n_k - 1$

For the convergence proof we need an analogous notion to the control polygon in the case of schemes refining control points. This notion is the control piecewise-ruled-surface, defined at refinement level k by

$$PR^k = \cup_{i=0}^{n_k-1}\{P = \lambda P_i + (1-\lambda)\tau_i^*(P_i), \quad P_i \in C_i^k, \lambda \in [0,1]\}$$

Fig. 3.5. Geometric refinement of curves. From left to right: initial curves, the refined curves after two and three refinement steps.

It is shown in Dyn, Elber & Itai (2008) that if the initial curves are simple, nonintersecting, and admissible for subdivision, then the sequence of control piecewise-ruled-surfaces $\{PR^k\}_{k \geq 0}$ is well defined, and converges in the Hausdorff metric to a set in $\mathbb{R}^3$. Also it is shown there that if the initial curves are sampled densely enough from a smooth surface, then the limit of the scheme approximates the surface.

From the computational point of view the refinements are executed only a small number of times (at most 5), so the "limit" is represented by the surface PR^k with $3 \leq k \leq 5$. All computations are done discretely. The curves are sampled at a finite number of points, and τ^* is computed by dynamical programming.

An example demonstrating the performance of this scheme on an initial set of curves, sampled from a smooth surface, is given in Figure 3.5.

References

C. de Boor (2001), *A Practical Guide to Splines*, Springer Verlag.

A. S. Cavaretta, W. Dahmen and C. A. Michelli (1991), 'Stationary subdivision', *Memoirs Amer. Math. Soc* **453**.

G. M. Chaikin (1974), 'An algorithm for high speed curve generation', *Computer Graphics and Image Processing* **3**, 346–349.

E. Cohen, T. Lyche and R. F. Riesenfeld (1980), 'Discrete B-splines and subdivision techniques', *Computer Graphics and Image Processing* **14**, 87–111.

I. Daubechies (1992), *Ten Lectures on Wavelets*, SIAM Publications.

I. Daubechies and J. C. Lagarias (1992), 'Two-scale difference equations II, local regularity, infinite products of matrices and fractals', *SIAM J. Math. Anal.* **23**, 1031–1079.

G. Deslauriers and S. Dubuc (1989), 'Symmetric iterative interpolation', *Constr. Approx.* **5**, 49–68.

N. Dyn (1992), 'Subdivision schemes in Computer-Aided Geometric Design', in *Advances in Numerical Analysis - Volume II, Wavelets, Subdivision Algorithms and Radial Basis Functions*, W. Light (ed.), Clarendon Press, Oxford, 36–104.

N. Dyn (2005), 'Three families of nonlinear subdivision schemes', in *Multivariate Approximation and Interpolation*, K. Jetter, D. Buhmann, W. Haussmann, R. Schaback, J. Stöckler (eds.), Elsevier, 23–38.

N. Dyn, G. Elber and U. Itai (2008), 'A subdivision scheme generating surfaces by repeated refinements of curves', School of Mathematical Sciences, Tel-Aviv University, work in progress.

N. Dyn, M. S. Floater and K. Hormann (2008), 'Four-point curve subdivision based on iterated chordal and centripetal parametrizations', to appear in *Computer Aided Geometric Design.*

N. Dyn, J. A. Gregory and D. Levin (1987), 'A four-point interpolatory subdivision scheme for curve design', *Computer Aided Geometric Design* **4**, 257–268.

N. Dyn, J. A. Gregory and D. Levin (1991), 'Analysis of uniform binary subdivision schemes for curve design', *Constr. Approx.* **7**, 127–147.

N. Dyn and D. Levin (1990), 'Interpolating subdivision schemes for the generation of curves and surfaces', in *Multivariate Approximation and Interpolation*, W. Haussmann, K. Jetter (eds.), Birkhäuser Verlag, 91–106.

N. Dyn and D. Levin (1995), 'Analysis of asymptotically equivalent binary subdivision schemes', *J. Math. Anal. Appl.* **193**, 594–621.

N. Dyn and D. Levin (2002), 'Subdivision schemes in geometric modelling', *Acta Numerica* **11**, 73–144.

M.S. Floater (2008), 'On the deviation of a parametric cubic spline from its data polygon' to appear in *Computer Aided Geometric Design.*

P. Grohs and J. Wallner (2008), 'Log-exponential analogues of univariate subdivision schemes in Lie groups and their smoothness properties', in *Approximation Theory XII*, M. Neamtu, L. L. Schumaker (eds.), Nashboro Press, 181–190.

J. Hechler, B. Mößner and U. Reif (2008), 'C^1-continuity of the generalized four-point scheme', FB Mathematics, TU-Darmstadt, preprint.

J. Lane, R. Riesenfeld (1980), 'A theoretical development for the computer generation and display of piecewise polynomial surfaces', *IEEE Trans. Pattern Anal. and Machine Intell.* **2**, 35–46.

D. Levin (1999), 'Using Laurent plynomial representation for the analysis of the nonuniform binary subdivision schemes', *Adv. Comp. Math.* **11**, 41–54.

M. Marinov, N. Dyn and D. Levin (2005), 'Geometrically controlled 4-point interpolatory schemes', in *Advances in Multiresolution for Geometric Modelling*, N. A. Dodgson, M. S. Floater, M. A. Sabin (eds.), Springer-Verlag, 301–315.

H. Prautzsch, W. Boehm and M. Paluszny (2002), *Bézier and B-spline Techniques*, Springer.

G. de Rahm (1956), 'Sur une courbe plane', *J. Math. Pures & Appl.* **35**, 25–42.

I. U. Rahman, I. Drori, V. C. Stodden, D. L. Donoho and P. Schröder (2005), 'Multiscale representations for manifold-valued data', *Multiscale Modeling and Simulation* **4**, 1201–1232.

J. Wallner (2006), 'Smoothness analysis of subdivision schemes by proximity', *Constr. Approx.* **24**, 289–318.

J. Wallner and N. Dyn (2005), 'Convergence and C^1 analysis of subdivision schemes on manifolds by proximity', *Computer Aided Geometric Design* **22**, 593–622.

J. Wallner, E. Nava Yazdani and P. Grohs (2007), 'Smoothness properties of Lie group subdivision schemes', *Multiscale Modeling and Simulation* **6**, 493–505.

G. Xie and T. P.-Y. Yu (2007), 'Smoothness equivalence properties of manifold-data subdivision schemes based on the projection approach', *SIAM J. Num. Anal.* **45**, 1200–1225.

G. Xie and T. P.-Y. Yu (2008a), 'Smoothness equivalence properties of general manifold-valued data subdivision schemes', *Multiscale Modeling and Simulation*, to appear

G. Xie and T. P.-Y. Yu (2008b), 'Approximation order equivalence properties of manifold-valued data subdivision schemes', Department of Mathematics, Drexel University, preprint.

4

Energy Preserving and Energy Stable Schemes for the Shallow Water Equations

Ulrik S. Fjordholm

Department of Mathematics
University of Oslo
P.O. Box 1053, Blindern, N–0316 Oslo, Norway
e-mail: ulriksf@ulrik.uio.no

Siddhartha Mishra

Centre of Mathematics for Applications (CMA)
University of Oslo
P.O. Box 1053, Blindern, N–0316 Oslo, Norway
e-mail: siddharm@cma.uio.no

Eitan Tadmor

Department of Mathematics, Institute for Physical Sciences & Technology
and Center of Scientific Computation and Mathematical Modeling
University of Maryland
MD 20742-4015, USA
e-mail: tadmor@cscamm.umd.edu

Abstract

We design energy preserving and energy stable schemes for the shallow water equations. A new explicit energy preserving flux is proposed and is compared with existing energy preserving fluxes. This new flux results in a considerable reduction in the computational cost. We add suitably discretized viscous terms to the energy preserving scheme in order to obtain an energy stable scheme that replicates the energy decay of the continuous problem. The addition of physical viscosity dramatically reduces the oscillations on resolved meshes. For computing on under-resolved meshes, we propose a new Roe-type numerical flux that adds diffusion in terms of energy variables to an energy preserving scheme. The resulting scheme is energy stable and as accurate as the Roe scheme, at a similar computational cost. The robustness of the energy preserving and energy stable schemes is demonstrated through several numerical experiments in both one and two space dimensions.

4.1 Introduction

Many interesting models in hydrology and oceanography involve flows where the horizontal scales of motion are much greater than the vertical scale. Such models include flows in lakes, rivers and irrigation channels and near-shore models in oceanography and climate modeling. These phenomena are often modeled by the shallow water equations,

$$h_t + (hu)_x + (hv)_y = 0,$$

$$(hu)_t + \left(hu^2 + \frac{1}{2}gh^2\right)_x + (huv)_y = \nu((hu_x)_x + (hu_y)_y),\qquad (4.1)$$

$$(hv)_t + (huv)_x + \left(hv^2 + \frac{1}{2}gh^2\right)_y = \nu((hv_x)_x + (hv_y)_y),$$

where h is the height of the fluid, u and v are the velocity components in the x and y directions, respectively, g is the constant acceleration due to gravity and ν is the eddy viscosity. The eddy viscosity is responsible for the transfer of energy to the smaller scales of the motion. The system is equipped with suitable initial and boundary conditions.

The eddy viscosity ν generally determines the smallest scale of the flow, and in most applications it is very small. It is common to assume that $\nu = 0$ and consider the inviscid form of the shallow water system,

$$h_t + (hu)_x + (hv)_y = 0,$$

$$(hu)_t + \left(hu^2 + \frac{1}{2}gh^2\right)_x + (huv)_y = 0,\qquad (4.2)$$

$$(hv)_t + (huv)_x + \left(hv^2 + \frac{1}{2}gh^2\right)_y = 0.$$

This is a system of conservation laws, which in general is of the form

$$U_t + F(U)_x + G(U)_y = 0. \qquad (4.3)$$

In our case, $U = [h, hu, hv]^\top$ is the vector of conserved quantities of mass and momentum and $F \equiv F(U) = [hu, hu^2 + \frac{1}{2}gh^2, huv]^\top$ and $G \equiv G(U) = [hv, huv, hv^2 + \frac{1}{2}gh^2]^\top$ are the fluxes in the x- and y-directions, respectively. The eigenvalues λ_i and μ_i of the Jacobians F' and G', respectively, are

$$\lambda_1 = u - \sqrt{gh}, \qquad\qquad \mu_1 = v - \sqrt{gh},$$

$$\lambda_2 = u, \qquad\qquad \mu_2 = v,$$
$$\lambda_3 = u + \sqrt{gh}, \qquad\qquad \mu_3 = v + \sqrt{gh}.$$

The corresponding eigenvectors are easily calculated; see LeVeque (2002).

Nonlinear systems of conservation laws of the form (4.3) arise in a wide variety of problems in elasticity, fluid dynamics and plasma physics (see Dafermos (2000) for details). The most striking feature about these equations is the fact that even smooth initial data can lead to solutions with discontinuities. This feature is exhibited even in the simplest case of a scalar conservation law in one space dimension. The presence of discontinuities forces us to consider solutions of (4.3) in the weak or averaged sense (see Dafermos (2000) and other standard textbooks for a definition).

Weak solutions are not necessarily unique, and to obtain uniqueness the equations must be augmented with additional admissibility criteria. These criteria assume the form of entropy conditions Dafermos (2000). Many systems of conservation laws are equipped with a convex function $E(U)$ and associated entropy flux functions $H = H(U)$ and $K = K(U)$ such that

$$\partial_{U_i} H(U) = \langle V, \partial_U F^{(i)} \rangle, \quad \partial_{U_i} K(U) = \langle V, \partial_U G^{(i)} \rangle \quad \text{for } i = 1, \ldots, n, \tag{4.4}$$

where $V := \partial_U E$ is the vector of *entropy variables*. As an immediate consequence of these identities, smooth solutions of (4.3) will satisfy the additional conservation law

$$E(U)_t + H(U)_x + K(U)_y = 0. \tag{4.5}$$

However, this identity is not valid at shocks and has to be modified accordingly. The entropy identity (4.5) transforms into the entropy *inequality* Dafermos (2000)

$$E(U)_t + H(U)_x + K(U)_y \leq 0, \tag{4.6}$$

in the sense of distributions. Scalar conservation laws are equipped with an infinite number of entropy/entropy-flux pairs, and this paves the way for a proof of existence, uniqueness and stability. However, systems of conservation laws in general do not possess an infinite number of entropy functions. This is a key difficulty in proving existence and stability results, particularly in the case of multi-dimensional systems.

However, many interesting systems like the Euler equations of gas dynamics and the magnetohydrodynamics (MHD) system of plasma physics are equipped with at least one physically relevant entropy function. The shallow water system (4.2) also possesses an entropy function, the *total energy*

$$E = \frac{1}{2} \left(hu^2 + hv^2 + gh^2 \right),$$

which is the sum of kinetic and gravitational potential energy. A direct calculation reveals that smooth solutions of (4.2) satisfy the energy conservation law

$$E_t + \left(\frac{1}{2} \left(hu^3 + huv^2 \right) + guh^2 \right)_x + \left(\frac{1}{2} \left(hu^2 v + hv^3 \right) + gvh^2 \right)_y = 0. \tag{4.7}$$

Integrating (4.7) over space, we obtain

$$\frac{d}{dt} \int_{\mathbb{R}^2} E \equiv 0. \tag{4.8}$$

Hence, the total energy of smooth solutions of (4.2) is conserved.

The above identity is valid only for smooth solutions of (4.2). Energy will be dissipated at shocks, and the precise rate of this dissipation can be explicitly calculated from the viscous form (4.1). The energy identity (4.7) in the presence of viscous terms takes the form,

$$E_t + \left(\frac{1}{2} \left(hu^3 + huv^2 \right) + guh^2 \right)_x + \left(\frac{1}{2} \left(hu^2 v + hv^3 \right) + gvh^2 \right)_y \tag{4.9}$$
$$= \nu \Big(u\big((hu_x)_x + (hu_y)_y \big) + v\big((hv_x)_x + (hv_y)_y \big) \Big).$$

Integrating this identity in space and integrating by parts we obtain the energy dissipation estimate

$$\frac{d}{dt} \int_{\mathbb{R}^2} E = -\nu \int_{\mathbb{R}^2} h \left(u_x^2 + u_y^2 + v_x^2 + v_y^2 \right). \tag{4.10}$$

As the height h is always positive, the right-hand side of the above identity is always non-positive, so we get the energy dissipation

$$\frac{d}{dt} \int_{\mathbb{R}^2} E \leq 0. \tag{4.11}$$

Thus, we recover the energy inequality that holds for weak solutions of (4.2). Furthermore, (4.10) gives an explicit rate for the dissipation of energy into smaller scales. Note that a bound on the total energy E automatically implies a bound on the L^2 norms of height and the velocity field, as the height is strictly positive. Hence, the energy estimate is also a statement of stability of solutions of the shallow water system.

However, energy estimates are not enough to obtain proofs of existence or uniqueness of solutions.

In the absence of existence results or explicit formulas for the solution of (4.3), numerical methods are the main tools in the study of these models. Numerical methods for systems of conservation laws have undergone extensive development in the last few decades, and the subject is in a fairly mature stage; see LeVeque (2002). The most popular methods are the so-called finite volume methods. For simplicity, we consider a uniform Cartesian mesh in $\mathbb{R}^2$ with mesh sizes Δx and Δy, respectively. We denote the nodes as $x_i := i\Delta x$ and $y_j := j\Delta y$ and a prototypical cell as $I_{ij} := [x_{i-\frac{1}{2}}, x_{i+\frac{1}{2}}) \times [y_{j-\frac{1}{2}}, y_{j+\frac{1}{2}})$. A standard cell-centered finite volume method consists of updating the cell averages,

$$U_{ij}(t) := \frac{1}{\Delta x \Delta y} \int_{I_{ij}} U(x, y, t) dx dy,$$

at each time level. For simplicity, we drop the time dependence of every quantity and write a standard finite volume scheme for (4.3) in the semi-discrete form as

$$\frac{d}{dt} U_{ij} = -\frac{1}{\Delta x} \left(F_{i+\frac{1}{2},j} - F_{i-\frac{1}{2},j} \right) - \frac{1}{\Delta y} \left(G_{i,j+\frac{1}{2}} - G_{i,j-\frac{1}{2}} \right), \quad (4.12)$$

where $F_{i\pm\frac{1}{2},j\pm\frac{1}{2}}$ and $G_{i\pm\frac{1}{2},j\pm\frac{1}{2}}$ are numerical fluxes at the cell-edges, consistent with the fluxes F and G, respectively[1]. The key step is the choice of the numerical fluxes. The numerical fluxes are computed using the neighboring cell-averages across the normal directions of cell-edges, and high-order accuracy in space is obtained with non-oscillatory reconstructions of point values from these cell-averages, e.g., Cockburn et al. (1998), LeVeque (2002) and the references therein. The time-integration is often performed with strong stability preserving Runge-Kutta methods, e.g., Gottlieb, Shu and Tadmor (2001). This standard framework has proved very successful in computing solutions of many interesting flow problems and is used extensively in practice.

However, very few rigorous stability results are obtained for finite volume schemes, particularly for systems of conservation laws. One of the reasons for the lack of stability is the failure to design schemes that satisfy a discrete form of the entropy inequality (4.6). Most finite volume

---

[1] Note that the differential fluxes $F(U), G(U)$ depend on the one conserved quantity U whereas numerical fluxes depend on two or more neighboring cell quantities, e.g., $F_{i+\frac{1}{2},j} = F(\ldots, U_{ij}, U_{i+1,j}, \ldots)$. This enables us to uniquely distinguish between differential fluxes such as $F_{ij} = F(U_{ij})$, and the corresponding numerical fluxes, e.g., $F_{i\pm\frac{1}{2},j}$ etc.

schemes (particularly at higher-order) do not necessarily satisfy such inequalities. This might result in numerical instabilities and the computation of incorrect solutions. Many finite volume schemes (in their first-order versions) do add enough numerical diffusion to dissipate entropy; however, the diffusion added is excessive and does not respect the rate of entropy diffusion in the continuous problem.

In the specific context of the shallow water equations (4.2), there are many different finite volume schemes in use. Most of these schemes do not preserve energy for smooth solutions (i.e., satisfy discrete versions of (4.7)), which can lead to significant loss of accuracy, particularly for problems involving large time scales. Few schemes will respect the energy balance (4.10) because entropy dissipation at shocks is too excessive. This can lead to both instabilities as well as numerical artifacts. See Arakawa (1966), Arakawa and Lamb (1977), Arakawa and Lamb (1981) for a detailed discussion on numerical effects of schemes that do not respect the energy balance in shallow water equations.

In the pioneering papers of Tadmor (1987) and (2003), the problem of *designing* finite volumes schemes which satisfy discrete versions of the entropy inequality (4.6) was tackled . The main feature of Tadmor (1987) was the design of an *entropy conservative* finite volume flux, i.e., consistent numerical fluxes $\widetilde{F}_{i\pm\frac{1}{2},j}$ and $\widetilde{G}_{i,j\pm\frac{1}{2}}$, which ensure that the numerical scheme (4.12) satisfies the entropy identity (4.5). Then, a novel entropy comparison principle is introduced in order to design an *entropy stable* scheme – a scheme that satisfies a discrete version of the entropy inequality. The idea is to compare the numerical diffusion of any given finite volume scheme with the diffusion of the entropy conservative scheme. A scheme that contains more diffusion than an entropy conservative scheme is entropy stable. Thus, one is able to investigate entropy stability using a comparison principle. Tadmor (1987) did not, however, contain explicit expressions for any interesting systems of conservation laws. A novel pathwise decomposition was introduced in Tadmor (2003) in order to obtain an explicit formula for the entropy conservative scheme. This approach was used in Tadmor and Zhong (2006) to compute solutions of the Euler equations, and in Tadmor and Zhong (2008) to approximate the shallow water equations. In both papers the authors used the explicit entropy conservative scheme of Tadmor (2003) to compute the solutions and used this scheme as a basis for a "faithful" discretization of the entropy (energy) balance.

Our aim in this paper is to consider energy preserving and energy stable discretizations of the shallow water system. We consider three different energy preserving finite volume schemes: The original entropy conservative scheme of Tadmor (1987), which can be explicitly integrated in the case of shallow water equations in one dimension; the pathwise explicit scheme of Tadmor (2003), Tadmor and Zhong (2008); and a novel explicit energy preserving scheme. We compare the three energy preserving schemes, and find that the new explicit entropy preserving scheme is both simpler to implement and computationally less expensive compared to the other two schemes. We use this new energy conservative scheme to design novel numerical diffusion operators that result in energy stable schemes. The schemes are implemented in a series of numerical experiments which illustrate their different features in the one-dimensional setup, in Section 4.2 and with the two-dimensional problems, in Section 4.3. Conclusions are drawn in Section 4.4.

4.2 The One-Dimensional Problem

For simplicity of the description, we start with the one-dimensional form of the inviscid shallow water system (4.2),

$$h_t + (hu)_x = 0,$$
$$(hu)_t + \left(hu^2 + \frac{1}{2}gh^2\right)_x = 0. \tag{4.13}$$

These equations are obtained from (4.2) simply by ignoring variation in the y-direction and setting the vertical velocity component $v = 0$.

The above equation is a form of the generic one-dimensional system of conservation laws,

$$U_t + F_x = 0, \tag{4.14}$$

with U the n-vector of unknowns and $F = F(U)$ the flux vector. Assume that an entropy/entropy-flux function pair (E, H) exists, so that

$$E(U)_t + H(U)_x = 0. \tag{4.15}$$

Define the vector of entropy variables as $V := \partial_U E$. For the one-dimensional shallow water system, the entropy function is given by the *energy*, $E = \frac{1}{2}(hu^2 + gh^2)$, which for smooth solutions satisfies

$$E_t + \left(\frac{1}{2}hu^3 + guh^2\right)_x = 0. \tag{4.16}$$

The vector V takes the form $V = \left[gh - \dfrac{u^2}{2}, u\right]^\top$. Define the *entropy potential* as $\Psi := \langle V, F \rangle - H$. A direct calculation shows that for the one-dimensional shallow water equations, Ψ is given by $\Psi = \frac{1}{2} g u h^2$.

Our aim is to design a finite volume scheme for (4.14) that satisfies a discrete form of the entropy identity (4.15). A finite volume scheme (in semi-discrete form) on a uniform mesh $x_i = i \Delta x$ is given by,

$$\frac{d}{dt} U_i = -\frac{1}{\Delta x} (F_{i+\frac{1}{2}} - F_{i-\frac{1}{2}}), \qquad (4.17)$$

where U_i is the cell average on $[x_{i-\frac{1}{2}}, x_{i+\frac{1}{2}})$ and $F_{i+\frac{1}{2}}$ is the numerical flux at the interface $x_{i+\frac{1}{2}}$.

As mentioned in the introduction, an arbitrary choice of a consistent numerical flux is not enough to satisfy a discrete version of the entropy identity (4.15). Instead, we follow the general procedure introduced in Tadmor (1987) to define an entropy preserving numerical flux. Here and below we use the following abbreviations

$$V_i = V(U_i), \qquad F_i = F(U_i), \qquad \Psi_i = \Psi(U_i),$$

$$[\![a_{i+\frac{1}{2}}]\!] := a_{i+1} - a_i, \qquad \overline{a}_{i+\frac{1}{2}} := \frac{1}{2}(a_i + a_{i+1}),$$

where $[\![a_{i+\frac{1}{2}}]\!]$ represents the jump of a across the interface at $x_{i+\frac{1}{2}}$.

Theorem 4.1 (Tadmor (1987)) *Consider the one-dimensional system of conservation laws* (4.14) *with entropy function* E, *entropy variables* V, *entropy flux* F *and potential* Ψ, *as defined above. Let* $\widetilde{F}_{i+\frac{1}{2}}$ *be a numerical flux, consistent with* (4.14), *that satisfies*

$$\langle [\![V_{i+\frac{1}{2}}]\!], \widetilde{F}_{i+\frac{1}{2}} \rangle = [\![\Psi_{i+\frac{1}{2}}]\!]. \qquad (4.18)$$

Then, the scheme (4.17) *satisfies the discrete entropy identity*

$$\frac{d}{dt} E(U_i(t)) = -\frac{1}{\Delta x}(\widetilde{H}_{i+\frac{1}{2}} - \widetilde{H}_{i-\frac{1}{2}}), \qquad (4.19)$$

with numerical entropy flux $\widetilde{H}_{i+\frac{1}{2}} := \langle \overline{V}_{i+\frac{1}{2}}, \widetilde{F}_{i+\frac{1}{2}} \rangle - \overline{\Psi}_{i+\frac{1}{2}}$. *Summing over all* i, *we end up with the conservation law* $\frac{d}{dt} \sum_i E(U_i(t)) \equiv 0$.

Hence, the finite volume scheme (4.17), (4.18) *is energy preserving.*

The proof of the above theorem can be found in Tadmor (1987). The theorem is very general and the key ingredient for obtaining entropy preserving fluxes is the condition (4.18).

Note that the condition (4.18) provides a single constraint for a flux to be entropy conservative. We will reserve the notation of $\widetilde{F}$ to distinguish such entropy conservative fluxes. Thus, for example, when $n = 1$ entropy conservative fluxes are uniquely determined as $\widetilde{F}_{i+\frac{1}{2}} = \left[\Psi_{i+\frac{1}{2}}\right] / \left[V_{i+\frac{1}{2}}\right]$. For general $n \times n$ systems, however, the choice of an entropy preserving flux in (4.14) is not unique for $n > 1$. Given this fact, we proceed to describe three different choices of entropy conservative fluxes for the 1D shallow water system (4.13).

4.2.1 Scheme I: Averaged Energy Conservative (AEC) Scheme

We start with an entropy conservative scheme that was first proposed in Tadmor (1987) for a general system (4.14). For $\xi \in [-1/2, 1/2]$, define the straight line

$$V_{i+\frac{1}{2}}(\xi) = \frac{1}{2}(V_i + V_{i+1}) + \xi(V_{i+1} - V_i).$$

Clearly, $V_{i+\frac{1}{2}}$ connects the vectors V_i and V_{i+1}. Next, we define the numerical flux as

$$\widetilde{F}_{i+\frac{1}{2}} = \int_{-1/2}^{1/2} F\left(V_{i+\frac{1}{2}}(\xi)\right) d\xi. \tag{4.20}$$

This flux represents the path integral of the flux along a straight line connecting two adjacent states in phase space. Clearly, this flux is consistent and it was shown in Tadmor (1987) that it is also entropy conservative. This follows from a straightforward calculation showing that (4.20) satisfies the identity (4.18).

In general, (4.20) cannot be evaluated explicitly in the phase space. But we will see that for the simple case of shallow water equations, the path integral can be explicitly computed in terms of the physical variables although the resulting formulas are still quite complicated. A long direct calculation of the integral in (4.20) for the shallow water system yields the following formulas:

$$\begin{aligned}
F^{(1)}_{i+\frac{1}{2}} &= \frac{h_i u_i}{3} + \frac{h_{i+1} u_{i+1}}{3} + \frac{h_i u_{i+1}}{6} + \frac{h_{i+1} u_i}{6} \\
&\quad - \frac{u_i^3}{24} - \frac{u_{i+1}^3}{24} + \frac{u_i u_{i+1}^2}{24} + \frac{u_{i+1} u_i^2}{24} \\
F^{(2)}_{i+\frac{1}{2}} &= \frac{1}{12} h_i u_i^2 + \frac{1}{12} h_{i+1} u_{i+1}^2 + \frac{1}{6} h_i u_{i+1}^2 + \frac{1}{6} h_{i+1} u_i^2 \\
&\quad + \frac{1}{4} h_i u_i u_{i+1} + \frac{1}{4} h_{i+1} u_i u_{i+1} + \frac{7g}{24} h_i^2 + \frac{7g}{24} h_{i+1}^2 \\
&\quad - \frac{g}{12} h_i h_{i+1} + \frac{1}{96} u_i^4 + \frac{1}{96} u_{i+1}^4 - \frac{1}{48} u_i^2 u_{i+1}^2.
\end{aligned} \tag{4.21}$$

A direct calculation verifies that this flux is indeed energy preserving. Note that the flux is symmetric in its arguments, and it is easy to check that it is consistent. The main problem with this flux is the complexity of the formulas: This 1D computation is rather specific and a similar calculation to obtain an explicit form of (4.20) for the two-dimensional shallow water system (4.2) becomes algebraically intractable. This necessitates the search for alternative fluxes satisfying (4.18).

4.2.2 Scheme II: Pathwise Energy Conservative (PEC) Scheme

An explicit solution of (4.18) was found in Tadmor (2003). This scheme was implemented for the shallow water system (4.2) in Tadmor and Zhong (2008), and we mention it here simply for the sake of completeness and comparison.

Consider the $n \times n$ system of conservation laws (4.14). Let $\{r_k\}, \{l_k\}$ for $k = 1, \ldots, n$ be any orthogonal eigensystem that spans $\mathbb{R}^n$. At an interface $x_{i+\frac{1}{2}}$, we have the two adjacent entropy variable vectors $V^{[0]} := V_i$ and $V^{[n]} := V_{i+1}$. Then, define the following paths:

$$V^{[0]} = V_i,$$
$$V^{[k]} = V^{[k-1]} + \langle [V_{j+\frac{1}{2}}], l_k \rangle r_k \qquad \text{for } k = 1, \ldots, n.$$

Note that $V^{[n]} = V_{i+1}$. We are replacing the straight line joining the two adjacent states in the flux (4.20) by a piecewise linear path that corresponds to the basis vectors. Now define, for $k = 1, 2, \ldots, n,$

$$\widetilde{F}^{[k]} = \frac{\Psi(V^{[k]}) - \Psi(V^{[k-1]})}{\langle [V_{j+\frac{1}{2}}], l_k \rangle} l_k. \qquad (4.22)$$

Then the PEC flux is given by

$$\widetilde{F}_{i+\frac{1}{2}} = \sum_{k=1}^{n} \widetilde{F}^{[k]}. \qquad (4.23)$$

To show that this flux is entropy conservative, multiply both sides of (4.23) with $[V_{i+\frac{1}{2}}]$ to get

$$\langle [V_{i+\frac{1}{2}}], \widetilde{F}_{i+\frac{1}{2}} \rangle = \sum_{k=1}^{n} \left(\Psi(V^{[k]}) - \Psi(V^{[k-1]}) \right)$$
$$= \Psi(V^{[n]}) - \Psi(V^{[0]}) = [\Psi_{i+\frac{1}{2}}].$$

Consistency of this flux has been shown in Tadmor (2003). The only remaining step in designing the flux is to specify the choice of the path, i.e., of the orthogonal eigensystem. Following Tadmor and Zhong (2008) and (2006), we take the path given by the eigenvectors of a Roe matrix corresponding to the shallow water system. Details of the implementation are provided in Tadmor and Zhong (2006). A significant modification is made in Remark 3.5 of Tadmor and Zhong (2006) to treat the case where $\langle [\![V_{j+\frac{1}{2}}]\!], l_k \rangle$ vanish, making the flux $\widetilde{F}^{[k]}$ in (4.22) singular; this case has been treated in Tadmor and Zhong (2006) and (2008).

The main feature of the flux (4.23) is its generality. It provides a recipe to construct entropy conservative schemes for any system. The main difficulty, though, is its computational cost, which is high because one has to evaluate the eigensystem and solve for the orthogonal system for every mesh point. Even though it has been implemented for both the shallow water and Euler equations, we would like to find a energy preserving flux which is computationally cheaper and easier to implement.

4.2.3 Scheme III: Explicit Energy Conservative (EEC) Scheme

Both of the above schemes can be explicitly written down and implemented, although at a high computational cost. Our aim is to design a much simpler scheme. To construct the entropy conservative flux at the interface $x_{i+\frac{1}{2}}$ we will use the following identity between jumps and averages,

$$[\![a_{i+\frac{1}{2}} b_{i+\frac{1}{2}}]\!] \equiv \overline{b}_{i+\frac{1}{2}} [\![a_{i+\frac{1}{2}}]\!] + [\![b_{i+\frac{1}{2}}]\!] \overline{a}_{i+\frac{1}{2}}. \tag{4.24}$$

In order to satisfy the entropy preserving constraint (4.18) for the special case of the shallow water equations (4.13) we use (4.24) to express the jumps across $x_{i+\frac{1}{2}}$ in terms of the jumps in the primitive variables h and u. We have

$$\left[\!\!\left[V^{(1)}_{i+\frac{1}{2}} \right]\!\!\right] = \left[\!\!\left[gh_{i+\frac{1}{2}} - \tfrac{1}{2} u^2_{i+\frac{1}{2}} \right]\!\!\right] = g [\![h_{i+\frac{1}{2}}]\!] - \overline{u}_{i+\frac{1}{2}} [\![u_{i+\frac{1}{2}}]\!],$$

$$\left[\!\!\left[V^{(2)}_{i+\frac{1}{2}} \right]\!\!\right] = [\![u_{i+\frac{1}{2}}]\!],$$

$$[\![\Psi_{i+\frac{1}{2}}]\!] = \frac{1}{2} g \left[\!\!\left[u_{i+\frac{1}{2}} h^2_{i+\frac{1}{2}} \right]\!\!\right] = g \overline{u}_{i+\frac{1}{2}} \overline{h}_{i+\frac{1}{2}} [\![h_{i+\frac{1}{2}}]\!] + \frac{g}{2} (\overline{h^2})_{i+\frac{1}{2}} [\![u_{i+\frac{1}{2}}]\!].$$

Writing down the desired flux componentwise as $\widetilde{F}_{i+\frac{1}{2}} = \left[\widetilde{F}^{(1)}_{i+\frac{1}{2}}, \widetilde{F}^{(2)}_{i+\frac{1}{2}} \right]$, inserting all the above quantities into (4.18) and then equating jumps in h and u, we get the following set of equations:

$$\widetilde{F}^{(1)}_{i+\frac{1}{2}} = \overline{h}_{i+\frac{1}{2}}\overline{u}_{i+\frac{1}{2}},$$

$$\widetilde{F}^{(2)}_{i+\frac{1}{2}} - \overline{u}_{i+\frac{1}{2}}\widetilde{F}^{(1)}_{i+\frac{1}{2}} = \frac{g}{2}(\overline{h^2})_{i+\frac{1}{2}}.$$

Solving the above equations, we get

$$\widetilde{F}^{(1)}_{i+\frac{1}{2}} = \overline{h}_{i+\frac{1}{2}}\overline{u}_{i+\frac{1}{2}},$$

$$\widetilde{F}^{(2)}_{i+\frac{1}{2}} = \overline{h}_{i+\frac{1}{2}}\left(\overline{u}_{i+\frac{1}{2}}\right)^2 + \frac{g}{2}(\overline{h^2})_{i+\frac{1}{2}}. \tag{4.25}$$

Thus, we can write down the energy preserving flux explicitly by expanding (4.25) as

$$\widetilde{F}^{(1)}_{i+\frac{1}{2}} = \left(\frac{h_i + h_{i+1}}{2}\right)\left(\frac{u_i + u_{i+1}}{2}\right),$$

$$\widetilde{F}^{(2)}_{i+\frac{1}{2}} = \left(\frac{h_i + h_{i+1}}{2}\right)\left(\frac{u_i + u_{i+1}}{2}\right)^2 + \frac{g}{2}\left(h_i^2 + h_{i+1}^2\right). \tag{4.26}$$

Clearly the flux (4.26) is energy preserving as well as consistent. It is symmetric and second-order accurate in space (as shown by a simple truncation error analysis). It is also extremely easy to code. Furthermore, numerical experiments will reveal that the flux is very cheap computationally and is robust. The above flux is similar in spirit, though different in details, to the entropy conservative flux for the Euler equations of gas dynamics designed in Roe (2006).

Remark 4.1 Both the AEC scheme (4.21) and the PEC scheme (4.23) were based on integrating the flux along a suitable path in the space of energy variables. The AEC scheme relied on a straight line path connecting the adjacent states whereas the PEC scheme was based on a piecewise straight line path parallel to eigenvectors of the Jacobian of the flux. A natural question arises — can the EEC scheme (4.26) be written as an energy preserving flux using integration along a suitable path in the phase space of energy variables? We were unable to obtain such a path; however, we believe it exists, and if so, it would be extremely interesting if it can be written down explicitly.

4.2.4 Time Stepping

The three energy conservative schemes above were formulated in the semi-discrete framework (4.17). We also need to describe the discrete time evolution in order to compute approximations to (4.13) and to this end, we choose two different time stepping schemes. First, note that all the three energy preserving schemes are central schemes and hence unstable when used together with the forward Euler method. Therefore, we need to use Runge-Kutta methods to stabilize the computations. Furthermore, it is advisable to use the strong-stability preserving (SSP) Runge-Kutta methods developed in Gottlieb, Shu and Tadmor (2001). Given a numerical flux F, let $\mathcal{L}(U_i^n)$ denote the net flux into grid cell i at time t_n, i.e.,

$$\mathcal{L}(U_i^n) := -\frac{1}{\Delta x}\left(F_{i+\frac{1}{2}} - F_{i-\frac{1}{2}}\right).$$

$\mathcal{L}(U_i^n)$ is precisely the right-hand side of the finite volume scheme (4.17). We use the following second-order SSP Runge-Kutta method of Gottlieb, Shu and Tadmor (2001), identified below as RK2:

$$\begin{aligned}
U_i^* &= U_i^n + \Delta t \mathcal{L}(U_i^n) \\
U_i^{**} &= U_i^* + \Delta t \mathcal{L}(U_i^*) \\
U_i^{n+1} &= \frac{1}{2}(U_i^n + U_i^{**}),
\end{aligned} \qquad \text{(RK2)}$$

and the third-order SSP Runge-Kutta method of Gottlieb, Shu and Tadmor (2001), denoted RK3:

$$\begin{aligned}
U_i^* &= U_i^n + \Delta t \mathcal{L}(U_i^n) \\
U_i^{**} &= \frac{3}{4}U_i^n + \frac{1}{4}U_i^* + \frac{\Delta t}{4}\mathcal{L}(U_i^*) \\
U_i^{n+1} &= \frac{1}{3}U_i^n + \frac{2}{3}U_i^{**} + \frac{2\Delta t}{3}\mathcal{L}(U_i^{**}).
\end{aligned} \qquad \text{(RK3)}$$

4.2.5 Energy Preserving Schemes: Numerical Experiments 1–2

We have three different energy preserving schemes for the shallow water system. We denote the finite volume scheme (4.17) with the energy conservative fluxes (4.21), (4.23) and (4.26) as AEC, PEC and EEC schemes, respectively. Our aim in this section is to compare these three schemes on a series of numerical experiments in one space dimension.

Numerical experiment #1: one-dimensional dam break

We start with a standard one-dimensional dam-break Riemann problem with initial data,

$$h(x,0) = \begin{cases} 2 & \text{if } x < 0, \\ 1.5 & \text{if } x > 0, \end{cases} \quad u(x,0) \equiv 0. \qquad (4.27)$$

In this problem, the initial data itself is discontinuous, and the exact solution consists of a left going rarefaction and a right going shock. We set the acceleration due to gravity to be $g = 1$. The computational domain is $[-1, 1]$ with 100 mesh points. We use RK2 time stepping scheme at a CFL number of 0.45, and we compute up to time $t = 0.4$. All the computations are performed with transparent Neumann type boundary conditions.

To begin with, we show the heights computed with the AEC, PEC and EEC schemes. The results are shown in Figure 4.1. In this figure, we also show a time history of the energy computed with all the three schemes.

As shown in Figure 4.1, all the three schemes are capturing the essential properties of the correct solution. The dam-break problem has a shock, and energy should be dissipated at a shock. Since the AEC, EEC and PEC schemes are energy preserving, the energy dissipation at the shock does not take place. Instead, the energy preserving schemes simply redistribute the energy into smaller scales in the form of oscillations trailing the shocks. This behavior is standard for energy (entropy) conservative schemes; see Tadmor and Zhong (2008) and the references therein. There is very little difference between the computed solutions.

Next we compute the behavior of the energy as a function of time. Figure 4.1 shows growth of energy of order 10^{-4} for all three schemes. The energy plots of the EEC and PEC schemes lie on top of each other, while the AEC scheme produces slightly more energy – at least initialy. This small-magnitude energy *growth*, which may appear puzzling at first sight, is solely due to time discretization. In fact, an RK2 time discretization of entropy conservative scheme will *produce* energy of order $\mathcal{O}(\Delta t)^3$, which explains the growth of energy as seen in Figure 4.1. We note in passing that fully discrete entropy conservative time-integration schemes are discussed in Tadmor (2003), LeFloch, Mercier and Rohde (2002).

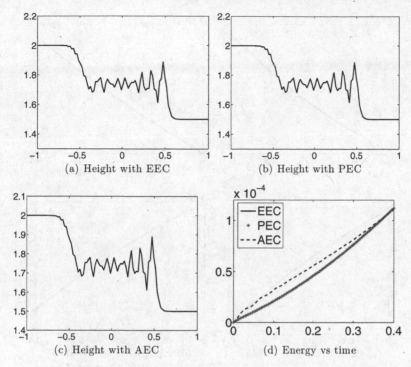

(a) Height with EEC

(b) Height with PEC

(c) Height with AEC

(d) Energy vs time

Fig. 4.1. Height h at $t = 0.4$ computed with the three energy conservative schemes on 100 mesh points with an RK2 method and a time history of energy computed with all the three schemes.

Effect of time stepping

The above comments need to be verified with more computations and variation of the time stepping routine and the CFL number. We therefore compare the semi-discrete EEC scheme which is discretized by the second- and third-order Runge-Kutta time stepping, RK2 and RK3, computed with CFL numbers 0.45 and 0.05. The energy history is computed and shown in Figure 4.2. This figure shows the effect time stepping schemes have on the energy balance. Comparing the RK2 results with two different CFL numbers reflects energy growth of order $\mathcal{O}(\Delta t)^3$. Thus, the energy growth of $\sim 10^{-7}$ corresponding to a CFL number 0.05 is three orders of magnitude smaller than the energy growth of $\sim 10^{-4}$ for a CFL number 0.45.

Next, we discuss the effect of using a higher-order time integration scheme. Observe that the third-order RK3 actually *dissipates* energy.

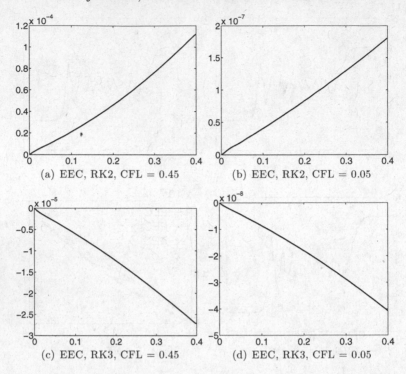

Fig. 4.2. A time history of energy E computed with the EEC scheme on 100 mesh points for two different RK schemes at two different CFL numbers.

Indeed, consideration of absolute stability regions implies that one needs at least third-order discretizations to enforce energy dissipation of entropy conservative scheme, e.g., Tadmor (2002). Moreover, energy dissipates at a rate of order $\mathcal{O}(\Delta t)^3$, ranging from $\sim 10^{-5}$ with CFL = 0.45 to $\sim 10^{-8}$ with CFL = 0.05. Similar results, which we omit, were found for the PEC and AEC schemes.

Effect of resolution: Numerical experiment #2

Another issue of interest for energy preserving schemes is the nature of their dispersive oscillations. To this end we compute the EEC-RK2 scheme with CFL number of 0.45 on four different meshes using 100, 400, 800 and 1600 mesh points, respectively. As shown in Figure 4.3, the solution becomes more and more oscillatory as the mesh is refined. Regardless of mesh size, the amplitude of the oscillations remains bounded and is of the order of the jump in the initial height. However, the fre-

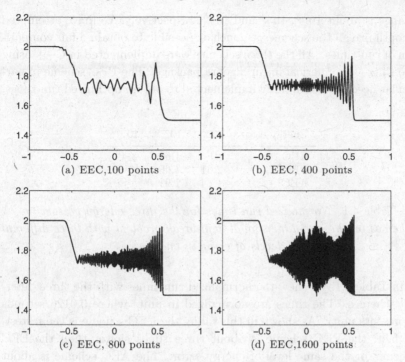

Fig. 4.3. Height h at time $t = 0.4$ computed with EEC scheme and RK2 at CFL 0.45 on four different meshes.

quency of the oscillations keeps increasing until a well-defined wave train is formed behind the shock. In fact, the frequency of the oscillations is at the scale of the mesh size. This wave-train is often referred to as a *modulation*, and is an interesting object of study, see e.g., Goodman and Lax (1988), Hou and Lax (1991) for more details). This behavior is expected as energy, which must be dissipated at shocks, is trapped by our energy preserving scheme, and non-linear dispersive effects redistributes this energy into the smallest resolvable scale on the computational grid, in the form of modulated high-frequencies.

Computational cost

Given the similarities in the behavior of the three energy preserving schemes, a key point is their computational cost. To compare the cost of these schemes we consider the above problem on a fixed mesh of 100 grid points and use the RK2 method of time integration. Then, we change the time step in three different computations so that the energy change is about 10^{-3}, 10^{-4} and 10^{-5}, respectively. We have attempted to optimize all the schemes as much as possible to obtain a fair comparison of run-times. All the three schemes were implemented in C++ using the Blitz++ numerical linear algebra package. The PEC scheme in particular needs to be carefully implemented to obtain optimized run-times.

Energy error	10^{-3}	10^{-4}	10^{-5}
EEC	1	1.69	3.49
PEC	2.61	4.68	10.47
AEC	1.14	1.91	-

Table 4.1. *Normalized run-times for the three energy preserving schemes on the one-dimensional dam-break problem with three different levels of error in energy.*

In Table 4.1, we show the normalized run-times with the three different schemes. The times are normalized in units where 0.0126 seconds represents unity. As shown in this table, the EEC scheme is the fastest. In fact, the PEC scheme is about three times slower than the EEC scheme for the same level of energy error. The AEC scheme is about 1.2 times slower than the EEC scheme. We were unable to obtain a time step small enough so that the AEC scheme gave an energy error of about 10^{-5}.

Summarizing, we see that all the three energy preserving schemes preserve energy to a satisfactory level. The energy dissipation/production due to explicit time stepping can be reduced by using smaller time steps and higher-order time integration methods. The presence of shocks in the computed solution results in high-frequency oscillations as energy is distributed into smaller scales. In terms of computational cost, the EEC scheme is the best, being about three times cheaper than the PEC scheme. This is expected as the PEC requires eigenvector decompositions at each cell interface. The EEC scheme, on the other hand, is explicit and requires only a few simple floating point operations.

4.2.6 Eddy Viscosity: Numerical Experiment 3

We obtain the one-dimensional form of the shallow water equations with eddy viscosity (4.1) by setting v and all change in y-direction to zero. The energy preservation of the inviscid form is no longer true in the context of (4.1) as energy will dissipate due to the viscous terms. The precise rate of energy dissipation is given by the one-dimensional form of the estimate (4.10). One of the key aims of designing energy preserving schemes is to obtain a faithful discretization of this energy dissipation balance.

We start by writing down the following scheme for the one-dimensional shallow water equations with eddy viscosity in semi-discrete form,

$$\frac{dU_i(t)}{dt} + \frac{1}{\Delta x}\left(\widetilde{F}_{i+\frac{1}{2}} - \widetilde{F}_{i-\frac{1}{2}}\right) = \frac{\nu}{\Delta x}\left(Q_{i+\frac{1}{2}} - Q_{i-\frac{1}{2}}\right). \qquad (4.28)$$

Here, $F_{i+\frac{1}{2}}$ is any energy preserving flux so that (4.18) holds, and $Q_{i+\frac{1}{2}}$ is a discretization of the viscous terms:

$$Q_{i+\frac{1}{2}} = \left[0, \overline{h}_{i+\frac{1}{2}}\left(\frac{u_{i+1} - u_i}{\Delta x}\right)\right]^\top. \qquad (4.29)$$

Note that (4.28) is a consistent discretization of (4.1) in one space dimension.

Lemma 4.1 *Consider the viscous form of the shallow water equations (4.1) in one space dimension. Let $E := \frac{1}{2}\left(hu^2 + gh^2\right)$ denote the energy and let $\widetilde{F}_{i+\frac{1}{2}}$ be any consistent, energy preserving numerical flux satisfying (4.18). Then, the solution U_i of the "eddy viscosity" scheme (4.28),(4.29) satisfies the discrete energy dissipation*

$$\frac{d}{dt}E(U_i(t)) = -\frac{1}{\Delta x}\left(\widetilde{H}_{i+\frac{1}{2}} - \widetilde{H}_{i-\frac{1}{2}}\right)$$
$$-\frac{\nu}{2}\left(\overline{h}_{i+\frac{1}{2}}\left(\frac{u_{i+1} - u_i}{\Delta x}\right)^2 + \overline{h}_{i-\frac{1}{2}}\left(\frac{u_i - u_{i-1}}{\Delta x}\right)^2\right),$$
$$(4.30)$$

where

$$\widetilde{H}_{i+\frac{1}{2}} := \left\langle \overline{V}_{i+\frac{1}{2}}, \ F_{i+\frac{1}{2}}\right\rangle - \overline{\Psi}_{i+\frac{1}{2}} - \overline{h}_{i+\frac{1}{2}}\,\overline{u}_{i+\frac{1}{2}}\frac{[\![u_{i+\frac{1}{2}}]\!]}{\Delta x}.$$

Summing over all i, we get the following energy dissipation estimate:

$$\frac{d}{dt}\sum_i E_i \equiv -\frac{\nu}{2}\sum_i(h_i + h_{i+1})\left(\frac{u_{i+1} - u_i}{\Delta x}\right)^2.$$

Hence the finite volume scheme (4.28) leads to a discrete form of the exact energy balance (4.10).

The proof follows by direct calculations with energy preserving fluxes; see Theorem 5.1 in Tadmor and Zhong (2008) for details.

Thus, any energy preserving flux together with a central discretization of the viscous terms as in (4.28) results in a faithful discretization of the energy balance for the viscous form of the shallow water equations. We will check this fact and study its implications by considering the following numerical experiment.

Numerical experiment with eddy viscosity

We consider the viscous form (4.1) in one space dimension together with the viscosity parameter $\nu = 0.01$ and the initial data given by the one-dimensional dam-break problem (4.27). Since the results with all the three energy preserving fluxes were so similar, we only report on the results obtained with the EEC scheme, together with a central discretization of viscous terms (4.28). Time integration is performed with an RK2 scheme with a CFL number that takes into account the viscous terms in the scheme. We choose the CFL number of 0.1 in subsequent computations. The results in Figure 4.4 show that the EEC scheme to-

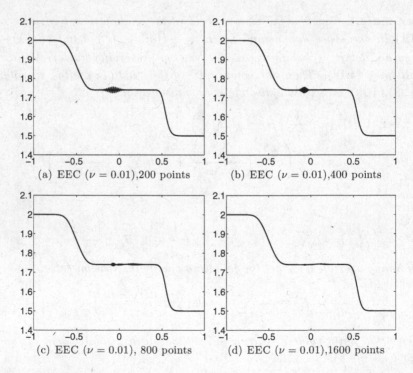

Fig. 4.4. Height h at time $t = 0.4$ computed with EEC scheme, RK2 at CFL 0.1 and central discretization of viscous terms (4.28) with eddy viscosity $\nu = 0.01$ on four different meshes.

gether with eddy viscosity results in a stable and robust approximation of the viscous shallow water equations. The right-going shock is dissipated as we are using a relatively large eddy viscosity $\nu = 0.01$. These results should be contrasted with plots from inviscid schemes shown in Figure 4.3. We see that the addition of the physical eddy viscosity leads to a dramatic reduction of the oscillations generated by the energy preserving scheme. Clearly, the addition of physical viscosity leads to an energy dissipation of the form (4.30), and the energy balance is faithfully discretized. However, we see from Figure 4.4 that small amplitude, high-frequency oscillations are still present on under-resolved meshes. As the mesh is refined, the amplitude of these oscillations is further reduced and the oscillations are hardly visible on a fine mesh of 1600 points. This behavior is consistent with expectations: the scales in the computed solution are given in terms of a *mesh Reynolds number* and we need to resolve them in order to obtain a solution without any oscillations. The mesh Reynolds number in this example appears to yield a mesh of roughly 1600 points to resolve all the scales in the viscous problem. An extreme illustration of the effect of eddy viscosity is obtained by comparing the approximate heights computed on 1600 mesh points with the EEC scheme and $\nu = 0$ (inviscid) and $\nu = 0.01$ (viscous) eddy viscosities. We consider the bottom right figures in Figure 4.3 and 4.4, respectively, and observe that the addition of viscosity dampens the oscillations dramatically.

4.2.7 Numerical Diffusion: Numerical Experiments 4–5

The previous section clearly illustrates the role of eddy viscosity in designing a stable numerical scheme for shallow water equations. The key observation is the need to resolve the viscous scales of the problem. The use of physical eddy viscosity is constrained by the fact that eddy viscosity in real physical applications is expected to be very small (smaller than 10^{-2}). This implies that we need to compute on very fine meshes in order to resolve all the viscous scales. Since this necessitates using large computational resources, we need to find an alternative way of designing stable numerical schemes for real world applications.

The standard way of designing stable schemes for problems with shocks in the finite volume framework is to add numerical diffusion. The numerical diffusion is built into the structure of numerical fluxes. In fact, standard finite volume fluxes $\widetilde{F}_{i+\frac{1}{2}}$ for the system (4.17) based on 3-point stencils can be written in the viscous form (consult Tadmor (1984))

$$F_{i+\frac{1}{2}} = F\left(U_i, U_{i+1}\right) = \frac{1}{2}\left(F(U_i) + F(U_{i+1})\right) - Q_{i+\frac{1}{2}}\left(V_{i+1} - V_i\right), \quad (4.31)$$

where $Q_{i+\frac{1}{2}}$ is a suitable numerical diffusion matrix coefficient. One of the fundamental results of Tadmor (1987) was to provide a criterion for whether the flux in (4.31) is entropy stable or not. We restate this result in the following lemma.

Lemma 4.2 (Tadmor (1987)) *Assume that the system of conservation laws* (4.14) *is equipped with an entropy/entropy flux pair* (E, F). *Let* $F_{i+\frac{1}{2}}$ *be a finite volume numerical flux consistent with* (4.14), *and let* $\widetilde{F}_{i+\frac{1}{2}}$ *be an entropy conservative numerical flux so that* (4.18) *holds. Let* $Q_{i+\frac{1}{2}}$ *and* $\widetilde{Q}_{i+\frac{1}{2}}$ *be the corresponding numerical diffusions associated with* $F_{i+\frac{1}{2}}$ *and* $\widetilde{F}_{i+\frac{1}{2}}$, *respectively. If*

$$Q_{i+\frac{1}{2}} \geq \widetilde{Q}_{i+\frac{1}{2}} \quad \forall\, i \quad\quad (4.32)$$

(in the sense that $Q_{i+\frac{1}{2}} - \widetilde{Q}_{i+\frac{1}{2}}$ *is a symmetric positive definite matrix), then the scheme*

$$\frac{dU_i(t)}{dt} = -\frac{1}{\Delta x}\left(F_{i+\frac{1}{2}} - F_{i-\frac{1}{2}}\right) \quad\quad (4.33)$$

is entropy stable, i.e., $\dfrac{d}{dt}\displaystyle\sum_i E(U_i(t)) \leq 0$.

In words — a finite volume flux, with *more* numerical viscosity than the entropy conservative flux, necessarily dissipates entropy and hence is stable. Several examples of entropy preserving fluxes were specified in Tadmor (1987) and in Tadmor (2003). We shall mention two in the context of the shallow water system (4.13).

(i) *Rusanov flux.* This energy stable flux takes the form,

$$F_{i+\frac{1}{2}}^{Rus} = \frac{1}{2}(F(U_i) + F(U_{i+1})) - \max_k \left\{|(\lambda_k)_i|, |(\lambda_k)_{i+1}|\right\}(U_{i+1} - U_i),$$
$$(4.34)$$

where $(\lambda_k)_i$ are the eigenvalues of $F'(U_i)$,

$$(\lambda_1)_i = u_i - \sqrt{gh_i}, \qquad (\lambda_2)_i = u_i + \sqrt{gh_i}.$$

Note that the diffusion is scaled by a local estimate of the speed of propagation.

(ii) *Roe flux.* This well-known flux is given by

$$F_{i+\frac{1}{2}}^{Roe1} = \frac{1}{2}\big(F(U_i) + F(U_{i+1})\big) - R_{i+\frac{1}{2}}|\Lambda_{i+\frac{1}{2}}|R_{i+\frac{1}{2}}^{-1}(U_{i+1} - U_i), \quad (4.35)$$

where

$$|\Lambda_{i+\frac{1}{2}}| = \text{diag}\{|(\lambda_1)_{i+\frac{1}{2}}|, |(\lambda_2)_{i+\frac{1}{2}}|\}$$

and

$$R_{i+\frac{1}{2}} = \begin{bmatrix} 1 & 1 \\ (\lambda_1)_{i+\frac{1}{2}} & (\lambda_2)_{i+\frac{1}{2}} \end{bmatrix}$$

for some suitable average states $U_{i+\frac{1}{2}}$. It was shown in Tadmor (1987) that the Roe flux, together with some appropriate entropy fix, is entropy stable. The choice of the parameters in the entropy fix was described in detail in Tadmor (2003).

It is well-known that the Rusanov flux (4.34) can be too diffusive for practical applications. The Roe flux (4.35) is much less diffusive and the shocks are captured sharply. In particular, choosing

$$h_{i+\frac{1}{2}} = \frac{h_i + h_{i+1}}{2}, \quad u_{i+\frac{1}{2}} = \frac{\sqrt{h_i}u_i + \sqrt{h_{i+1}}u_{i+1}}{\sqrt{h_i} + \sqrt{h_{i+1}}},$$

as the average states in (4.35) ensures that isolated single shocks are resolved exactly. However, the Roe flux lacks entropy stability and we refer to Tadmor (2003) for a discussion on entropy stable modifications of the Roe flux.

Here we propose a novel choice of a numerical diffusion flux for (4.13), in the spirit of a Roe flux, which is energy stable . The resulting numerical flux is designed in two steps. In the first step, we will replace the central part of the flux, $(f(U_i) + f(U_{i+1}))/2$, with the energy preserving fluxes satisfying (4.18); we can choose any of the three energy preserving fluxes for the shallow water equation. The next step is to design a numerical diffusion operator. Here, we need the following simple lemma.

Lemma 4.3 *Consider the shallow water equations in one dimension, together with energy $E = (gh^2 + hu^2)/2$ and the energy variables $V = \left[gh - \frac{u^2}{2}, u\right]^\top$. Let the values U_i, U_{i+1} across an interface be given.*

(i) *We have the identity*

$$[U_{i+\frac{1}{2}}] = (\overline{U_V})_{i+\frac{1}{2}}[V_{i+\frac{1}{2}}], \quad (4.36)$$

where

$$(\overline{U_V})_{i+\frac{1}{2}} = \frac{1}{g} \begin{bmatrix} 1 & \overline{u}_{i+\frac{1}{2}} \\ \overline{u}_{i+\frac{1}{2}} & (\overline{u^2})_{i+\frac{1}{2}} + g\overline{h}_{i+\frac{1}{2}} \end{bmatrix}$$

and $\overline{h}_{i+\frac{1}{2}}$ and $\overline{u}_{i+\frac{1}{2}}$ are the arithmetic averages across the interface.

(ii) *Define the scaled matrix of right eigenvectors of $F'(U_{i+\frac{1}{2}})$ for some averaged state $U_{i+\frac{1}{2}}$:*

$$R_{i+\frac{1}{2}} = \frac{1}{\sqrt{2g}} \begin{bmatrix} 1 & 1 \\ (\lambda_1)_{i+\frac{1}{2}} & (\lambda_2)_{i+\frac{1}{2}} \end{bmatrix}. \tag{4.37}$$

Then we have

$$R_{i+\frac{1}{2}} R_{i+\frac{1}{2}}^{\top} = (U_V)_{i+\frac{1}{2}}, \tag{4.38}$$

where U_V is the symmetric positive definite change of variables matrix

$$(U_V)_{i+\frac{1}{2}} = \frac{1}{g} \begin{bmatrix} 1 & u_{i+\frac{1}{2}} \\ u_{i+\frac{1}{2}} & u_{i+\frac{1}{2}}^2 - gh_{i+\frac{1}{2}} \end{bmatrix}$$

evaluated at the same average state.

The proof of the above identities follows by a straightforward calculation which we omit. Note that part (i) of the lemma above provides an appropriate average state at which the jump of the conservative variables can be expressed in terms of the jump of the energy variables. This state is not the arithmetic average of the conservative variables, but rather of the primitive variables. Part (ii) provides a suitable scaling for the eigenvectors.

This lemma can be used to design a suitable Roe-type diffusion operator. We consider the usual Roe diffusion, $Q_{i+\frac{1}{2}}^{Roe1}$ in (4.35), evaluated at the average states in (4.36) and use Lemma 4.3 to obtain,

$$
\begin{aligned}
Q_{i+\frac{1}{2}}^{Roe1} \llbracket U_{i+\frac{1}{2}} \rrbracket &= R_{i+\frac{1}{2}} |\Lambda_{i+\frac{1}{2}}| R_{i+\frac{1}{2}}^{-1} \llbracket U_{i+\frac{1}{2}} \rrbracket, \\
&= R_{i+\frac{1}{2}} |\Lambda_{i+\frac{1}{2}}| R_{i+\frac{1}{2}}^{-1} \overline{(U_V)}_{i+\frac{1}{2}} \llbracket V_{i+\frac{1}{2}} \rrbracket, \quad \text{(by (4.36))} \\
&= R_{i+\frac{1}{2}} |\Lambda_{i+\frac{1}{2}}| R_{i+\frac{1}{2}}^{-1} R_{i+\frac{1}{2}} R_{i+\frac{1}{2}}^{\top} \llbracket V_{i+\frac{1}{2}} \rrbracket, \quad \text{(by (4.38))} \\
&= R_{i+\frac{1}{2}} |\Lambda_{i+\frac{1}{2}}| R_{i+\frac{1}{2}}^{\top} \llbracket V_{i+\frac{1}{2}} \rrbracket.
\end{aligned}
$$

These formal calculations suggest that we can choose a numerical diffusion operator in terms of the energy variables with proper scaling of the eigenvectors and evaluation of the eigenvalues and eigenvectors at an appropriate state.

Now, let $\widetilde{F}_{i+\frac{1}{2}}$ be any energy preserving numerical flux (4.18). We introduce the following Roe-type flux:

$$F_{i+\frac{1}{2}}^{ERoe1} = F^{ERoe1}(U_i, U_{i+1}) := \widetilde{F}_{i+\frac{1}{2}} - \frac{1}{2}R_{i+\frac{1}{2}}|\Lambda_{i+\frac{1}{2}}|R_{i+\frac{1}{2}}^{\top}[V_{i+\frac{1}{2}}],$$
$$(4.39)$$

where $R_{i+\frac{1}{2}}$ is as defined in (4.37), and

$$|\Lambda_{i+\frac{1}{2}}| = \mathrm{diag}\left\{|\overline{u}_{i+\frac{1}{2}} - \sqrt{g\overline{h}_{i+\frac{1}{2}}}|, \ |\overline{u}_{i+\frac{1}{2}} + \sqrt{g\overline{h}_{i+\frac{1}{2}}}|\right\}.$$

$F_{i+\frac{1}{2}}^{ERoe1}$ is clearly consistent. It differs from the usual Roe flux (4.35) in that the central term is an energy preserving flux, and that the diffusion is given in terms of energy variables rather than conservative variables. We refer to $F_{i+\frac{1}{2}}^{ERoe1}$ as *entropic Roe flux*: indeed, the associated scheme is entropy stable as quantified in the following theorem.

Theorem 4.2 *Let U_i be the solution of the entropic Roe scheme*

$$\frac{dU_i(t)}{dt} = -\frac{1}{\Delta x}\left(F_{i+\frac{1}{2}}^{ERoe1} - F_{i-\frac{1}{2}}^{ERoe1}\right). \qquad (4.40)$$

Then the following discrete energy estimate holds:

$$\frac{d}{dt}(E(U_i)(t)) = -\frac{1}{\Delta x}(\widetilde{H}_{i+\frac{1}{2}} - \widetilde{H}_{i-\frac{1}{2}})$$
$$-\frac{1}{4\Delta x}\left\langle [V_{i+\frac{1}{2}}], R_{i+\frac{1}{2}}|\Lambda_{i+\frac{1}{2}}|R_{i+\frac{1}{2}}^{\top}[V_{i+\frac{1}{2}}]\right\rangle \qquad (4.41)$$
$$-\frac{1}{4\Delta x}\left\langle [V_{i-\frac{1}{2}}], R_{i-\frac{1}{2}}|\Lambda_{i-\frac{1}{2}}|R_{i-\frac{1}{2}}^{\top}[V_{i-\frac{1}{2}}]\right\rangle,$$

where

$$\widetilde{H}_{i+\frac{1}{2}} := \left\langle \overline{V}_{i+\frac{1}{2}}, \widetilde{F}_{i+\frac{1}{2}}\right\rangle - \overline{\Psi}_{i+\frac{1}{2}} + \frac{1}{2}\left\langle \overline{V}_{i+\frac{1}{2}}, R_{i+\frac{1}{2}}|\Lambda_{i+\frac{1}{2}}|R_{i+\frac{1}{2}}^{\top}[V_{i+\frac{1}{2}}]\right\rangle.$$

Summing over all i, we get the following energy dissipation estimate:

$$\frac{d}{dt}\sum_i E_i = -\frac{1}{2\Delta x}\sum_i \left\langle [V_{i+\frac{1}{2}}], R_{i+\frac{1}{2}}|\Lambda_{i+\frac{1}{2}}|R_{i+\frac{1}{2}}^{\top}[V_{i+\frac{1}{2}}]\right\rangle.$$

In particular, since the matrix $R|\Lambda|R^{\top}$ is symmetric non-negative definite, the scheme (4.40) is energy stable.

The proof of this theorem is a straightforward generalization of the proof of entropy stability with numerical diffusion in entropy variables (consult Tadmor (1987),(2003)): one multiplies both sides of (4.33) by V_i and summation by parts while taking into account the energy preserving flux and the special form of the diffusion matrix in (4.39) yields the energy estimate (4.41).

Remark 4.2 The idea of defining the numerical diffusion operator for finite volume fluxes in terms of entropy variables is not new but was proposed in Tadmor (1986),(1987) and subsequently, was used in Hughes Franca and Mallet (1986), Khalfallah and Lerat (1989) and others. The specific structure of the diffusion operator in (4.39), however, is novel for the shallow water equations. The flux (4.39) is less dissipative than the Rusanov flux and does not need an entropy fix like the Roe flux.

Numerical experiments with numerical diffusion

We present a series of one-dimensional numerical experiments of schemes with added numerical diffusion. We will test both the standard Rusanov and Roe schemes (4.34) and (4.35), and compare the results with the new entropy stable Roe-type scheme (4.39). The latter employs an energy preserving flux $\widetilde{F}_{i+\frac{1}{2}}$ and we chose to use here the EEC flux (4.26). The resulting entropic scheme (4.39), (4.26) is denoted *ERoe*. We start with the one-dimensional dam-break problem with initial data (4.27) and domain $[-1, 1]$. We compute the approximate solutions with the Roe, Rusanov and ERoe schemes on a uniform mesh with 100 mesh points and plot the approximate heights in Figure 4.5. As shown in this figure,

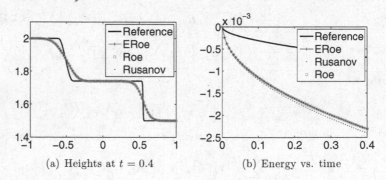

(a) Heights at $t = 0.4$ (b) Energy vs. time

Fig. 4.5. Solutions computed with the Roe, Rusanov and ERoe schemes for 1D dam break problem with 100 mesh points. Left: Height at $t = 0.4$, Right: Energy vs. time.

the three schemes behave as expected. The Rusanov scheme is slightly more diffusive than the Roe-type schemes, with the right-going shock being slightly more smeared. However, the difference is not much. Both the Roe and ERoe schemes compute the solution quite well at this coarse mesh resolution. The difference between the two schemes is negligible for this problem. Figure 4.5(b) shows the energy history, and we observe that both the Roe and ERoe schemes dissipate energy in an identical manner. The Rusanov scheme dissipates more energy than either of the two Roe-type schemes. All three schemes are energy-stable, in that they diffuse more energy than the physical solution.

The above numerical experiment illustrated that the new ERoe scheme is robust and very similar in behavior to the standard Roe scheme. The ERoe scheme, however, was proved to be energy stable, thus providing an entropy fix for the Roe scheme. The following two numerical experiments will illustrate these points.

Numerical experiment #4: A dam-break problem

We consider a different one-dimensional dam-break problem for the shallow water equations with initial data

$$h(x, 0) = \begin{cases} 15 & \text{if } x < 0, \\ 1 & \text{if } x > 0, \end{cases} \qquad u(x, 0) \equiv 0 \qquad (4.42)$$

in the computational domain $[-1, 1]$, and we set $g = 1$. The difference between the above initial data and the one in experiment #1 in (4.27) is the large initial jump in height. The exact solution still consists of a right going shock and a left going rarefaction, separated by a region of a constant height of 5. We compute the solutions with the Roe and ERoe schemes on a mesh of 100 points up to time $t = 0.15$. We also compute the Rusanov scheme and a reference solution on a finer mesh of 3200 points. As shown in Figure 4.6, the ERoe scheme provides a good approximation to the exact solution. The Roe scheme approximates the shock equally well; however it produces a spurious steady shock at approximately $x = 0$. The magnitude of this shock is about 2.5 and it is entirely unphysical. This behavior of spurious waves generated by the Roe scheme is well known (see LeVeque (2002)), and the scheme needs to be entropy fixed, as in, e.g., Tadmor (2003). This is a key difference between the Roe and ERoe schemes. They seem to have the same resolution, but the Roe scheme (at least in its non-entropy fixed version) is unstable in some cases.

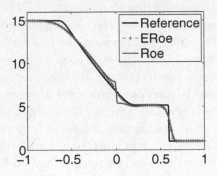

Fig. 4.6. Heights computed at time $t = 0.4$ with the Roe and ERoe schemes for numerical experiments with 100 mesh points.

Numerical experiment #5: An expansion problem

Another well documented issue with the standard Roe scheme is its lack of positivity, i.e., it may produce negative heights. The ERoe scheme might also suffer from this problem. We present an experiment where the Roe scheme leads to negative heights, whereas the ERoe scheme retains positivity. We consider initial data

$$h(x,0) \equiv 1, \quad u(x,0) = \begin{cases} -4 & \text{if} \quad x < 0, \\ 4 & \text{if} \quad x > 0. \end{cases} \tag{4.43}$$

The computational domain is $[-1, 1]$ with transparent boundary conditions and $g = 1$. The initial velocity is chosen such that the fluid is pushed away from the center of the domain in both directions, leading to the formation of an almost dry zone, with very low values of height in the center of the domain. We compute the solutions with the Roe and ERoe schemes for a uniform mesh of 100 mesh points. The results from Figure 4.7 show that the ERoe scheme approximates the solution quite well and retains positivity of the height in the near-dry zone around $x = 0$. However, the Roe scheme fails in this case due to negative heights which form at $t \sim 0.006$ (we therefore show the computed height just before this failure). This illustrates how the ERoe stabilizes the positivity failure of the Roe scheme; we do not, however, claim that the ERoe is positivity preserving, and one might expect it to fail if we increase the velocity in (4.43).

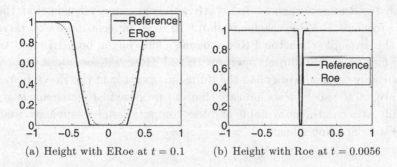

(a) Height with ERoe at $t = 0.1$ (b) Height with Roe at $t = 0.0056$

Fig. 4.7. Solutions computed with the Roe, and ERoe schemes for 1D expansion problem with 100 mesh points. Left: Height at $t = 0.4$ with ERoe, Right: Height at $t = 0.01$ with Roe.

Computational costs

The above examples illustrate that the ERoe is quite robust and energy stable. It is as accurate as the standard Roe scheme and is more stable with respect to energy and positivity. A natural question is whether this increased stability comes at a price. Hence, we consider the standard one-dimensional dam-break problem with initial data (4.27) and compute the solutions with Roe, ERoe and Rusanov schemes. We compute a reference solution with the Rusanov scheme on a fine mesh of 3200 mesh points and consider L^1 errors in height with respect to this reference solution. Next, we compute with three different mesh resolutions and find the meshes on which each of the schemes will yield a relative error of 1, 0.5 and 0.1 percent, respectively. The run times are then normalized such that unit time corresponds to 0.0205 seconds. We present the normalized run times with all the three schemes in Table 4.2. We see

Relative Energy error	1	0.5	0.1
Rusanov	1.05	8.24	203.41
Roe	1.15	8.43	208.29
ERoe	1	7.36	171.7

Table 4.2. *Normalized run-times for the Rusanov, Roe and ERoe schemes on the one-dimensional dam-break problem with three different levels of relative error in height.*

from the table that the ERoe scheme has the lowest computational cost for the same level of error in all the three different error levels. Surprisingly the Rusanov scheme is less costly and hence more efficient than the Roe scheme. The Roe scheme is about 15 to 20 percent more expensive in this example than the ERoe scheme. This might be attributed to the fact that the diffusion operator in the ERoe scheme might involve less computational work than the diffusion operator in the Roe scheme. Hence, all the above tests indicate that our proposed ERoe scheme is actually as accurate, more stable and less computationally expensive than the standard Roe scheme.

4.2.8 Second-order Extensions: Numerical Experiment 6

The ERoe flux (4.39) has two parts: A second-order symmetric energy preserving part and a numerical diffusion operator. The numerical diffusion is first-order accurate in space and hence, the overall accuracy of the ERoe scheme is restricted to first-order. This leads to smeared shocks, as we observed in the previous numerical experiments. It is straightforward to extend this scheme to obtain second-order accuracy in space. For this purpose, we follow the fairly standard approach and replace the piecewise-constant cell averages U_i in (4.33) with a non-oscillatory piecewise linear reconstruction, e.g., Kurganov and Tadmor (2002), of the form

$$p_i(x) = U_i + \frac{U_i'}{\Delta x}(x - x_i). \tag{4.44}$$

The numerical derivative U_i' is given by

$$U_i' := \text{minmod}\left(U_{i+1} - U_i, \ \frac{1}{2}(U_{i+1} - U_{i-1}), \ U_i - U_{i-1}\right), \tag{4.45}$$

where minmod is the function

$$\text{minmod}(a, b, c) := \begin{cases} \text{sgn}(a) \min\{|a|, |b|, |c|\}, & \text{if } \text{sgn}(a) = \text{sgn}(b) = \text{sgn}(c) \\ 0, & \text{otherwise.} \end{cases}$$

The reconstruction is fairly standard, and we can use other limiters for the numerical derivatives. Let

$$U_i^E = p_i(x_{i+\frac{1}{2}}), \quad U_i^W = p_i(x_{i-\frac{1}{2}}).$$

The second-order version of the ERoe flux (4.39) is then

$$F_{i+\frac{1}{2}}^{ERoe2} = \widetilde{F}(U_i^E, U_{i+1}^W) - \frac{1}{2}R|\Lambda|R^\top(V_{i+1}^W - V_i^E), \tag{4.46}$$

where $\widetilde{F}$ is any energy preserving flux (4.18), V_i^E, V_i^W are the energy variables evaluated at U_i^E and U_i^W, and the matrices R and Λ are the corresponding matrices in (4.36) and (4.38) with respect to the reconstructed values U_i^E and U_{i+1}^W. The resulting finite volume scheme with this flux is formally second-accurate in space. However, we can no longer prove that it is energy stable like the first-order version. Instead we will test with this second-order flux in our next numerical experiment.

Numerical experiment #6: Second-order computation of dam-break problem

We consider the standard one-dimensional dam-break problem with initial data (4.27) and consider the first-order (4.39) and second-order (4.46) versions of the ERoe scheme. We compute heights on a uniform mesh of 100 mesh points. Both schemes are integrated in time using a standard second-order RK2 method with a CFL number of 0.45, and we show the computed heights at time $t = 0.4$ in Figure 4.8. This figure

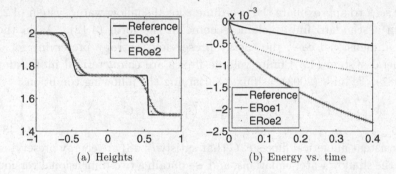

(a) Heights   (b) Energy vs. time

Fig. 4.8. Solutions computed with the first and second-order versions of the ERoe scheme with 100 mesh points.

clearly shows the effect of a higher order of accuracy. The shock and rarefaction are both resolved to a much better extent with the second-order scheme. Furthermore, the energy dissipation with the second-order scheme is much less than that of the first-order scheme. The second-order scheme seems to be energy stable in this example although we were unable to obtain a proof of this. Even higher order extensions can be performed by using fairly standard ENO and WENO type reconstructions.

4.3 The Two-Dimensional Problem

The analysis and numerics presented in Section 4.2 for the one-dimensional problem can be easily extended to the two-dimensional shallow water equations (4.2). We will be brief in the exposition, as most of the details are similar to the one-dimensional case. To begin with, the energy variables for the two-dimensional form of the equations (4.2) and the associated potentials are given by

$$V = \left[gh - \frac{u^2 + v^2}{2}, u, v \right]^\top, \quad \Psi = \frac{1}{2} guh^2, \quad \Phi = \frac{1}{2} gvh^2, \qquad (4.47)$$

where Ψ and Φ are the energy potentials in the x- and y-directions, respectively.

4.3.1 Energy Preserving Schemes: Numerical Experiment 7

We seek to approximate the two-dimensional shallow water system (4.2) with a standard finite volume scheme of the form (4.12). As in the one-dimensional case, the energy preserving (entropy preserving for a general system (4.3)) finite volume fluxes are characterized in Tadmor (1987), Tadmor (2003), as fluxes satisfying the following conditions:

$$\left\langle \left[V_{i+\frac{1}{2},j} \right], \widetilde{F}_{i+\frac{1}{2},j} \right\rangle = \left[\Psi_{i+\frac{1}{2},j} \right], \quad \left\langle \left[V_{i,j+\frac{1}{2}} \right], \widetilde{G}_{i,j+\frac{1}{2}} \right\rangle = \left[\Phi_{i,j+\frac{1}{2}} \right].$$
$$(4.48)$$

Consistent numerical fluxes F, G that satisfy (4.48) are energy preserving for the shallow water equations, and we obtain a two-dimensional version of Theorem 4.1 (we skip the details and refer the reader to Tadmor (1987) for details).

In the one-dimensional case, we obtained three different schemes satisfying (4.18). The flux of the form (4.20) can be similarly defined for the two-dimensional case. However, we failed to explicitly compute the averages in phase space and obtain a simple explicit flux as we did in the one-dimensional case. This is largely due to the increased complexity of the resulting algebraic expressions. Hence, we do not have a flux analogous to the AEC flux (4.21) for the two-dimensional case.

The pathwise approach of Tadmor (2003) can be easily extended to cover the two-dimensional case, and has been done so in Tadmor and Zhong (2008). We skip the details of the flux and refer the reader to Algorithm 3.1 in Tadmor and Zhong (2008) for the description of this flux. The pathwise flux is a natural generalization of the PEC flux (4.23), and we continue referring to the resulting scheme as the PEC scheme.

It is straightforward to extend the explicit differencing based EEC flux (4.26) to the two-dimensional case. We carry out the steps indicated in the derivation of (4.26) for the two-dimensional case, and obtain the following entropy conservative flux:

$$\widetilde{F}_{i+\frac{1}{2},j} = \begin{bmatrix} \overline{h}_{i+\frac{1}{2},j}\overline{u}_{i+\frac{1}{2},j} \\ \overline{h}_{i+\frac{1}{2},j}(\overline{u}_{i+\frac{1}{2},j})^2 + \frac{g}{2}(\overline{h^2})_{i+\frac{1}{2},j} \\ \overline{h}_{i+\frac{1}{2},j}\overline{u}_{i+\frac{1}{2},j}\ \overline{v}_{i+\frac{1}{2},j} \end{bmatrix}, \qquad (4.49a)$$

and

$$\widetilde{G}_{i,j+\frac{1}{2}} = \begin{bmatrix} \overline{h}_{i,j+\frac{1}{2}}\overline{v}_{i,j+\frac{1}{2}} \\ \overline{h}_{i,j+\frac{1}{2}}\overline{u}_{i,j+\frac{1}{2}}\ \overline{v}_{i,j+\frac{1}{2}} \\ \overline{h}_{i,j+\frac{1}{2}}(\overline{v}_{i,j+\frac{1}{2}})^2 + \frac{g}{2}(\overline{h^2})_{i,j+\frac{1}{2}} \end{bmatrix}. \qquad (4.49b)$$

It is easy to check that the above fluxes satisfy (4.48) and are consistent. This flux is also trivial to implement in code as we simply need to evaluate averages. It is symmetric in its arguments and reduces to (4.26) for the one-dimensional case. We will denote the resulting scheme with (4.49) as the EEC scheme.

Both the PEC scheme and EEC scheme are energy preserving for the two-dimensional shallow water equations. We compare their numerical behavior in the following numerical experiment.

Numerical Experiment #7: 2D cylindrical dam-break

We consider the shallow water equations (4.2) in the domain $[-1,1] \times [-1,1]$ with initial data

$$h(x,y,0) = \begin{cases} 2, & \text{if } \sqrt{x^2 + y^2} < 0.5 \\ 1, & \text{otherwise} \end{cases}, \quad u(x,y,0) \equiv 0. \qquad (4.50)$$

The initial data represents the breaking of a cylindrical dam, and the problem has radial symmetry. We set $g = 1$. The exact solution consists of a circular shock moving out and a rarefaction moving inward. We compute with both EEC and PEC schemes on a uniform 50×50 mesh. The computations are performed with an RK2 time integration routine and a CFL number of 0.45.

As shown in Figure 4.9, the energy preserving schemes show very little difference in the numerical results. As in the one-dimensional cases, there are oscillations for both schemes as the energy is redistributed into smaller scales, and dispersive effects dominate due to the lack of any diffusive mechanism. We also show the energy errors in time in Figure 4.10. The results from Figure 4.10 show that the energy errors for

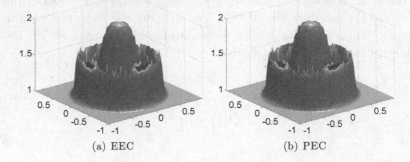

(a) EEC   (b) PEC

Fig. 4.9. Approximate heights for the cylindrical dam-break problem at time $t = 0.2$ computed on a uniform 50×50 mesh with both EEC and PEC schemes.

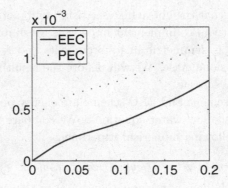

Fig. 4.10. Energy vs. time for the 2-d cylindrical dam-break problem and EEC, PEC schemes for a 50×50 mesh and RK2 with CFL 0.45.

both schemes are quite low for this extremely coarse mesh and second-order time integration with a moderately high CFL number. The results are consistent with those observed for the one-dimensional problem in Section 4.2. However, it seems that the PEC scheme generates a larger error in energy of about 2-3 times more than the EEC scheme.

The errors in energy are purely due to the time integration. As in one dimension, we observe a considerable reduction of the energy error by decreasing the time step. A sample computation is presented in Figure 4.11, where the energy errors generated with the EEC scheme and an RK2 method for two different CFL numbers is shown. The results in Figure 4.11 show that halving the CFL number (i.e., halving the time step) reduces the energy error by a factor of eight. Thus, the energy errors behave like $\mathcal{O}(\Delta t)^3$, as in the one-dimensional case. As in the one-

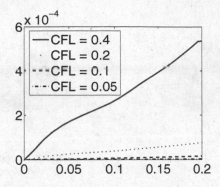

Fig. 4.11. Energy vs. time for the 2-d cylindrical dam-break problem and EEC scheme for a 100×100 mesh and RK2 with four different CFL numbers.

dimensional case, going to an RK3 time integration will further reduce the energy decay.

As stated before, energy preserving schemes lead to oscillations near shocks as there is no dissipative mechanism. In the one-dimensional case, these oscillations increased in frequency forming a modulated behavior as the mesh was refined. The same effects hold true in the two-dimensional case. We illustrate this behavior by presenting the approximate heights computed with the EEC scheme for a 100×100 mesh and a 200×200 mesh with RK2 time-integration in Figure 4.12. The figure clearly shows that increasing the mesh resolution results in oscillations of higher frequency although the magnitude of oscillations remains bounded. This is clear evidence of the dispersive behavior of the schemes in the absence of diffusion.

Computational costs

There are minor differences in the quality of the solutions obtained with the EEC and PEC schemes in two dimensions. The main difference lies in the simplicity of the EEC scheme and its low computational cost. As in one dimension, we illustrate the efficiency of the EEC scheme vis-à-vis the PEC scheme by fixing a uniform 50×50 mesh and lowering the time step to obtain energy errors of $10^{-3}, 10^{-4}$ and 10^{-5}, respectively. The resulting run times are then shown in Table 4.3. All the times are normalized so that unit time is taken to be 0.0617 seconds. As shown in Table 4.3, the PEC scheme is about 5 to 6 times more expensive than the EEC scheme. This is to be expected, as the PEC scheme was about 2.5 to 3 times more expensive in one dimension (see Table 4.1). The EEC

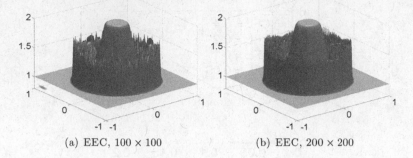

(a) EEC, 100×100 (b) EEC, 200×200

Fig. 4.12. Approximate heights for the cylindrical dam-break problem at time $t = 0.4$ computed on a uniform 100×100 and 200×200 meshes with the EEC and PEC scheme.

Energy error	10^{-3}	10^{-4}	10^{-5}
EEC	1	2.18	5.07
PEC	4.79	11.5	30.47

Table 4.3. *Normalized run-times for the two energy preserving schemes on the two-dimensional cylindrical dam-break problem with three different levels of error in energy.*

scheme provides a considerable speed up if one is interested in higher dimensional computations.

4.3.2 Eddy Viscosity: Numerical Experiment #8

By a simple generalization of (4.29), the energy preserving schemes can be used together with a central discretization of the viscous terms to obtain a scheme that is energy stable and with energy dissipating at a rate dictated by the viscous terms in (4.1). Without going into details, we observe that we will obtain a discrete form of (4.10) (a two-dimensional generalization of (4.30)). Thus, combining energy preserving schemes with a central discretization yields a faithful discretization of the continuous energy decay estimate.

We test this contention on a numerical example by considering the two-dimensional cylindrical dam-break problem (4.50) with eddy viscosity $\nu = 0.01$ and show the results in Figure 4.13. As expected, the presence of eddy viscosity serves to dramatically reduce the oscillations in the energy preserving schemes. Since the resulting solution has very low

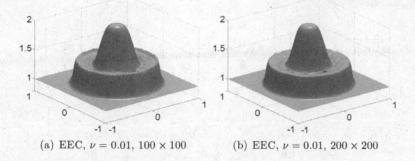

(a) EEC, $\nu = 0.01$, 100×100 (b) EEC, $\nu = 0.01$, 200×200

Fig. 4.13. Approximate heights for the cylindrical dam-break problem with eddy viscosity $\nu = 0.01$ at time $t = 0.4$ computed on uniform 100×100 and 200×200 meshes with the EEC scheme.

amplitude oscillations, it appears that the reasonably coarse 200×200 mesh is enough to resolve all the viscous scales. This is similar to what we observed in one dimension. Using an energy preserving scheme together with central viscous terms leads to an energy stable discretizations of the system, and the oscillations obtained for the inviscid problem are damped on a resolved mesh when one adds eddy viscosity.

4.3.3 Numerical Diffusion: Numerical Experiment #9

As in the one-dimensional case, we can use the energy preserving schemes to design suitable numerical diffusion operators leading to energy stable discretizations of the inviscid shallow water system (4.2), or the under-resolved viscous form (4.1). We imitate the strategy used in designing the energy stable Roe-type flux (4.39) in Section 4.2 and present its two-dimensional generalization. We start with the following lemma, which is a generalization of Lemma 4.3.

Lemma 4.4 *Consider the shallow water equations in two dimensions (4.2). Let the values $U_{i,j}, U_{i+1,j}, U_{i,j+1}$ across an interface be given.*
[i] *We have the identities*

$$\left[\!\left[U_{i+\frac{1}{2},j}\right]\!\right] = \overline{(U_V)}_{i+\frac{1}{2},j}\left[\!\left[V_{i+\frac{1}{2},j}\right]\!\right], \qquad \left[\!\left[U_{i,j+\frac{1}{2}}\right]\!\right] = \overline{(U_V)}_{i,j+\frac{1}{2}}\left[\!\left[V_{i,j+\frac{1}{2}}\right]\!\right],$$
$$(4.51)$$

where

$$\overline{(U_V)}_{i+\frac{1}{2},j} := \frac{1}{g} \begin{bmatrix} 1 & \overline{u}_{i+\frac{1}{2},j} & \overline{v}_{i+\frac{1}{2},j} \\ \overline{u}_{i+\frac{1}{2},j} & \overline{u}^2_{i+\frac{1}{2},j} + g\overline{h}_{i+\frac{1}{2},j} & \overline{u}_{i+\frac{1}{2},j}\,\overline{v}_{i+\frac{1}{2},j} \\ \overline{v}_{i+\frac{1}{2},j} & \overline{u}_{i+\frac{1}{2},j}\,\overline{v}_{i+\frac{1}{2},j} & \overline{v}^2_{i+\frac{1}{2},j} + g\overline{h}_{i+\frac{1}{2},j} \end{bmatrix}$$

and

$$\overline{(U_V)}_{i,j+\frac{1}{2}} := \frac{1}{g} \begin{bmatrix} 1 & \overline{u}_{i,j+\frac{1}{2}} & \overline{v}_{i,j+\frac{1}{2}} \\ \overline{u}_{i,j+\frac{1}{2}} & \overline{u}^2_{i,j+\frac{1}{2}} + g\overline{h}_{i,j+\frac{1}{2}} & \overline{u}_{i,j+\frac{1}{2}}\,\overline{v}_{i,j+\frac{1}{2}} \\ \overline{v}_{i,j+\frac{1}{2}} & \overline{u}_{i,j+\frac{1}{2}}\,\overline{v}_{i,j+\frac{1}{2}} & \overline{v}^2_{i,j+\frac{1}{2}} + g\overline{h}_{i,j+\frac{1}{2}} \end{bmatrix}.$$

[ii] *Define the scaled matrices of right eigenvectors of* $F'(U_{i+\frac{1}{2},j})$ *and* $G'(U_{i,j+\frac{1}{2}})$ *for averaged states* $U_{i+\frac{1}{2},j}$ *and* $U_{i,j+\frac{1}{2}}$:

$$R^x_{i+\frac{1}{2},j} = \frac{1}{\sqrt{2g}} \begin{bmatrix} 1 & 0 & 1 \\ \overline{u}_{i+\frac{1}{2},j} - \sqrt{g\overline{h}_{i+\frac{1}{2},j}} & 0 & \overline{u}_{i+\frac{1}{2},j} + \sqrt{g\overline{h}_{i+\frac{1}{2},j}} \\ \overline{v}_{i+\frac{1}{2},j} & \sqrt{g\overline{h}_{i+\frac{1}{2},j}}\,\overline{v}_{i+\frac{1}{2},j} \end{bmatrix},$$

$$R^y_{i,j+\frac{1}{2}} = \frac{1}{\sqrt{2g}} \begin{bmatrix} 1 & 0 & 1 \\ \overline{u}_{i,j+\frac{1}{2}} & -\sqrt{g\overline{h}_{i,j+\frac{1}{2}}} & \overline{u}_{i,j+\frac{1}{2}} \\ \overline{v}_{i,j+\frac{1}{2}} - \sqrt{g\overline{h}_{i,j+\frac{1}{2}}} & 0 & \overline{v}_{i,j+\frac{1}{2}} + \sqrt{g\overline{h}_{i,j+\frac{1}{2}}} \end{bmatrix}.$$

$$(4.52)$$

Then we have

$$R^x_{i+\frac{1}{2},j}\left(R^x_{i+\frac{1}{2},j}\right)^\top = \overline{(U_V)}_{i+\frac{1}{2},j}, \qquad R^y_{i,j+\frac{1}{2}}\left(R^y_{i,j+\frac{1}{2}}\right)^\top = \overline{(U_V)}_{i,j+\frac{1}{2}}.$$

$$(4.53)$$

The proof follows from direct calculations. Using this lemma, we follow Section 4.2.7 and define the following Roe-type flux:

$$F^{ERoe}_{i+\frac{1}{2},j} := \widetilde{F}_{i+\frac{1}{2},j} - \frac{1}{2}R^x_{i+\frac{1}{2},j}|\Lambda^x_{i+\frac{1}{2},j}|\left(R^x_{i+\frac{1}{2},j}\right)^\top [V_{i+\frac{1}{2},j}],$$

$$G^{ERoe}_{i,j+\frac{1}{2}} := \widetilde{G}_{i,j+\frac{1}{2}} - \frac{1}{2}R^y_{i,j+\frac{1}{2}}|\Lambda^y_{i,j+\frac{1}{2}}|\left(R^y_{i,j+\frac{1}{2}}\right)^\top [V_{i,j+\frac{1}{2}}],$$

$$(4.54)$$

where $\widetilde{F}_{i+\frac{1}{2},j}, \widetilde{G}_{i,j+\frac{1}{2}}$ are any pair of consistent, energy preserving fluxes satisfying (4.48), V is the vector of energy variables, $R^x_{i+\frac{1}{2},j}$ and $R^y_{i,j+\frac{1}{2}}$ are defined in (4.52) and

$$\left|\Lambda^x_{i+\frac{1}{2},j}\right| = \text{diag}\left\{\left|\overline{u}_{i+\frac{1}{2},j} - \sqrt{g\overline{h}_{i+\frac{1}{2},j}}\right|, \left|\overline{u}_{i+\frac{1}{2},j}\right|, \left|\overline{u}_{i+\frac{1}{2},j} + \sqrt{g\overline{h}_{i+\frac{1}{2},j}}\right|\right\},$$

$$\left|\Lambda^y_{i,j+\frac{1}{2}}\right| = \text{diag}\left\{\left|\overline{v}_{i,j+\frac{1}{2}} - \sqrt{g\overline{h}_{i,j+\frac{1}{2}}}\right|, \left|\overline{v}_{i,j+\frac{1}{2}}\right|, \left|\overline{v}_{i,j+\frac{1}{2}} + \sqrt{g\overline{h}_{i,j+\frac{1}{2}}}\right|\right\}.$$

The above flux is clearly consistent. A simple generalization of Theorem 4.2 shows that the finite volume scheme based on fluxes (4.54) is energy stable. We skip the details of this estimate as it is a straightforward generalization of (4.41).

We denote the two-dimensional entropic Roe-type scheme with energy preserving fluxes (4.49), (4.54) as the ERoe scheme and compare it with standard Rusanov and Roe schemes. We consider the two-dimensional cylindrical dam-break problem (4.50) and compute with the standard Roe and ERoe schemes on a uniform 100×100 mesh. Figure 4.14 shows

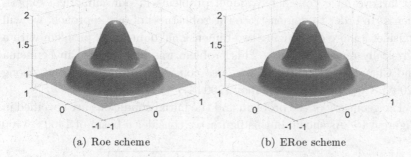

(a) Roe scheme   (b) ERoe scheme

Fig. 4.14. Approximate heights for the cylindrical dam-break problem at time $t = 0.2$ computed on a uniform 100×100 mesh with the Roe and ERoe schemes.

that both the standard Roe and ERoe schemes approximate the solution rather well, but with smearing at the outward shock and the rarefaction. As with one-dimensional dam-break problems, the differences between the two schemes are very small. However, as in Section 4.2, we can find several examples where the ERoe scheme is stable, whereas the standard Roe scheme is unstable.

As in Section 4.2.7, we have computed the cost with each scheme, and we present the normalized run-times (unit time is set to 0.0475 seconds) with respect to relative errors in height in Table 4.4. As shown in this table, the Roe and ERoe schemes have approximately the same computational cost and are considerably more efficient than the Rusanov scheme. Given the fact that the ERoe scheme is as accurate, has the same computational cost as the Roe scheme but is provably energy-stable, it seems preferable to the Roe scheme.

Relative height error (in %)	4	2	1
Rusanov	1.59	32.1	404.21
Roe	1	20.99	229.47
ERoe	1.03	21.17	231.57

Table 4.4. *Normalized run-times for the Rusanov, Roe and ERoe schemes on the two-dimensional cylindrical dam-break problem with three different levels of relative error in height.*

Numerical experiment #9: A physical dam-break problem

So far, we have considered model problems in our numerical experiments. In order to demonstrate the robustness of our approach, we will consider a more challenging two-dimensional dam-break problem with a physically realistic set up. This problem was first studied in Fennema and Chaudhery (1990) and was also considered in Tadmor and Zhong (2008), Chertock and Kurganov (2004).

The geometry of the problem and the initial conditions are specified in Figure 4.15. As shown in this figure, we consider a basin of 1400×1400

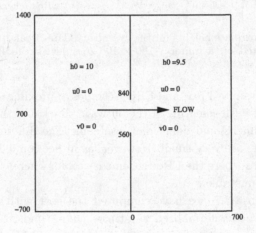

Fig. 4.15. Set-up and initial conditions of the physical dam-break problem

m^2 with a dam in the middle. The walls of the basin are solid and frictionless and the bottom is assumed to be flat. The walls are reflective with initial water level at 10 m and tail water level of 9.5 m. At $t = 0$, the central part of the dam fails and water is released downstream through a breach, as shown in Figure 4.15. We set acceleration due to gravity to $9.8ms^{-2}$. The boundary treatment is similar to the one used in Tadmor and Zhong (2008). We compute approximate solutions of this problem with three different schemes: the energy preserving EEC scheme, the EEC scheme together with eddy viscosity of $\nu = 10 m^2 s^{-1}$ and the energy stable ERoe scheme. We used a uniform 100×100 mesh with a second-order RK2 method for time integration at a CFL number of 0.45. The results are displayed in Figure 4.16. The results are along expected lines.

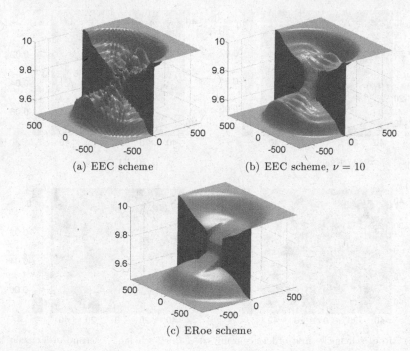

(a) EEC scheme   (b) EEC scheme, $\nu = 10$

(c) ERoe scheme

Fig. 4.16. Approximate heights for the physical dam-break problem at time $t = 50$ computed on a uniform 100×100 mesh with EEC scheme together with eddy viscosity and with the ERoe scheme.

The EEC schemes produce oscillations, but the basic flow features are computed quite accurately. In particular, the circular shock wave is resolved quite sharply.

The addition of eddy viscosity reduces the oscillations considerably. However, some oscillations are still present, indicating an under-resolved computation. When we combine the EEC fluxes (4.49) with the Roe-type entropy variables based numerical diffusion operator (4.54) to obtain the ERoe scheme, we observe that all the oscillations are removed. However, the shocks and other wave fronts are smeared. These results are very similar to those obtained in Tadmor and Zhong (2008), Chertock and Kurganov (2004) and confirm the robustness of our approach.

4.3.4 Numerical Experiment #10: Advection of Vorticity

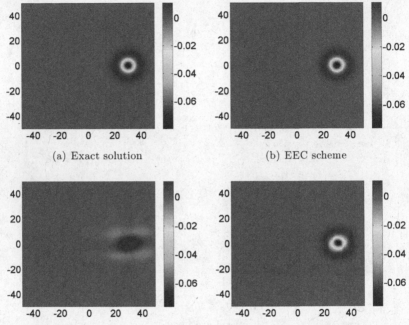

(a) Exact solution

(b) EEC scheme

(c) ERoe1 scheme — first-order accuracy (d) ERoe2 scheme — second-order accuracy

Fig. 4.17. Discrete vorticity at time $t = 100$ on a 200×200 mesh for numerical experiment 6: Vortex advection with EEC and ERoe schemes.

All the numerical experiments presented so far involved the formation and propagation of shocks. Energy preserving schemes led to oscillations near the shock on account of energy transfer to the small scales in the problem. We needed to introduce either eddy viscosity or numerical diffusion to remove these oscillations. Another interesting object of study, particularly for the two-dimensional form of the shallow water equations (4.2), is the vorticity. We define the vorticity as $\omega = v_x - u_y$, with u and v being the velocities. It is well known (Arakawa and Lamb (1977)) that the vorticity satisfies the following advection equation,

$$\omega_t + (u\omega)_x + (v\omega)_y = 0. \tag{4.55}$$

This identity is valid only for the smooth solutions of (4.2). Many papers, e.g., Arakawa and Lamb (1981), have dealt with the question of designing numerical schemes that satisfy a discrete version of the vorticity advection (4.55). The vorticity errors generated by a scheme discretizing (4.2) are considered a key component of its overall performance, particularly for meteorological applications. It was speculated in Tadmor and Zhong (2008) that energy preserving schemes for the shallow water equations might generate large vorticity errors. Hence, we test the energy preserving EEC scheme in a numerical experiment dealing with the advection of a vortex.

We adapt a standard test case (see Ismail and Roe (2005) and references therein) for the isentropic Euler equations to the shallow water equations. One can check that the following functions,

$$h(x,y,t) = 1 - \frac{c_1^2}{4c_2 g}e^{2f}$$
$$u(x,y,t) = M + c_1(y - y_0)e^f$$
$$v(x,y,t) = -c_1(x - x_0 - Mt)e^f,$$

where

$$f = f(x,y,t) = -c_2\left((x - x_0 - Mt)^2 + (y - y_0)^2\right),$$

are a smooth solution to the shallow water equations (4.2) for any choice of constants M, c_1, c_2, x_0 and y_0. For our numerical experiment, we consider the above solution at time $t = 0$ as our initial data and choose $M = 0.5$, $g = 1$, perturbation coefficients $(c_1, c_2) = (-0.04, 0.02)$ and starting position $(x_0, y_0) = (-20, 0)$. The exact solution is a vortex moving at a constant velocity in the x-direction. This test case has been considered in many papers in the literature and standard schemes have been found to generate unacceptably large vorticity errors. We compute on a domain $[-50, 50] \times [-50, 50]$ and show the approximate solutions in Figure 4.17.

We compute the standard discrete vorticity given by

$$\omega_{i,j} = \frac{v_{i+1,j} - v_{i-1,j}}{2\Delta x} - \frac{u_{i,j+1} - u_{i,j-1}}{2\Delta y},$$

and plot the discrete vorticity on a uniform 200×200 mesh with EEC and ERoe schemes at time $t = 100$. The time integration is performed with an RK2 method at a CFL number of 0.45. The vorticity of the exact solution is plotted for comparison. As shown in the figure, the EEC scheme does a remarkable job of resolving the solution at this fairly coarse mesh with $\Delta x = \Delta y = 0.5$ and a long period of time with final time $t = 100$. There are no oscillations whatsoever, and the shape of the vortex is retained. This should be contrasted with the oscillations that a EEC scheme generates near shocks. Similarly, the performance of the EEC scheme is considerably better than the ERoe scheme. As seen in Figure 4.17(c), the structure of the vortex is destroyed in the first-order ERoe scheme. Such behavior is expected from standard schemes (Ismail and Roe (2005)). A possible explanation lies in the order of accuracy, as the EEC scheme is formally second-order accurate, whereas the ERoe1 scheme is restricted to first-order of accuracy. We extended the ERoe scheme to second-order accuracy by applying minmod limiters, as in Section 4.2.8; the result of ERoe2 is shown in Figure 4.17(d). The rendition is now much more accurate, but the increased accuracy comes at the cost of small spurious oscillations that make the vortex appear non-symmetric.

We believe that the order alone does not explain the good performance of the EEC scheme. One reason for its performance may lie in the preservation of energy, illustrated in Figure 4.18. The figure shows that the energy error is very low, and this may be one of the reasons that the scheme preserves the structure of the vortex.

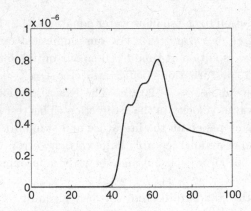

Fig. 4.18. Energy vs. time for EEC scheme in numerical experiment 6 with 200×200 mesh and RK2 method with $CFL = 0.45$.

This experiment illustrates that the EEC scheme is very robust for smooth solutions and needs no eddy viscosity or numerical diffusion for stability. Furthermore, it advects vorticity without destroying the vortical structures in the problem.

4.4 Conclusion

We considered the shallow water equations in one and two space dimensions. Our aim was to design finite volume schemes for these equations that either preserve energy or dissipate it in the right way. We considered three different energy preserving schemes in one dimension – the PEC scheme proposed in Tadmor (2003), Tadmor and Zhong (2008), an explicit evaluation of the scheme proposed in Tadmor (1987), and a new explicit energy preserving scheme, inspired by the entropy preserving scheme for the Euler equations in Roe (2006). The new scheme, called the EEC scheme, is very easy to implement and is robust. We compared the three schemes in numerical experiments and observed that the PEC and EEC schemes behave in a similar manner. The main advantage of the EEC scheme over the PEC scheme is its low computational cost: about three times faster in one dimension and about six times faster in two dimensions. The low cost coupled with the ease of implementation makes the EEC scheme an ideal energy preserving scheme for the shallow water equations.

The energy preserving schemes lead to oscillations at the mesh scale due to the absence of diffusive mechanisms. Addition of eddy viscosity by using suitable discretization of the viscous terms leads to energy stable discretizations of the shallow water equations. Eddy viscosity dramatically reduces the oscillations, particularly on resolved meshes. However, oscillations might remain on under-resolved meshes.

Computations on under-resolved meshes or for the inviscid version of the equations require adding suitable artificial diffusion operators. We proposed a novel numerical flux of the Roe type that is based on using an energy preserving symmetric flux together with a numerical diffusion based on energy variables. The diffusion operators need to be scaled suitably, and the resulting scheme is energy-stable without any additional fixes. The resulting scheme, termed an ERoe scheme, is very robust. It has the same accuracy and a similar computational cost as the standard Roe scheme, but is more stable.

All the above points are demonstrated in a series of numerical experiments in both one and two spatial dimensions. We believe that this approach of using energy (entropy) preserving schemes together with suitable physical or numerical diffusions operators is a practical, cost-effective and stable approach for computing flows. We will aim to extend this approach to more complicated models like the Euler equations of gas dynamics, incompressible flows and equations of magneto-hydrodynamics in forthcoming papers.

Acknowledgment. The work on this paper was started when S. M. visited the Center of Scientific Computation and Mathematical Modelling (CSCAMM) and he thanks CSCAMM and all its members for the excellent hospitality and facilities. The research of E. T. was supported in part by NSF grant 07-07949 and ONR grant N00014-91-J-1076.

References

A. Arakawa (1966), 'Computational design for long-term numerical integration of the equations of fluid motion: Two-dimensional incompressible flow', *J. Comput. Phys.* **1**, 119–143.

A. Arakawa and V. Lamb (1977), 'Computational design of the basic dynamical process of the UCLA general circulation model', *Meth. Comput. Phys.* **17**, 173–265.

A. Arakawa and V. Lamb (1981), 'A potential enstrophy and energy conserving scheme for the shallow water equations', *Mont. Weat. Rev.* **109**, 18–36.

A. Chertock and A. Kurganov (2004), 'On a hybrid finite-volume-particle method', *M2AN Math. Model. Num. Anal.* **38**, 1971–1991.

B. Cockburn, C. Johnson, C.-W. Shu and E. Tadmor (1998), *Advanced Numerical Approximation of Nonlinear Hyperbolic Equations*, C.I.M.E. course held in Cetraro, Italy, June 1997 (A. Quarteroni ed.), Lecture notes in Mathematics **1697**, Springer Verlag.

C. Dafermos (2000), *Hyperbolic Conservation Laws in Continuum Physics*, Springer, Berlin.

R. Fennema and M. Chaudhry (1990), 'Explicit methods for the 2d-transient free surface flows' *J. Hydraul. Eng. ASCE.* **116**, 1013–1034.

S. Gottlieb, C.-W. Shu and E. Tadmor (2001), 'High order time discretizations with strong stability properties', *SIAM. Review* **43**, 89–112.

J. Goodman and P. Lax (1988), 'On dispersive-diffusive schemes I', *Comm. Pure. Appl. Math.* **41**, 591–613.

T. Hou and P. Lax (1991), 'Dispersive approximations in fluid dynamics', *Comm. Pure. Appl. Math.* **44**, 1–40.

T. Hughes, L. Franca and M. Mallet (1986), 'A new finite element formulation for CFD I: Symmetric forms of the compressible Euler and Navier-Stokes equations and the second law of thermodynamics', *Comp. Meth. Appl. Mech. Eng.* **54**, 223–234.

F. Ismail and P. Roe (2005), 'Towards a vorticity preserving second order finite volume scheme solving the Euler equations', *Proc. of AIAA conference on CFD*, Toronto, Canada.

K. Khalfallah and A. Lerat (1989), 'Correction d'entropie pour des schémas numériques approchant un système hyperbolique', *C. R. Acad. Sci. Paris Sér. II* **308**, 815–820.

A. Kurganov and E. Tadmor (2002), 'New high resolution central schemes for non-linear conservation laws and convection-diffusion equations', *J. Comput. Phys* **160**, 241–282.

P. LeFloch, J. Mercier and C. Rohde (2002), 'Fully discrete entropy conservative schemes of arbitraty order', *SIAM J. Numer. Anal.* **40**, 1968–1992.

R. LeVeque (2002), *Finite Volume Methods for Hyperbolic Problems*, Cambridge University Press, Cambridge.

P. Roe (2006), 'Entropy conservative schemes for Euler equations', talk at *HYP 2006, Lyon, France*, unpublished. Lecture available from http://math.univ-lyon1.fr/~hyp2006.

E. Tadmor (1984), 'Numerical viscosity and entropy conditions for conservative difference schemes', *Math. Comp.* **43**, 369–381.

E. Tadmor (1986), 'Entropy conservative finite element schemes' in *Numerical Methods for Compressible Flows - Finite Difference Element and Volume Techniques*, Proc. annual meeting of the AMSE (T. E. Tezduyar and T. J. R. Hughes, eds.), v. **78** , 149–158.

E. Tadmor (1987), 'The numerical viscosity of entropy stable schemes for systems of conservation laws, I.', *Math. Comp.* **49**, 91–103.

E. Tadmor (2002), 'From semi-discrete to fully discrete: stability of Runge-Kutta schemes by the energy method. II.', in *Collected Lectures on the Preservation of Stability under Discretization*, Lecture Notes Colorado State Univ. Conference, Fort Collins, CO, 2001 (D. Estep and S. Tavener, eds.), Proc. Appl. Math. **109**, SIAM, 25–49.

E. Tadmor (2003), 'Entropy stability theory for difference approximations of nonlinear conservation laws and related time-dependent problems', *Acta Numerica* **12**, 451–512.

E. Tadmor and W. Zhong (2006), 'Entropy stable approximations of Navier-Stokes equations with no artificial numerical viscosity', *J. Hyperbolic. Differ. Equ.* **3**, 529–559.

E. Tadmor and W. Zhong (2008), 'Energy preserving and stable approximations for the two-dimensional shallow water equations', in *Mathematics and Computation: A Contemporary View*, Proc. of the third Abel symposium, Ålesund, Norway, Springer, 67–94.

5

Pathwise Convergence of Numerical Schemes for Random and Stochastic Differential Equations

A. Jentzen, P. E. Kloeden and A. Neuenkirch

Institut für Mathematik
Johann Wolfgang Goethe Universität
D-60054 Frankfurt am Main
Germany
e-mail: {jentzen,kloeden,neuenkirch}@math.uni-frankfurt.de

Abstract

Itô stochastic calculus is a mean-square or L_2 calculus with different transformation rules to deterministic calculus. In particular, in view of its definition, an Itô integral is not as robust to approximation as the deterministic Riemann integral. This has some critical implications for the development of effective numerical schemes for stochastic differential equations (SDEs). Higher order numerical schemes for both strong and weak convergences have been derived using stochastic Taylor expansions, but most proofs in the literature assume the uniform boundedness of all relevant partial derivatives of the coefficient functions, which is highly restrictive. Problems also arise if solutions are restricted to certain regions, e.g. must be positive and the coefficient functions involve square roots.

Pathwise convergence is an alternative to the strong and weak convergences of the literature. It arises naturally in many important applications and, in fact, numerical calculations are carried out pathwise. Pathwise convergent numerical schemes for both SDEs and random ordinary differential equations (RODEs) will be discussed here. In particular, we see that a strong Taylor scheme for SDEs of order γ converges pathwise with order $\gamma - \epsilon$ for every arbitrarily small $\epsilon > 0$ under the usual assumptions. We also introduce modified Taylor schemes which converge pathwise in a similar way, even when the coefficient derivatives are not uniformly bounded or are restricted to certain regions. In particular, these schemes can be applied without difficulty to volatility models in finance.

The relationship between SDEs and RODEs via Doss-Sussman transformations will also be considered and new classes of numerical schemes for RODEs will be presented. Finally, extensions to stochastic and random partial differential equations will be mentioned briefly.

5.1 Introduction

The inclusion of noise terms in differential equations dates back to the the early 1900s when Einstein provided an explanation for the physical phenomenon of Brownian motion. At the same time Langevin was interested in the actual motion of the particles and formulated noisy differential equations of the form

$$\frac{dx}{dt} = a(x) + b(x)\eta_t, \tag{5.1}$$

while Bachelier proposed similar equations to model the evolution of share prices.

Many technical difficulties arose, for mathematicians at least, since the noise process η_t was meant to be Gaussian white noise and it took a good half century before K. Itô (in fact, independently at about the same time, I. Gikhman and W. Döblin, too) could develop a stochastic calculus which allowed a rigorous formulation and mathematical development of a theory of *stochastic differential equations* (SDEs).

Itô stochastic calculus is a mean-square or L_2 calculus with different transformation rules to deterministic calculus. In particular, the integrand function in an Itô integral is always evaluated at the left end point of a discretization subinterval rather than at an arbitrary point as in the deterministic Riemann integral, which means that it is not as robust to approximation. This has some critical implications for the development of effective numerical schemes for SDEs.

In many physical applications the noise often has a wide band rather than white spectrum, i.e., it is a Δ-correlated stationary Gaussian process $\eta_t^{(\Delta)}$ with a Gaussian white noise limit as $\Delta \to 0$. In this case the noisy differential equation (5.1) is in fact an ordinary differential equation (ODE)

$$\frac{dx}{dt} = a(x) + b(x)\,\eta_t^{(\Delta)}, \tag{5.2}$$

which can be handled pathwise by the methods of deterministic calculus. What kind of SDE might arise in the limit as $\Delta \to 0$ has been investigated intensively, see e.g. Wong & Zakai (1965). More generally, a *random ordinary differential equation* (RODE) is formulated pathwise as an ODE

$$\frac{dx}{dt} = f(\zeta_t, x) \tag{5.3}$$

where ζ_t is a stochastic process. RODEs seem to have had a shadow existence to SDEs, but have been around for as long, if not longer, than SDEs and have many important applications, see e.g. Bunke (1972), Soong (1973) and Arnold (1997). Although the rules of deterministic calculus apply pathwise to them, it is important to note that the vector field function in (5.3) is at most Hölder continuous in time like the driving stochastic process ζ_t and thus lacks the smoothness needed to justify the error analysis of traditional numerical methods for ODEs. Such methods can be used but will attain at best a low convergence order, so new higher order numerical schemes must be derived for RODEs.

In this article we briefly summarize the existing theory of numerical methods for SDEs and then show how some rather strong assumptions on the SDE coefficients used in the literature can be relaxed if pathwise rather than mean-square convergence is used. We also show how numerical schemes can be modified so that the solutions remain in certain prescribed regions of admissibility, e.g. the preservation of the boundary domain for SDEs with square root coefficients. Some important relationships between RODEs and SDEs will be indicated and recent work on higher order numerical schemes derived for RODEs will also be mentioned. Finally, these new schemes will be applied with the method of lines to random and stochastic partial differential equations (RPDE, SPDE).

5.2 Numerical Approximation of Itô SDEs

For simplicity we consider an Itô SDE in $\mathbb{R}^d$

$$dX_t = a(X_t)\, dt + \sum_{j=1}^{m} b_j(X_t) dW_t^j, \qquad t \in [0, T], \tag{5.4}$$

with drift and diffusion coefficients a, $b_j : \mathbb{R}^d \to \mathbb{R}^d$ $(j = 1, \ldots, m)$ which do not depend explicitly on the time variable t. Here $W_t = (W_t^1, \ldots, W_t^m)$, $t \geq 0$, is an m-dimensional Wiener process (also called a Brownian motion) and superscripts label components of vectors. Such an Itô SDE is only symbolic for an Itô stochastic <u>integral</u> equation

$$X_t = X_0 + \int_0^t a(X_s)\,ds + \sum_{j=1}^m \int_0^t b_j(X_s)dW_s^j, \qquad (5.5)$$

where the drift integral is pathwise a Riemann integral and the others are Itô stochastic integrals.

Due to differences in the deterministic and stochastic calculi, traditional numerical schemes for ODEs are either inconsistent or, at best, converge with a very low order when adapted to SDEs like (5.4). New types of numerical schemes are thus necessary and can be derived systematically by means of stochastic Taylor expansions in integral form; see the monograph of Kloeden & Platen (1992), and also Milstein (1995), for a thorough development of the theory.

Consider a partition $0 = t_0 < t_1 < \cdots < t_{N_T} = T$ of the interval $[0, T]$ with step sizes $\Delta_n := t_{n+1} - t_n > 0$ and maximum step size $\Delta := \max_n \Delta_n$. Let $Y_n^{(\Delta)}$ be an approximation generated by some numerical schemes of X_{t_n} for a solution X_t of an SDE (5.4). In the literature *average error criteria* with deterministic constants are usually considered:

Weak approximation of order β if

$$|\mathbb{E}\phi(X_T) - \mathbb{E}\phi(Y_{N_T})| \le K_{\phi,T}\,\Delta^\beta$$

for smooth test functions $\phi : \mathbb{R}^d \to \mathbb{R}$;

Strong approximation of order γ (usually just $p = 1$ or 2) if

$$\left(\mathbb{E}\sup_{n=0,\ldots,N_T}|X_n - Y_n|^p\right)^{1/p} \le K_{p,T}\,\Delta^\gamma.$$

Itô–Taylor numerical schemes for both types of convergences are based on appropriate Itô–Taylor expansions (see chapter 5 in Kloeden & Platen (1992)). These involve:

(1) *Differential operators*

$$L^0 = \sum_{k=1}^d a^k \frac{\partial}{\partial x^k} + \frac{1}{2}\sum_{k,l=1}^d \sum_{j=1}^m b^{k,j} b^{l,j} \frac{\partial^2}{\partial x^k \partial x^l}, \quad L^j = \sum_{k=1}^d b^{k,j}\frac{\partial}{\partial x^k},$$

where a^k, $b^{k,j}$ are the k-th components of a and b_j, $j = 1,\ldots,m$.

(2) *Multi-indices*

$$\mathcal{M}_m = \left\{\alpha = (j_1,\ldots,j_l) \in \{0,1,2,\ldots,m\}^l : l \in \mathbb{N}\right\} \cup \{\emptyset\}$$

with $l(\alpha)$ the length of α, $n(\alpha)$ the number of zero entries of α, and $\emptyset$ the multi-index of length 0.

(3) *Iterated integrals and coefficient functions*:

$$I_\alpha(s,t) = \int_s^t \cdots \int_s^{\tau_2} dW_{\tau_1}^{j_1} \ldots dW_{\tau_l}^{j_l}, \qquad f_\alpha(x) = L^{j_1} \cdots L^{j_{l-1}} b^{j_l}(x)$$

with the notation $dW_t^0 = dt$, $b^0 = a$.

The *Itô-Taylor scheme of strong order* $\gamma = \frac{1}{2}, 1, \frac{3}{2}, 2, \ldots$,

$$Y_{n+1}^\gamma = Y_n^\gamma + \sum_{\alpha \in \mathcal{A}_\gamma^{(s)} \setminus \{\emptyset\}} f_\alpha(Y_n^\gamma) \cdot I_\alpha(t_n, t_{n+1}) \tag{5.6}$$

uses multi-indices in the hierarchical set

$$\mathcal{A}_\gamma^{(s)} = \{\alpha \in \mathcal{M}_m : l(\alpha) + n(\alpha) \le 2\gamma \text{ or } l(\alpha) = n(\alpha) = \gamma + 1/2\},$$

whereas the *Itô-Taylor scheme of weak order* $\beta = 1, 2, 3, \ldots$,

$$Y_{n+1}^\beta = Y_n^\beta + \sum_{\alpha \in \mathcal{A}_\beta^{(w)} \setminus \{\emptyset\}} f_\alpha(Y_n^\beta) \cdot I_\alpha(t_n, t_{n+1}) \tag{5.7}$$

uses multi-indices in the hierarchical set $\mathcal{A}_\beta^{(w)} = \{\alpha \in \mathcal{M}_m : l(\alpha) \le \beta\}$.

Example 5.1 For the scalar Itô SDE

$$dX_t = a(X_t)\, dt + b(X_t)\, dW_t$$

the *Euler-Maruyama scheme*

$$Y_{n+1} = Y_n + a(Y_n)\, \Delta_n + b(Y_n)\, \Delta W_n,$$

where $\Delta W_n = I_{(1)}(t_n, t_{n+1}) = W_{t_{n+1}} - W_{t_n}$, is the Itô-Taylor scheme of strong order $\gamma = \frac{1}{2}$ and the Itô-Taylor scheme of weak order $\beta = 1$ with the hierarchical sets $\mathcal{A}_{\frac{1}{2}}^{(s)} = \mathcal{A}_1^{(w)} = \{(0), (1)\}$. Note that the index (0) occurs for different reasons in the two hierarchical sets. For the scalar SDE the *Milstein scheme*

$$Y_{n+1} = Y_n + a(Y_n)\, \Delta_n + b(Y_n)\, \Delta W_n + \frac{1}{2} b(Y_n) b'(Y_n) \left[(\Delta W_n)^2 - \Delta_n\right]$$

is the Itô-Taylor scheme of strong order $\gamma = 1$ with the hierarchical set $\mathcal{A}_1^{(s)} = \{(0), (1), (1,1)\}$. It also has weak order $\beta = 1$, but is not the Itô-Taylor scheme of this weak order.

Here we have used the coefficient functions $f_{(0)} = a$, $f_{(1)} = b$, $f_{(1,1)} = bb'$ and the iterated integrals

$$I_{(0)}(t_n, t_{n+1}) = \int_{t_n}^{t_{n+1}} dW_s^0 = \Delta_n, \tag{5.8}$$

$$I_{(1)}(t_n, t_{n+1}) = \int_{t_n}^{t_{n+1}} dW_s^1 = \Delta W_n^1, \tag{5.9}$$

and

$$I_{(1,1)}(t_n, t_{n+1}) = \int_{t_n}^{t_{n+1}} \int_{t_n}^{s} dW_\tau^1 \, dW_s^1 = \frac{1}{2} \left[(\Delta W_n^1)^2 - \Delta_n \right]. \tag{5.10}$$

Remark 5.1 There are usually no simple expressions like (5.10) for multiple stochastic integrals involving different independent Wiener processes. How to approximate such integrals is a major issue in stochastic numerics. The difficulty and cost of approximating stochastic integrals of higher multiplicity restricts the practical usefulness of higher order strong schemes, see Kloeden & Platen (1992), Kloeden (2002) and in particular Gaines & Lyons (1994, 1997). However, the work of Wiktorsson (2001) on the simulation of the second iterated integral is very promising.

Remark 5.2 The Itô-Taylor schemes are the "basic" schemes for the weak and strong approximation of stochastic differential equations. Based on these schemes, numerous other numerical methods as Runge-Kutta methods and multi-step methods have been constructed in the last years. See e.g. Kloeden & Platen (1992), Milstein (1995), Burrage *et al.* (2004), Milstein & Tretyakov (2004) and the references therein.

While the Itô-Taylor schemes are developed from the Itô-Taylor expansion of the solution of the SDE, the use of the exponential Lie series of the solution SDE leads to stochastic Lie group numerical schemes, see e.g. Castell & Gaines (1995) and Malham & Wiese (2008).

Proofs in the literature of the above convergence orders, e.g. in the monographs Kloeden & Platen (1992) and Milstein (1995), assume that the coefficient functions f_α in the Itô-Taylor schemes are *uniformly bounded* on $\mathbb{R}^d$, i.e., the partial derivatives of appropriately high order of the SDE coefficient functions $a, b_1, \ldots, b_m$ are *uniformly bounded* on $\mathbb{R}^d$. This assumption is <u>not satisfied</u> for many SDEs in important applications such as the Duffing-van der Pol oscillator with multiplicative noise

$$dX_t^1 = X_t^2 \, dt, \tag{5.11}$$
$$dX_t^2 = \left[-X_t^1 + \beta X_t^2 - (X_t^1)^3 - (X_t^1)^2 X_t^2 \right] dt + \sigma X_t^2 \, dW_t$$

with linear and cubic terms (recall that superscripts label vector components, so that powers are indicated by enclosing their arguments in parentheses), the Fisher-Wright equation in biomathematics

$$dX_t = \left[\kappa_1 (1 - X_t) - \kappa_2 X_t \right] dt + \sqrt{X_t(1 - X_t)} \, dW_t \tag{5.12}$$

with $\kappa_1, \kappa_2 \geq 0$ and $X_t \in [0,1]$, and the Cox-Ingersoll-Ross model in mathematical biology and finance

$$dV_t = \kappa \left(\lambda - V_t \right) dt + \theta \sqrt{V_t} \, dW_t \tag{5.13}$$

with $\kappa, \lambda, \theta \geq 0$ and $V_t \geq 0$.

One way to overcome this problem is to restrict attention to SDEs with special dynamical properties such as ergodicity, e.g. by assuming that the coefficients satisfy certain dissipativity and nondegeneracy conditions, see Mattingly *et al.* (2002), Higham *et al.* (2002) and Milstein & Tretyakov (2005). This yields the appropriate order estimates without bounded derivatives of coefficients. However, several type of SDEs and in particular SDEs with square root coefficients remain a problem. Many numerical schemes do not preserve the domain of the solution of the SDE and hence may crash when implemented, which has led to various ad hoc modifications to prevent this happening.

5.3 Pathwise Convergence

By the *pathwise convergence* of an approximate solution we mean that

$$\sup_{n=0,\ldots,N_T} \left| X_{t_n}(\omega) - Y_n^{(\Delta)}(\omega) \right| \to 0 \qquad \text{as} \quad \Delta \to 0$$

for (at least) almost all $\omega \in \Omega$, where Ω is the sample space of the underlying probability space $(\Omega, \mathcal{F}, \mathbb{P})$.

This is interesting because numerical calculations of the approximating random variable $Y_n^{(\Delta)}$ are carried out path by path. Moreover, the modern theory of random dynamical systems, which deals with random attractors and stochastic bifurcations is of pathwise nature, see Arnold (1997). In addition, the solutions of some SDEs are non-integrable, i.e., $\mathbb{E}|X_t| = \infty$ for some $t \geq 0$, so strong or weak convergent approximation is not possible.

We should not forget that Itô calculus is an L_2 or a mean-square calculus and not a pathwise calculus. Nevertheless some results for pathwise approximation of SDEs are known. For example, in 1983 Talay showed that the Milstein scheme SDE with a scalar Brownian motion has the pathwise error estimate

$$\sup_{n=0,\ldots,N_T} \left| X_{t_n}(\omega) - Y_n^{(\Delta)}(\omega) \right| \leq K_{\epsilon,T}^{(M)}(\omega) \Delta^{\frac{1}{2}-\epsilon},$$

for all $\epsilon > 0$ and almost all $\omega \in \Omega$, i.e., is pathwise of order $\frac{1}{2} - \epsilon$. Later Gyöngy (1998) and Fleury (2005) showed that the Euler–Maruyama scheme has the same pathwise convergence order. Note that the error constants here depend on ω, so they are in fact random variables. The nature of their statistical properties is an interesting question, of which less is known theoretically so far and which requires further investigations. Some empirical distributions of the random error constants are plotted in Figures 5.2 and 5.4 below. Given that the sample paths of a Wiener process are Hölder continuous with exponent $\frac{1}{2} - \epsilon$ one may ask: *Is the convergence order $\frac{1}{2} - \epsilon$ "sharp" for pathwise approximation?* The answer is no! In fact, Kloeden & Neuenkirch (2007) have shown recently that an arbitrary pathwise convergence order is possible.

Theorem 5.1 *Under classical assumptions the Itô–Taylor scheme of strong order $\gamma > 0$ converges pathwise with order $\gamma - \epsilon$ for all $\epsilon > 0$, i.e.,*

$$\sup_{n=0,\ldots,N_T} |X_{t_n}(\omega) - Y_n^\gamma(\omega)| \leq K_{\epsilon,T}^\gamma(\omega) \cdot \Delta^{\gamma-\epsilon}$$

for almost all $\omega \in \Omega$.

Thus, for example, the Milstein scheme has pathwise order $1 - \epsilon$ rather than the lower order $\frac{1}{2} - \epsilon$ obtained in Talay (1983) (which was a consequence of the proof used there).

The proof of Theorem 5.1 is based on the Burkholder–Davis–Gundy inequality

$$\mathbb{E} \sup_{s \in [0,t]} \left| \int_0^s X_\tau \, dW_\tau \right|^p \leq C_p \cdot \mathbb{E} \left| \int_0^t (X_\tau)^2 \, d\tau \right|^{p/2}$$

and a Borel–Cantelli argument in the following:

Lemma 5.1 *Let $\gamma > 0$ and $c_p \geq 0$ for $p \geq 1$. If $\{Z_n\}_{n \in \mathbb{N}}$ is a sequence of random variables with*

$$(\mathbb{E}|Z_n|^p)^{1/p} \leq c_p \cdot n^{-\gamma}$$

for all $p \geq 1$ and $n \in \mathbb{N}$, then for each $\epsilon > 0$ there exists a finite non-negative random variable K_ϵ such that

$$|Z_n(\omega)| \leq K_\epsilon(\omega) \cdot n^{-\gamma + \varepsilon} \qquad a.s.$$

for all $n \in \mathbb{N}$.

5.3.1 SDEs without uniformly bounded coefficients

Using a localization argument, Jentzen *et al.* (2008a) showed that Theorem 5.1 remains true if the SDE coefficients satisfy $a, b_1, \ldots, b_m \in C^{2\gamma+1}(\mathbb{R}^d; \mathbb{R}^d)$, i.e., they do not necessarily have uniformly bounded derivatives. So, this extension of Theorem 5.1 applies to the Duffing-van der Pol oscillator with multiplicative noise (5.11). Whether mean-square convergence rates under these assumptions can be derived, remains an open problem. Compare e.g. Higham *et al.* (2002), where the standard convergence rates of several Euler-type methods are recovered under the assumptions that the drift coefficient has a polynomial behaviour and satisfies a one-sided Lipschitz condition, while the diffusion coefficient satisfies a global Lipschitz condition. For SDEs with a discontinuous but monotone increasing drift (such as a Heaviside function) and additive noise the mean-square convergence of the Euler method has been derived in Halidas & Kloeden (2008).

Numerical Example I   Consider the Duffing-van der Pol oscillator with multiplicative noise (5.11) with $\beta = 3$, $\sigma = 2$ and initial conditions $X_0^1 = X_0^2 = 1$ for $T = 1$.

Numerical Example II Consider the empirical distributions for the random error constants of the Euler-Maruyama and Milstein schemes applied to the SDE

$$dX_t = -(1+X_t)(1-(X_t)^2)\, dt + (1-(X_t)^2)\, dW_t, \qquad t \in [0,1], \quad X_0 = 0.$$

for $N = 10^4$ sample paths.

5.3.2 SDEs on restricted regions

The Fisher–Wright SDE (5.12) and the Cox–Ingersoll–Ross SDE (5.13) have square-root coefficients, which require the solutions to remain in

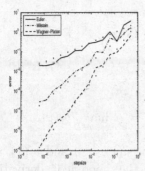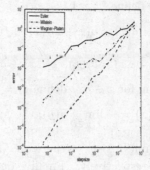

Fig. 5.1. Pathwise maximum error vs. stepsize for two sample paths for the Euler scheme $(-)$, the Milstein scheme $(-\cdot-)$ and the Wagner–Platen scheme $(--)$.

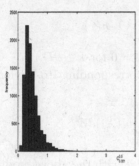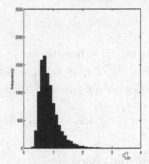

Fig. 5.2. Empirical distribution of $K_{0.001}^{0.5}$ and $K_{0.001}^{1.0}$ (sample size: $N = 10^4$).

the region where the expression under the square-root is non-negative. However, numerical iterations may leave this restricted region, in which case the algorithm will terminate.

One possibility to avoid this problem is to use appropriately modified Itô–Taylor schemes, see Jentzen *et al.* (2008a). So consider the SDE

$$dX_t = a(X_t)\,dt + \sum_{j=1}^{m} b_j(X_t)dW_t^j, \qquad t \in [0, T], \qquad (5.14)$$

which takes values in a domain $D \subseteq \mathbb{R}^d$ and suppose that the SDE coefficients $a, b_1, \ldots, b_m$ are r-times continuously differentiable on D and that SDE (5.14) has a unique strong solution. Define $E := \{x \in \mathbb{R}^d : x \notin \overline{D}\}$. Then choose auxiliary functions $f, g_1, \ldots, g_m \in C^s(E; \mathbb{R}^d)$ for

$s \in \mathbb{N}$ and define $(j = 1, \ldots, m)$

$$\widetilde{a}(x) = a(x) \cdot \mathbf{1}_D(x) + f(x) \cdot \mathbf{1}_E(x), \qquad x \in \mathbb{R}^d, \qquad (5.15)$$

$$\widetilde{b}_j(x) = b_j(x) \cdot \mathbf{1}_D(x) + g_j(x) \cdot \mathbf{1}_E(x), \qquad x \in \mathbb{R}^d. \qquad (5.16)$$

In addition, for $x \in \partial D$ define $(j = 1, \ldots, m)$

$$\widetilde{a}(x) = \lim_{y \to x;\, y \in D} \widetilde{a}(y), \qquad \widetilde{b}_j(x) = \lim_{y \to x;\, \in y \in D} \widetilde{b}_j(y), \qquad (5.17)$$

if these limits exist. Otherwise, define $\widetilde{a}(x) = 0$ and $\widetilde{b}_j(x) = 0$ for $x \in \partial D$, respectively. Finally, define the "modified" derivative of a function $h : \mathbb{R}^d \to \mathbb{R}^d$ by $(l = 1, \ldots, d)$

$$\partial_{x^l} h(x) = \frac{\partial}{\partial x^l} h(x), \quad x \in D \cup E, \qquad (5.18)$$

and for $x \in \partial D$ define

$$\partial_{x^l} h(x) = \lim_{y \to x;\, y \in D} \partial_{x^l} h(x) \qquad (5.19)$$

if this limit exists; otherwise set $\partial_{x^l} h(x) = 0$ for $x \in \partial D$.

A *modified Itô–Taylor* scheme is the corresponding Itô-Taylor scheme for the SDE with modified coefficients

$$dX_t = \widetilde{a}(X_t)\, dt + \sum_{j=1}^m \widetilde{b}_j(X_t) dW_t^j, \qquad (5.20)$$

using differential operators $\widetilde{L}^0, \widetilde{L}^1, \ldots, \widetilde{L}^m$ with the above modified derivatives. Note that this method is well defined as long as the coefficients of the equation are $(2\gamma - 1)$-times differentiable on D and the auxiliary functions are $(2\gamma - 1)$-times differentiable on E. The purpose of the auxiliary functions is twofold: to obtain a well defined approximation scheme and to "reflect" the numerical scheme back to D, once it has left D. In particular, the auxiliary functions can always be chosen affine or even constant.

It was shown in Jentzen *et al.* (2008a) that Theorem 5.1 adapts to modified Itô-Taylor schemes for SDEs on domains in $\mathbb{R}^d$.

Theorem 5.2 *Assume that* $\widetilde{a}, \widetilde{b}_1, \ldots, \widetilde{b}_m \in C^{2\gamma+1}(D; \mathbb{R}^d) \cap C^{2\gamma-1}(E; \mathbb{R}^d)$ *and let* $Y_n^{mod,\gamma}$ *be the modified Itô-Taylor scheme for* $\gamma = \frac{1}{2}, 1, \frac{3}{2}, \ldots$.

Then for all $\epsilon > 0$ *there exists a finite, non-negative random variable* $K_{\gamma,\epsilon}^{f,g}$ *such that*

$$\sup_{n=0,\ldots,N_T} \left| X_{t_n}(\omega) - Y_n^{mod,\gamma}(\omega) \right| \le K_{\gamma,\epsilon}^{f,g}(\omega) \cdot \Delta^{\gamma-\epsilon}$$

for almost all $\omega \in \Omega$ *and all* $n = 1, \ldots, N_T$.

Note that the convergence rate does not depend on the choice of the auxiliary functions, but the random constant in the error bound clearly does.

Example 5.2 The Fisher–Wright SDE (5.12) has the property that if $\min\{\kappa_1, \kappa_2\} \geq \frac{1}{2}$ and $X_0 \in (0, 1)$, then

$$\mathbb{P}(X_t \in (0, 1) \text{ for all } t \geq 0) = 1.$$

However, iterates of standard Itô–Taylor numerical schemes may leave $[0, 1]$, so we will use a modified numerical scheme with extended coefficients outside of $[0, 1]$ defined by

$$\text{auxiliary drift}: \quad f(x) = \kappa_1(1 - x) - \kappa_2 x, \qquad x \notin [0, 1]$$

$$\text{auxiliary diffusion}: \quad g(x) = 0, \qquad x \notin [0, 1]$$

and appropriately defined coefficients at the boundary points $x \in \{0, 1\}$. By Theorem 5.2 the modified Itô–Taylor scheme of order γ converges pathwise with order $\gamma - \epsilon$. This is illustrated in Numerical Example III for the Euler–Maruyama and Milstein schemes.

Numerical Example III Consider the Fisher–Wright SDE (5.12) with parameters $\kappa_1 = 0.5$ and $\kappa_2 = 1$, the initial value $X_0 = 0.1$ and final time $T = 1$.

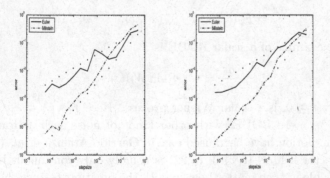

Fig. 5.3. Pathwise maximum error vs. stepsize for two sample paths for the Euler scheme $(-)$ and the Milstein scheme $(-\cdot-)$.

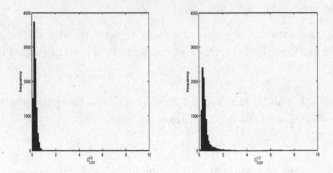

Fig. 5.4. Empirical distribution of $K^{0.5}_{0.001}$ and $K^{1.0}_{0.001}$ (sample size: $N = 10^4$).

5.4 Random Ordinary Differential Equations (RODEs)

Let $\zeta_t, t \geq 0$, be an m-dimensional stochastic process and suppose that $f : \mathbb{R}^m \times \mathbb{R}^d \to \mathbb{R}^d$ is smooth. Then

$$\frac{dx}{dt} = f(\zeta_t, x) \qquad (5.21)$$

is a *random ordinary differential equation* on $\mathbb{R}^d$ driven by the noise process ζ in $\mathbb{R}^m$, i.e., (5.21) is pathwise the ordinary differential equation (ODE) on $\mathbb{R}^d$

$$\frac{dx}{dt} = F_\omega(t, x) := f(\zeta_t(\omega), x), \qquad \omega \in \Omega. \qquad (5.22)$$

A simple example of a scalar RODE is

$$\frac{dx}{dt} = -x + \sin W_t(\omega), \qquad (5.23)$$

where $W_t, t \geq 0$, is a scalar Wiener process. Here $f(z, x) = -x + \sin z$ and $d = m = 1$. RODEs with other kinds of noise such as fractional Brownian motion have been used e.g. in Garrido-Atienza *et al.* (2008).

The vector field $F_\omega(t, x)$ in (5.22) is usually only continuous but not differentiable in t, no matter how smooth the function f, since the paths of the driving stochastic process ζ are usually at most Hölder continuous. This has important implications for the efficient numerical solution of RODEs.

Random ordinary differential equation occur in many applications and, as the wideband noise example (5.2) in the introduction shows, they may even be more realistic than SDEs with their idealized noise, which are then just a convenient limit, cf. Wong & Zakai (1965) and Godin & Molchanov (2007). Moreover, RODEs with Wiener processes can be rewritten as stochastic differential equations and hence all the results of Section 5.3 can be applied to them. For example, the scalar RODE (5.23) can be rewritten as the 2-dimensional SDE

$$d \begin{pmatrix} X_t \\ Y_t \end{pmatrix} = \begin{pmatrix} -X_t + \sin Y_t \\ 0 \end{pmatrix} dt + \begin{pmatrix} 0 \\ 1 \end{pmatrix} dW_t.$$

On the other hand, any finite dimensional SDE can be transformed to a RODE. In the case of commutative noise this is the famous Doss-Sussmann result (Doss (1977), Sussmann (1978)), which was generalized to all SDEs in recent years by Imkeller & Lederer (2001, 2002). It is easily illustrated for a scalar SDE with additive noise: The equation

$$dX_t = f(X_t) \, dt + dW_t$$

is equivalent to the RODE

$$\frac{dz}{dt} = f(z + O_t) + O_t \tag{5.24}$$

where $z(t) := X_t - O_t$, $t \geq 0$ and O_t, $t \geq 0$, is the stochastic stationary Ornstein-Uhlenbeck process satisfying the linear SDE

$$dO_t = -O_t \, dt + dW_t. \tag{5.25}$$

To see this, subtract integral versions of both SDEs and substitute to obtain

$$z(t) = z(0) + \int_0^t [f(z(s) + O_s) + O_s] \, ds.$$

Then, by continuity and the fundamental theorem of deterministic calculus, it follows that z is pathwise differentiable.

In particular, we can use deterministic calculus pathwise for RODEs. This greatly facilitates the investigation of dynamical behaviour and other qualitative properties of RODEs. For example, suppose that f satisfies a one-sided dissipative Lipschitz condition ($L > 0$),

$$\langle x - y, f(x) - f(y) \rangle \leq -L|x - y|^2, \qquad x, y \in \mathbb{R}^d.$$

Then, for any two solutions z_1 and z_2 of the RODE (5.24),

$$\frac{d}{dt}|z_1(t) - z_2(t)|^2 = 2\left\langle z_1(t) - z_2(t), \frac{dz_1}{dt} - \frac{dz_2}{dt} \right\rangle$$

$$= 2\langle z_1(t) - z_2(t), f(z_1(t) + O_t) - f(z_2(t) + O_t)\rangle$$

$$\leq -2L\,|z_1(t) - z_2(t)|^2$$

from which it follows that pathwise

$$|z_1(t) - z_2(t)|^2 \leq e^{-2Lt}|z_1(0) - z_2(0)|^2 \to 0 \quad \text{as } t \to \infty.$$

Then from the theory of random dynamical systems, Arnold (1997), there exists a pathwise asymptotically stable stochastic stationary solution. For an application to synchronization in the presence of noise see Caraballo & Kloeden (2005) and Caraballo *et al.* (2008).

5.4.1 Numerical schemes for RODEs

We can solve RODEs pathwise as ODEs with Runge-Kutta schemes, but these schemes do <u>not</u> attain their traditional order, since the vector field $F_\omega(t, x)$ in (5.22) is not smooth enough in t. For example, let ζ be pathwise Hölder continuous of order $\frac{1}{2}$. Then the Euler scheme

$$Y_{n+1}(\omega) = (1 - \Delta_n)\,Y_n(\omega) + \zeta_{t_n}(\omega)\,\Delta_n$$

for the RODE

$$\frac{dx}{dt} = -x + \zeta_t(\omega),$$

attains the pathwise order $\frac{1}{2}$. One can do better, however, by using the pathwise *averaged Euler scheme*

$$Y_{n+1}(\omega) = (1 - \Delta_n)\,Y_n(\omega) + \int_{t_n}^{t_{n+1}} \zeta_t(\omega)\,dt,$$

which was proposed by Grüne & Kloeden (2001). It attains the pathwise order 1 provided the integral is approximated with Riemann sums

$$\int_{t_n}^{t_{n+1}} \zeta_t(\omega)\,dt \approx \sum_{j=1}^{J_{\Delta_n}} \zeta_{t_n + j\delta}(\omega)\,\delta$$

with step size $\delta \approx \Delta_n^2$ and $\delta \cdot J_{\Delta_n} = \Delta_n$. In fact, this was done more generally in Grüne & Kloeden (2001) for RODEs with an affine structure, i.e., of the form

$$\frac{dx}{dt} = f(x) + G(x)\zeta_t, \tag{5.26}$$

where $f : \mathbb{R}^d \to \mathbb{R}^d$ and $G : \mathbb{R}^d \to \mathbb{R}^d \times \mathbb{R}^m$. The explicit averaged Euler scheme then reads

$$Y_{n+1} = Y_n + [f(Y_n) + G(Y_n) I_n] \Delta_n, \tag{5.27}$$

where

$$I_n(\omega) := \frac{1}{\Delta_n} \int_{t_n}^{t_{n+1}} \zeta_s(\omega) \, ds. \tag{5.28}$$

For the general RODE (5.21) this suggests that one should pathwise average the vectorfield, i.e.,

$$\frac{1}{\Delta_n} \int_{t_n}^{t_{n+1}} f(\zeta_s(\omega), Y_n(\omega)) \, ds,$$

which is computationally expensive even for low dimensional systems. An alternative is to use the averaged noise within the vector field, which leads to the explicit *averaged Euler scheme*

$$Y_{n+1} = Y_n + f(I_n, Y_n) \Delta_n. \tag{5.29}$$

A systematic derivation of higher order numerical schemes for RODEs based on this idea using generalized Taylor expansions and approximated noise integrals has been carried out by Jentzen & Kloeden (2007, 2008a). See also Carbonell *et al.* (2005) for the local linearization method.

It is well known from the theory of classical Runge-Kutta schemes for ODEs that an implicit scheme is required for the stable integration of an ODE obtained from the spatial discretization of a parabolic PDE. In fact, implicit schemes which are B-stable (see e.g. Hairer & Wanner (1991)) i.e., preserve the structure of merging trajectories of an ODE with a dissipative one-sided Lipschitz condition, are even better. (Recall that no explicit or linear-implicit Runge-Kutta scheme is ever B-stable.) Since RODEs are generalizations of ODEs this applies equally well to RODEs. Jentzen & Kloeden (2008b) introduced the *implicit averaged Euler scheme* (IAES)

$$Y_{n+1} = Y_n + f(I_n, Y_{n+1}) \Delta_n \tag{5.30}$$

and the *implicit averaged midpoint scheme* (IAMS)

$$Y_{n+1} = Y_n + f\left(I_n, \frac{1}{2}(Y_n + Y_{n+1})\right) \Delta_n \tag{5.31}$$

for the RODE (5.21). Both of these numerical schemes are B-stable and have pathwise convergence orders $\min(2\theta, 1)$ and 2θ, respectively, where θ is the Hölder exponent of the noise process ζ.

5.5 Stochastic and Random Partial Differential Equations

As with RODEs and SDEs, we also distinguish between random and stochastic partial differential equations. For simplicity we restrict attention to parabolic reaction-diffusion type PDE on a bounded spatial domain $\mathcal{D}$ in $\mathbb{R}^d$ with smooth boundary $\partial\mathcal{D}$ and assume a Dirichlet boundary condition. In particular, we consider *random PDEs* of the form

$$\frac{\partial u}{\partial t} = \Delta u + f(\zeta_t, u), \qquad u\big|_{\partial\mathcal{D}} = 0, \tag{5.32}$$

with a stochastic process ζ_t, $t \geq 0$, (possibly infinite dimensional) which we interpret and analyze pathwise as a deterministic PDE. We also consider *stochastic PDEs* of the form

$$dU = [\Delta U + f(U)]\, dt + g(U)\, dW, \qquad U\big|_{\partial\mathcal{D}} = 0, \tag{5.33}$$

where W is an infinite dimensional Wiener process of the form

$$W(t, x) = \sum_{j=1}^{\infty} c_j W_t^j \phi_j(x), \qquad t \geq 0, x \in \mathbb{R}^d,$$

with mutually independent scalar Wiener processes W_t^j, $t \geq 0$, $j \in \mathbb{N}$; here the ϕ_j are a basis system in, e.g. $L_2(\mathcal{D})$ formed by the eigenfunctions of the Laplace operator on $\mathcal{D}$ with Dirichlet boundary conditions. Note that when the $c_j \equiv 1$ for all $j \in \mathbb{N}$ and $d = 1$, then W is called a Brownian or Wiener sheet, which is rough in both space and time. As for SDEs the theory of SPDEs is a mean-square theory and requires Itô stochastic calculus, see e.g. Krylov & Rozovskij (1982) and DaPrato & Zabcyzk (1992). The theory is complicated by different types of solutions and function spaces depending on the spatial regularity of the driving noise process.

The Doss-Sussmann theory is not as well developed for SPDEs as for SDEs, but in simple cases we can transform an SPDE to an RPDE. For example, the SPDE (5.33) with such additive noise, i.e.,

$$dU = [\Delta U + f(U)]\, dt + dW, \qquad U\big|_{\partial\mathcal{D}} = 0,$$

is equivalent to the RPDE

$$\frac{\partial v}{\partial t} = \Delta v + f(v + \widehat{O}_t) + \widehat{O}_t$$

with $v(t) = U_t - \dot{O}_t$, $t \geq 0$, where $\widehat{O}_t$, $t \geq 0$ is the (infinite-dimensional) Ornstein-Uhlenbeck stochastic stationary solution of the linear SPDE

$$dU = [\Delta U - U]\,dt + dW, \qquad U\big|_{\partial \mathcal{D}} = 0. \tag{5.34}$$

5.5.1 Numerical methods

All of the difficulties encountered in solving deterministic PDEs numerically reoccur with RPDEs and SPDEs plus more due to the noise, in particular due to the roughness of the noise sample paths and the need to compute many sample paths if strong or weak error criteria are used. RPDEs are less computationally intensive, if a pathwise error criterion is used. In both cases, the theory is in its early stages, see e.g. Gyöngy & Nualart (1995), Grecksch & Kloeden (1996), Davie & Gaines (2000), Hausenblas (2002, 2003), Gyöngy & Krylov (2003), Gyöngy & Millet (2005), Müller-Gronbach & Ritter (2007), and much remains to be done. The temporal convergence rate is often very low when the noise is very rough, such as a Wiener sheet process, although a higher order is possible when the noise is spatially smoother or finite dimensional.

Standard spatial discretization methods such as the method of lines (spatial difference quotients) and the Galerkin or finite element methods (basis function expansions) lead to high dimensional RODEs or SDEs, which are then approximated through temporally discretized numerical schemes. Like their deterministic counterparts, such systems are stiff and require specially constructed stable numerical schemes, which are usually implicit, for their efficient computation such as the B-stable implicit averaged Euler (5.30) and implicit averaged midpoint (5.31) schemes introduced in Jentzen & Kloeden (2008b).

Figure 5.5 here shows a sample path of each of the implicit averaged Euler scheme (left) and explicit Euler scheme (right) for the 9 dimensional RODE obtained by the method of lines with 10 subintervals applied to the random PDE with a scalar Ornstein–Uhlenbeck process,

$$\frac{\partial u}{\partial t} = \frac{\partial^2 u}{\partial x^2} - u - (u + O_t)^3 \tag{5.35}$$

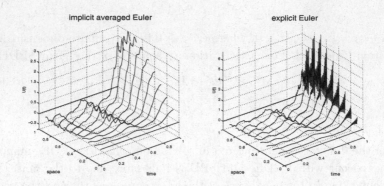

Fig. 5.5. Pathwise approximation of the random PDE (5.35) with a scalar Ornstein–Uhlenbeck process with the method of lines and the implicit averaged (left) and explicit (right) Euler schemes (see text).

on the interval $0 \leq x \leq 1$ with Dirichlet boundary condition and zero initial condition. This RPDE is equivalent to the SPDE with additive noise

$$dU_t = \left[\frac{\partial^2}{\partial x^2} U_t - U_t - U_t^3 \right] dt + dW_t \qquad (5.36)$$

on the interval $0 \leq x \leq 1$ with Dirichlet boundary condition with a scalar Wiener process, where $u(t) := U_t - O_t$.

In concluding, we mention that the stochastic Taylor expansion for finite dimensional SDEs does not generalize directly to SPDEs, but an alternative expansion in a Hilbert space to arbitrarily high strong order is possible for parabolic SPDEs using the smoothening properties of the semi-group operator, see Jentzen (2008).

Acknowledgement Supported by the DFG-Project *"Pathwise Numerics and Dynamics of Stochastic Evolution Equations"*

References

L. Arnold (1997), *Random Dynamical Systems*, Springer, Berlin.

H. Bunke (1972), *Gewöhnliche Differentialgleichungen mit zufälligen Parametern*, Akademie-Verlag, Berlin.

K. Burrage, P. M. Burrage and T. Tian (2004), 'Numerical methods for strong solutions of stochastic differential equations: An overview', *Proc. R. Soc. Lond., Ser. A, Math. Phys. Eng. Sci.* **460**, 373–402.

T. Caraballo and P. E. Kloeden (2005), 'The persistence of synchronization under environmental noise', *Proc. R. Soc. Lond., Ser. A, Math. Phys. Eng. Sci.* **461**, 2257–2267.

T. Caraballo, P. E. Kloeden and A. Neuenkirch (2008), 'Synchronization of systems with linear noise', *Stochastics & Dynamics* **8**, 139–154.

F. Carbonell, J. C. Jimenez, R. J. Biscay and H. de la Cruz (2005), 'The local linearization method for numerical integration of random differential equations', *BIT* **45**, 1–14.

F. Castell and J. Gaines (1995), 'An efficient approximation method for stochastic differential equations by means of the exponential Lie series', *Math. Comput. Simul.* **38**, 13–19.

G. Da Prato and G. Zabczyk (1992), *Stochastic Equations in Infinite Dimensions*, Cambridge University Press, Cambridge.

A. M. Davie and J. G. Gaines (2000), 'Convergence of numerical schemes for the solution of parabolic stochastic partial differential equations', *Math. Comput.* **70**, 123–134.

H. Doss (1977), 'Liens entre équations différentielles stochastiques et ordinaires', *Ann. Inst. Henri Poincaré, Nouv. Sér., Sect. B* **13**, 99–125.

G. Fleury (2005), 'Convergence of schemes for stochastic differential equations', *Prob. Engineering Mech.* **21**, 35–43.

J. G. Gaines and T. J. Lyons (1994), 'Random generation of stochastic area integrals', *SIAM J. Appl. Math.* **54**, 1132–1146.

J. G. Gaines and T. J. Lyons (1997), 'Variable step size control in the numerical solution of stochastic differential equations', *SIAM J. Appl. Math.* **57**, 1455–1484.

M. J. Garrido-Atienza, P. E. Kloeden and A. Neuenkirch (2008), 'Discretization of stationary solutions of stochastic systems driven by fractional Brownian motion', *J. Appl. Math. Optim.*, submitted.

Y. A. Godin and S. Molchanov (2007), 'Approximation of random dynamical systems with discrete time by stochastic differential equations: I. Theory', *Random Oper. Stoch. Eqns.* **15**, 205–222.

W. Grecksch and P. E. Kloeden (1996), 'Time-discretized Galerkin approximations of parabolic stochastic PDEs', *Bull. Austral. Math. Soc.* **54**, 79–84.

L. Grüne and P. E. Kloeden (2001), 'Pathwise approximation of random ordinary differential equations', *BIT* **12**, 711–721.

I. Gyöngy (1998), 'A note on Euler's approximations', *Potential Anal.* **8**, 205–216.

I. Gyöngy and N. Krylov (2003), 'On the splitting-up method and stochastic partial differential equations', *Ann. Probab.* **31**, 564–591.

I. Gyöngy and A. Millet (2005), 'On discretization schemes for stochastic evolution equations', *Potential Anal.* **23**, 99–134.

I. Gyöngy and D. Nualart (1997), 'Implicit scheme for stochastic parabolic partial differential equations driven by space-time white noise', *Potential Anal.* **7**, 725–757.

E. Hairer and G. Wanner (1991), *Solving Ordinary Differential Equations II*, Springer, Berlin.

N. Halidias and P. E. Kloeden (2008), 'A note on the Euler-Maruyama scheme for stochastic differential equations with a discontinuous monotone drift coefficient', *BIT* **48**, 51–59.

E. Hausenblas (2002), 'Numerical analysis of semilinear stochastic evolution equations in Banach spaces', *J. Computat. Appl. Math.* **147**, 485–516.

E. Hausenblas (2003), 'Approximation of semilinear stochastic evolution equations', *Potential Anal.* **18**, 141–186.

D. J. Higham, X. Mao and A. M. Stuart (2002), 'Strong convergence of Euler-type methods for nonlinear stochastic differential equations', *SIAM J. Num. Anal.* **40**, 1041–1063.

P. Imkeller and C. Lederer (2001), 'On the cohomology of flows of stochastic and random differential equations', *Probab. Theory Related Fields* **120**, 209–235.

P. Imkeller and C. Lederer (2002), 'The cohomology of stochastic and random differential equations, and local linearization of stochastic flows', *Stoch. Dyn.* **2**, 131–159.

A. Jentzen (2008), *Numerical Solution of Stochastic Partial Differential Equations*, Dissertation, Institut für Mathematik, Johann Wolfgang Goethe Universität, Frankfurt am Main.

A. Jentzen and P. E. Kloeden (2007), 'Pathwise convergent higher order numerical schemes for random ordinary differential equations', *Proc. R. Soc. Lond., Ser. A, Math. Phys. Eng. Sci.* **463**, 2929–2944.

A. Jentzen and P. E. Kloeden (2008), 'Pathwise Taylor schemes for random ordinary differential equations', *BIT*, to appear.

A. Jentzen and P. E. Kloeden (2008), 'Stable time integration of spatially discretized random and stochastic PDEs', submitted.

A. Jentzen, P. E. Kloeden and A. Neuenkirch (2008), 'Pathwise approximation of stochastic differential equations on domains: Higher order convergence rates without global Lipschitz coefficients', *Numer. Math.*, to appear

A. Jentzen, A. Neuenkirch and A. Rößler (2008), 'Runge-Kutta type schemes for random ordinary differential equations', preprint.

P. E. Kloeden (2002), 'The systematic derivation of higher order numerical methods for stochastic differential equations', *Milan J. Math.* **70**, 187–207.

P. E. Kloeden and A. Neuenkirch (2007), 'The pathwise convergence of approximation schemes for stochastic differential equations', *LMS J. Comp. Math.* **10**, 235–253.

P. E. Kloeden and E. Platen (1992), *The Numerical Solution of Stochastic Differential Equations*, Springer, Berlin.

N. V. Krylov and B. L. Rozovskij (1981), 'Stochastic evolution equations', *J. Sov. Math.* **16**, 1233–1277.

S. J. A. Malham and A. Wiese (2008), 'Stochastic Lie group integrators', *SIAM J. Sci. Comput.* **30**, 597–617.

J. C. Mattingly, A. M. Stuart and D. J. Higham (2002), 'Ergodicity for SDEs and approximations: locally Lipschitz vector fields and degenerate noise', *Stochastic Processes Appl.* **101**, 185–232.

G. N. Milstein (1995), *Numerical Integration of Stochastic Differential Equations*, Kluwer, Dordrecht.

G. N. Milstein and M. V. Tretjakov (2004), *Stochastic Numerics for Mathe-*

matical Physics, Springer, Berlin.

G. N. Milstein and M. V. Tretjakov (2005), 'Numerical integration of stochastic differential equations with nonglobally Lipschitz coefficients', *SIAM J. Numer. Anal.* **43**, 1139–1154.

T. Müller-Gronbach and K. Ritter (2007), 'Lower bounds and nonuniform time discretization for approximation of stochastic heat equations', *Found. Comput. Math.* **7**, 135–181.

T. T. Soong (1973), *Random Differential Equations in Science and Engineering*, Academic Press, New York.

H. J. Sussmann (1978), 'On the gap between deterministic and stochastic differential equations', *Ann. Probab.* **6**, 19–41.

D. Talay (1983), 'Résolution trajectorielle et analyse numérique des équations différentielles stochastiques', *Stochastics* **9**, 275–306.

M. Wiktorsson (2001), 'Joint characteristic function and simultaneous simulation of iterated Itô integrals for multiple independent Brownian motions', *Ann. Appl. Probab.* **11**, 470–487.

E. W. Wong and M. Zakai (1965), 'On the relation between ordinary and stochastic differential equations', *Internat. J. Engineering Sci.* **3**, 213–229.

6

Some Properties of the Global Behaviour of Conservative Low-Dimensional Systems

Carles Simó

Departament de Matemàtica Aplicada i Anàlisi
Universitat de Barcelona, Gran Via, 585
Barcelona 08007, Spain
e-mail: carles@maia.ub.es

Abstract

When studying a dynamical system from a global viewpoint, there are only few theoretical tools at our disposal. This is especially true if we want to describe all aspects of the dynamics with a reasonable amount of detail. A combination of analytic, symbolic and numerical tools, together with qualitative and topological considerations, can give a reasonably good description. Furthermore it is possible to derive paradigmatic models which can be analysed theoretically and allow us to study pieces of the dynamics. It is also important to know the relevance of different phenomena. Are they confined to a narrow domain of the phase space or to a tiny region of the parameter space or do they really play a significant role? Several theoretical/numerical tools are presented, and applied to different problems in celestial mechanics, unfolding of singularities and other problems. This is part of a project aimed towards understanding finite-dimensional systems in a global way. To avoid technicalities we shall assume that all maps and flows considered in this paper are analytic.

6.1 Introduction

Many properties are known for low-dimensional conservative systems, like Area-Preserving or Measure-Preserving Maps (APM, MPM) or systems which can be reduced to them as 2-degrees of freedom Hamiltonian systems and volume-preserving 3D flows. Most of these properties have a local character, either around a fixed point, around a given orbit, like a periodic orbit or a homoclinic orbit, or around an invariant curve or torus. However it is extremely relevant to study global properties, when different invariant objects like manifolds and tori and their relative positions play a role. A combination of theoretical tools and computer-assisted studies allow us to grasp some global properties. Still many fundamental questions remain unanswered. This is specially true when the dimension increases.

The purpose of this paper is to present some theoretical facts and numerical tools which can be of help in global studies. Accordingly we first present a short sample of problems, mainly in Celestial Mechanics, but also in bifurcation theory, in one of the paradigmatic models, the Hénon conservative map, and also references to similar problems in fluid mechanics. Then, in Section 6.3, some analytic tools are recalled, like KAM theory, splitting of separatrices and several return maps. Section 6.4 is devoted to describing a few numerical tools, like efficient long-time integrations, dynamics indicators and a simple method to test if some initial data gives rise to an invariant curve. In Section 6.5 additional information is given on the Hénon conservative map. We close with a short list of open problems.

All the studied systems are non-integrable. It is quite rare for a realistic problem to be integrable, although one can have interesting integrable approximations. Theoretical criteria for deciding if a Hamiltonian system is integrable fall into different classes. For the more algebraic aspects and a summary of other methods, applications and open problems one can see Morales et al. (2007)

A large part of this paper is based on work in progress with different authors. Even if this can be considered as a preliminary report, we hope it will be found useful by the reader. This text coincides, essentially, with the presentation given at the FoCM08 conference.

When mentioning different colours in the figures, we refer to the electronic version. In the printed version they are replaced by grey tones. The electronic version also offers the possibility of magnifying figures to check details. Most of plots have been produced in high resolution.

6.2 A Sample of Problems

6.2.1 The Restricted Three-Body Problem: Asteroids, Outer Comets, Satellites

The RTBP is a useful simple model for studying the dynamics of a massless particle under the gravitational attraction of two massive bodies (the primary S and the secondary J) assumed to move in circular orbits around their centre of mass. Using a rotating coordinate system (the synodical system) which fixes the positions of the main bodies and suitable units, the Hamiltonian is

$$H = \frac{1}{2}(p_x^2 + p_y^2 + p_z^2) + yp_x - xp_y - \frac{1-\mu}{r_1} - \frac{\mu}{r_2},$$

where $r_1 = ((x-\mu)^2 + y^2 + x^2)^{1/2}$, $r_2 = ((x-\mu+1)^2 + y^2 + z^2)^{1/2}$ and $\mu = m_J/(m_S + m_J)$, where x, y, z are Cartesian coordinates, p_x, p_y, p_z the conjugated momenta and m_S, m_J the masses of the main bodies. The energy is related to the classical Jacobi integral C as $H = -0.5(C - \mu(1-\mu))$. See, e.g., Simó (1997, 1998). Despite its simplicity it gives relevant practical information for the dynamics of satellites around a planet, for asteroids and comets, and for the design of space missions.

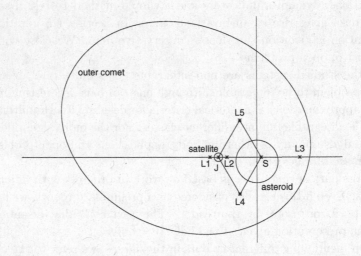

Fig. 6.1. The RTBP with the location of the libration points L_1 to L_5 and orbits of an asteroid, an outer comet and a satellite around a planet.

The system has 5 equilibrium points, denoted as libration or Euler-Lagrange points, see Figure 6.1. Several questions can be raised, such as

1) At a given level of energy when are the orbits of asteroids, outer comets, satellites bounded (no escape, no collision)? If an orbit is close to elliptic, then up to which eccentricity e is some stability found?
2) The triangular points $L_{4,5}$, are close to the so-called Trojan asteroids.

In the 2D case these points are linearly stable for $\mu \in [0, \mu_1]$, where

$$\mu_s = \omega_{\text{short}}/\omega_{\text{long}} = s, \ s \in \mathbb{N}, \ \omega_{\text{short,long}} = \left[(1 \pm (1 - 27\mu(1-\mu))^{1/2})/2\right]^{1/2}$$

$\omega_{\text{short}}, \omega_{\text{long}}$ being the frequencies at the $L_{4,5}$. Nonlinear stability has been proved, in the planar problem, for $\mu \in [0, \mu_1] \setminus \{\mu_2, \mu_3\}$. In the 3D case *practical stability*, see Giorgilli et al. (1989), has been proved in the absence of resonances. Here *practical* means that the changes in the orbit are below some tolerance for very large intervals of time (e.g., the age of the solar system) if motion starts close to $L_{4,5}$.

But the following question remains: Up to which distance do we have some kind of stability? Figure 6.2 provides a partial answer. To scan a reasonable set of initial conditions we consider motion starting at (x, y, z) with zero synodical velocity. An escape of the vicinity of, say, L_5 is defined as a too close approach to S or J ($d < 0.1, 0.01$), a too large distance to S ($d > 2$) or a projection on (x, y) with $y < -0.2$. Let (r, α, z) be the initial data in cylindrical coordinates. Long-time integration has been used, starting on a fine grid in (r, α, z), to detect non-escaping points, with special care on the boundary points (a "quasi-boundary" in the 3D case). Note that for a fixed value of μ the initial points are on different levels of H.

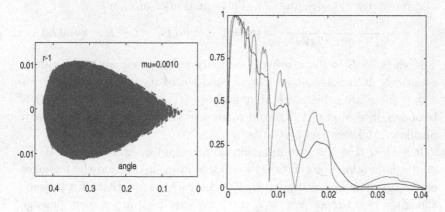

Fig. 6.2. Stability around $L_{4,5}$ in the RTBP. Left: domain of stability for the planar problem and $\mu \approx \mu_{\text{Sun-Jupiter}}$. Right: Relative size of the stability domain in the 2D case (blue curve) and the 3D one (red curve, preliminary results) as a function of μ. See text for details.

On the left part the set of non-escaping points for $\mu = 0.001$ is plotted on $(r - 1, \alpha/(2\pi))$ variables. Theoretically one can prove that for μ small the measure is $\mathcal{O}(\sqrt{\mu})$. On the right side we display the abundance of subsisting points, as a function of μ, normalized so that the maximal value is 1. The maximum occurs for $\mu \approx 0.0014$ in the 2D case and for $\mu \approx 0.0017$ in the 3D one. In the 2D case it is clear the role of the 2:1 and 3:1 resonances which occur for μ_2, μ_3, but one can clearly see the role of the 4:1, 5:1, 6:1,... resonances for smaller values of μ, which decrease the measure of non-escaping points. This effect is not so strong in the 3D case. For values of μ for which almost all (or all) initial points escape in the 2D case, there exist points in the space which are in invariant tori. In fact, for most of the μ values the region with a larger abundance of stable points corresponds to inclinations between 20° and 40°. This agrees with the experimental results on the abundance of Trojans as a function of the inclination. Furthermore we stress that tiny stable regions appear for $\mu > \mu_1$. The point itself loses stability, but some tori remain around it.

6.2.2 The Sitnikov Problem

This is a classical and popular toy problem which describes the motion of a massless particle on the vertical axis under the attraction of two bodies, with mass equal to $1/2$, moving in an horizontal plane in ellipses of eccentricity e, see Figure 6.3. The equation of motion is

$$\ddot{z} = -\frac{z}{(z^2 + r(t)^2/4)^{3/2}}, \quad r(t) = 1 - e\cos(E), \quad t = E - e\sin(E).$$

The variable E is the eccentric anomaly and last equality is Kepler's equation. It is relevant as a first example of *oscillatory motion*: for $e > 0$ there are orbits such that $\limsup z(t) = \infty$ while $\liminf z(t)$ is bounded. See Moser (1973) and references therein to works by Chazy, Sitnikov, Alekseev and McGehee.

It is clear that the system is autonomous and integrable if $e = 0$. To study the motion one can take as suitable Poincaré section the passages of the massless body through $z = 0$. Let $v = dz/dt$; then if a passage though $z = 0$ occurs at $t = t_k$ with velocity $|v_k|$ and a next passage occurs, the Poincaré map is defined as $\mathcal{P}(|v_k|, t_k) = (|v_{k+1}|, t_{k+1})$, which can be represented in polar coordinates (t is defined mod 2π). As $|v| = 0$ is a fixed point of the flow $\mathcal{P}$ is not defined on it. But $\mathcal{P}$ can be extended by continuity and then $|v| = 0$ can be seen as an elliptic fixed point of $\mathcal{P}$, except at tiny intervals in e where the linear character is of hyperbolic with reflection type, that is, the eigenvalues are λ, λ^{-1} with $\lambda < -1$.

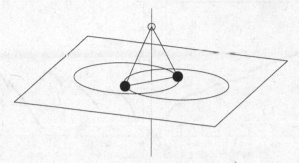

Fig. 6.3. Sketch of the motion of the bodies in Sitnikov's problem.

This follows from the analysis of the linear behaviour at the equilibrium point, given by the solutions of the Hill's equation of Ince's type

$$d\xi/dE = (1 - e\cos(E))\eta, \ d\eta/dE = -8\xi/(1 - e\cos(E))^2 \,.$$

If $e = 0$ the orbits with $|v_0| < 2$ are bounded.

Let us pass to *global aspects*: Up to which distance is it possible to find rotational invariant curves (*r.i.c.*) of $\mathcal{P}$? That is, curves which can be seen as a graph $|v| = g(t)$ and, therefore, confine the motion. It is clear that this depends on e. Figure 6.4 shows some results which depend on the value of t where we compute $g(t)$. Again using long-time integration and other tools, see Section 6.4, the set of non-escaping orbits has been computed. On the left we plot the maximal value of $|v|$ at $t = \pi$ as a function of e. One can easily see some jumps on the value of $|v|$. A modified plot is shown on the right: we have scaled $|v|$ and added a suitable function to enhance the jumps. It is natural to conjecture the existence of infinitely many jumps (most of them are tiny).

At the bottom of the left hand side plot one can see values ranging from 1 to 7. They correspond to the values of e for which the *r.i.c.* which were confining islands of periods $1, 2, \ldots, 7$ break down and points sitting on a domain of *confined chaos* are free to escape. Other smaller jumps correspond to the breakdown of *r.i.c.* which confined islands with rotation numbers of the form $p/q \in \mathbb{Q}$.

Another point to stress is that the values of e at which the jumps occur *do not* correspond, in general, to the values of e at which bifurcations occur at $|v| = 0$. These are different phenomena.

In fact the Sitnikov problem is an example of a one degree of freedom

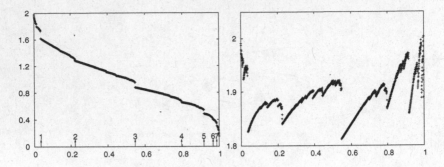

Fig. 6.4. Left: Approximate maximal value of $|v|$ at $t = E = \pi$ as a function of e. Right: Idem but plotting $|v|/(1-e)^{1/4} + \sqrt{e}$ vs e.

(1 d.o.f.) Hamiltonian system depending periodically on time. See Broer et al. (2004) for similar studies in a different context..

6.2.3 Bifurcations: The Hopf-Saddle-Node Conservative Unfolding

At a HSN bifurcation of a flow in $\mathbb{R}^3$ the eigenvalues are $0, \pm i$. Normal Forms (NF), formally rotationally invariant, can be computed and unfolded generically. The volume-preserving case helps to understand other more general unfoldings. From the different generic cases which can appear, the most interesting ones give rise, after a suitable blow up, to a limit behaviour as displayed in Figure 6.5. On the left we display the dynamics of the NF (to any order) which has 2 fixed points (south and north poles, S and N). Both the z axis and the $\mathbb{S}^2$ sphere shown are invariant. The sphere coincides with $W^u(N) = W^s(S)$ and the axis with branches of $W^s(N)$ and $W^u(S)$ which also match. Dynamics on the sphere spirals away from N and enters S also spiraling. On the right side we show the reduction of the NF to a 2D flow. It has 2 saddles and an additional fixed point. Tuning parameters one can obtain an heteroclinic connection of the saddles. The additional fixed point is, generically, a focus, but in the volume-preserving case it is a centre and then it is surrounded by periodic orbits. By rotation of the right hand side around the vertical axis one obtains the left plot. The centre fixed point becomes a periodic orbit of the NF of the 3D flow and the planar periodic orbits become 2D tori.

But the true dynamics can differ largely from the one given by the

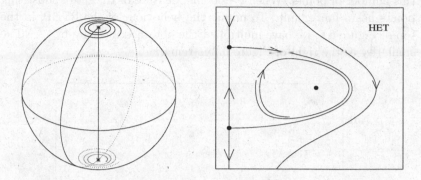

Fig. 6.5. An sketch of the dynamics of the Normal Form in an interesting case of a HSN bifurcation. Left: a 3D view. Right: the reduction to a planar flow in a generic case (non-volume-preserving) for parameters for which an heteroclinic connection occurs.

NF when we go away from the limit case. The coincidence of manifolds to any order implies that the splitting of the 2D manifolds and the distance between the 1D ones are exponentially small in the distance, in the parameter space, to the bifurcation. See Dumortier et al. (2009).

As a concrete example which is representative enough, we consider the Michelson system, which appears also as the equation for traveling waves in the Kuramoto-Shivasinski PDE:

$$x''' = -1 + ax' + x^2,\, a > 0.$$

It is convenient to use as parameter $\varepsilon = (-a)^{-3}$. The NF to any order has connecting 2D and 1D invariant manifolds of $(\pm 1, 0, 0)$. The real system has splitting exponentially small in ε. The variables $y = x'$ and $z = x''$ can be introduced and clearly the flow preserves volume.

As Poincaré section one has taken passages through $z = 0$ with $z' > 0$. Scaling of y by the factor $2\sqrt{-a}$ allows us to reduce the Poincaré section to a domain which is, approximately, the half unit disk (it is exactly the half disk when $\varepsilon \to 0$). Initial points are taken on the domain $(y, x) \in [-1, 0] \times [-1.05, 1.05]$. Transversality is lost along a curve which plays almost no role. Note that the Poincaré map preserves a measure whose density is proportional to $|z'|$ and, hence, it is absolutely continuous with respect to the Lebesgue measure on the (y, x) plane.

Starting on the given domain the points in a fine equispaced grid which do not escape are stored. As a simple measure we have taken this number of points. When $\varepsilon \to 0$ the measure of the set of subsisting points has a finite limit. To mimic the behaviour of the RTBP, in the top of Figure 6.6, we have multiplied the obtained measure by $\sqrt{\varepsilon}$. The similarity to the right of Figure 6.2 is remarkable.

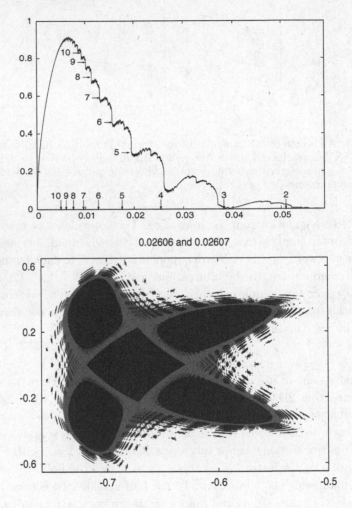

Fig. 6.6. Some results for the Michelson system. Top: A measure of the domain of confined motion, see text. The marks from 2 to 10 at the bottom part of the plot denote the bifurcations at E of islands with rotation number 1:2,...,1:10. The marks from 5 to 10 along the curve denote the places at which *r.i.c.* confining these island are broken. Bottom: In blue points which do not escape for $\varepsilon = \varepsilon_2 = 0.02607$. In red (and also the points in red hidden by the ones in blue) the non-escaping points for $\varepsilon = \varepsilon_1 = 0.02606$. This transition occurs at the jump associated with period 4 in the top plot.

In the bottom of Figure 6.6 we show the changes in the set of non-escaping points which occur for some small changes in ε. Concretely, we have taken the values $\varepsilon_1 = 0.02606$ and $\varepsilon_2 = 0.02607$. For ε_1 appears a large stable region, bounded by a *r.i.c.* which encloses the islands of period 4, and many chains of islands outside this curve. These are the points in red in the plot. Note that the points in blue are hiding points in red. Beyond period-4 elliptic points inside the islands, one can guess the existence of a period-4 hyperbolic orbit. The transversal homoclinic intersection of its manifolds creates a chaotic zone, but the points on it can not escape because of the *r.i.c.*

Changing to ε_2 the curves surrounding the chaotic zone break down. Hence points in the chaotic zone can (and do) escape. The non-escaping points for ε_2 are plotted in blue. One can see a central area around the fixed elliptic point, with very sharp boundaries and the big period-4 islands. These are surrounded by many chains of tiny islands. And also chains of islands can be seen surrounding the set of 4 islands. The major islands of these chains have only minor changes with respect to the islands found for ε_1.

6.2.4 A Problem in Fluid Dynamics: The Rayleigh-Bénard Model

This model describes the convection forced by a difference of temperature between the bottom and top boundaries. We have considered a cube with conducting lateral walls. The equations describing the velocity field V and the variation of the temperature with respect to a linear profile θ are

$$Pr^{-1}\left(\frac{\partial V}{\partial t} + Ra^{1/2}(V \cdot \nabla)V\right) - \nabla^2 V - Ra^{1/2}\theta\, e_z + \nabla p = 0,$$

$$\frac{\partial \theta}{\partial t} + Ra^{1/2}(V \cdot \nabla)\theta - \nabla^2\theta - Ra^{1/2}w = 0, \qquad \nabla \cdot V = 0,$$

where Pr and Ra denote the Prandtl and Rayleigh numbers, respectively. The boundary conditions are $V = \theta = 0$ at $|x| = |y| = |z| = 1/2$. Using a Galerkin method with 10976 basic functions we have computed the bifurcation diagram of steady-state solutions, see Puigjaner et al. (2008). Then, for some of the branches, the velocity field has been integrated. In that case we are interested in the mixing properties of the volume-preserving flow. Hence, the domain of interest is the complement of the invariant tori. Methods similar to the ones used for the

Michelson problem (i.e., Poincaré sections) combined with computation of Lyapunov exponents, allow us to evaluate how the mixing changes as a function of Ra, for fixed Pr and along different branches of the bifurcation diagram, see Simó et al. (2008a).

6.2.5 A Paradigmatic Example: The Hénon Map

Any quadratic area preserving map reduces to (see Hénon (1969))

$$H_c : \begin{pmatrix} x \\ y \end{pmatrix} \to \begin{pmatrix} \bar{x} \\ \bar{y} \end{pmatrix} = \begin{pmatrix} c(1 - x^2) + 2x + y \\ -x \end{pmatrix} \qquad x, y, c \in \mathbb{R}.$$

With this normalization fixed points are located at H$= (-1, 1)$ (hyperbolic) and at E$= (1, -1)$, elliptic if $c \in [0, 2]$ with Tr$(DH_c(E))=2 - 2c$. A Normal Form analysis proves nonlinear stability if $c \in (0, 2) \setminus \{1.5\}$. It appears typically in generic non-hyperbolic area-preserving maps, e.g., in many parts of the phase space. Despite its trivial character many of its properties are still not known in detail, mainly concerning its global aspects. It serves as a paradigm for many phenomena, see Simó–Vieiro (2009). A typical phase portrait, the abundance of non-escaping points and an example after the breakdown of *r.i.c.* around period-3 islands are shown in the top of Figure 6.7. Observe the self-similar character and the jump discontinuities and compare with the results displayed in Figures 6.2, 6.3 and 6.6. Compare the bottom part with the bottom of Figure 6.6. Similar patterns appear for any other kind of islands when confining outer curves break down. The choice of low periods in the plots is to prevent the appearance of small chaotic zones.

6.3 Analytic Tools

We present some ideas on a few concrete topics.

6.3.1 KAM Theory: Conditions for Applicability

A twist map on an annulus is a map given in polar coordinates as

$$T_0 : (r, \theta) \mapsto (r + \alpha(r))$$

and such that $d\alpha/dr \neq 0$. It is obvious that all the circles $r = r_0$ are invariant under T_0. We can consider perturbations like

$$T_\varepsilon : (r, \theta) \mapsto (r + \varepsilon f_1(r, \theta, \varepsilon), \theta + \alpha(r) + \varepsilon f_2(r, \theta, \varepsilon))$$

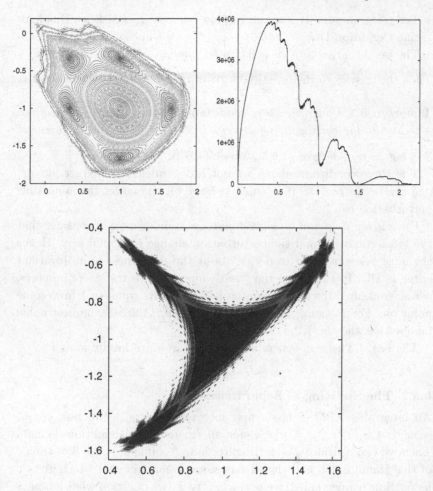

Fig. 6.7. Top left: Phase portrait for $c = 0.75$, shortly after the period-5 bifurcation at E. Outside that domain, points escape to infinity, close to W_H^u. Top right: number of non-escaping points on a grid $10^{-3} \times 10^{-3}$. Computations done by a combination of methods. Bottom: Subsisting points before (red) and after (blue) the destruction of *r.i.c.* around the period-3 islands.

with f_1, f_2 such that T_ε is still APM. A natural question is which curves subsist under the perturbation. Assume

- $d\alpha/dr \neq 0$ for T, the twist condition, i.e., the frequency changes with amplitude,

- The rotation number ρ of an invariant curve of T_0 satisfies a Diophantine Condition DC, i.e., $|q\rho - p| > c|q|^{-\tau}$, for some $c > 0, \tau \geq 1$,
- The perturbation $\|\varepsilon f\|$ is sufficiently small.

Then the answer to the question is given by

Theorem 6.1 *Under the above conditions T_ε has a r.i.c. with the given ρ ε-close to the corresponding invariant curve of T_0 (KAM Theorem).*

See, e.g., Arnol'd–Avez (1967), Moser (1973).

The three conditions above are not independent: the larger $|d\alpha/dr|$ and the better the DC (i.e., small τ, large c), the larger the admissible perturbation can be.

For a given map and on a given narrow annulus we can consider that the twist condition and the perturbation change in a mild way. Hence the most relevant point to decide about the robustness of an invariant curve is DC. In this sense the "best" numbers are the *noble numbers*, whose continued fraction expansion has quotients equal to 1 from some point on. For instance, in the range $[0.13579, 0.135791]$ the best noble number has the cfe $[7, 2, 1, 2, 1, 11, 1, 1, 1, \ldots]$.

The KAM Theorem generalizes to maps in a product of annuli.

6.3.2 The Splitting of Separatrices

An integrable APM T has a first integral, that is, a function $g(x, y)$ such that $(g(T(x, y)) = g(x, y)$ for all (x, y) in the definition domain. Then we can consider g as a Hamiltonian. Assume T is the flow time ε of this Hamiltonian and that it has some separatrix $\gamma(t)$ which gives a homoclinic connection. If we perturb T to T_ε we question what happens to the separatrix: does it subsist or does it split? Generically it splits, but the splitting can be hard to detect if ε is small. This is due to the *exponentially small character of the splitting*.

A first indication of this character is given by an application of Neishtadt's Theorem, see Neishtadt (1984).

Theorem 6.2 *Given $z' = \varepsilon f(z, t, \varepsilon)$ with f analytic with respect to $z \in K$, a compact in $\mathbb{C}^m$, 2π-periodic and continuous in t, bounded in ε, there exists a change $z = h(w, t, \varepsilon)$, with the same regularity, such that the new v.f. is $w' = \varepsilon g(w, \varepsilon) + r(w, t, \varepsilon)$ with $\|r\| < \exp(-d/\varepsilon)$, $d > 0$.*

To derive bounds on the splitting of an APM using this result it is enough to consider a suspension of the map using a periodic v.f. (any regularity with respect to t is enough). If the map is APM the suspension can be chosen to be a non-autonomous Hamiltonian and then one can apply the theorem to obtain g autonomous and Hamiltonian with an exponentially small remainder. See Broer et al. (1996) for examples of the application of this methodology.

The previous theorem can be extended to the case of f analytic quasi-periodic in t with frequencies satisfying a DC, see Simó (1994).

To obtain sharper bounds assume (Fontich–Simó (1990)),

- $\mathcal{H}(q, p)$ is a one d.o.f. (analytic) Hamiltonian with an homoclinic orbit $\gamma(t)$ to an hyperbolic point H,
- σ is the smallest absolute value of the imaginary part of the singularities of γ,
- T_ε is an analytic APM: $T_\varepsilon = \varphi_\varepsilon^{\mathcal{H}} + \mathcal{O}(\varepsilon^2)$,
- $h = \ln(\lambda(\varepsilon))$, where $\lambda(\varepsilon)$ is the dominant eigenvalue of T_ε at the hyperbolic point H_ε.

Theorem 6.3 *Under the assumptions above, for all $\delta > 0$ there exists $N(\delta)$ such that the size of the splitting is bounded by $N(\delta) \exp(-(2\pi\sigma - \delta)/h)$.*

The size of the splitting can be measured by the angle at an homoclinic point on a fixed domain, by the area of the lobes or by some other invariant, like the homoclinic invariant, see Gelfreich (1999), Gelfreich–Lazutkin (2001), Gelfreich–Simó (2008).

Generically the asymptotic size of the splitting as $\varepsilon \to 0$ is like

$$ah^r \exp(-2\pi\sigma/h)(1+o(1)) \text{ or } ah^r \exp(-2\pi\sigma/h) \left(\sum_i \cos(g_i/h + \xi_i) + o(1) \right).$$

Let us now consider the *inner and outer splittings* at a resonance. Let T be a twist map and $r_{m/n}$ an invariant circle with rotation number $\rho = \alpha(r_{m/n})/(2\pi) = m/n$ (a resonant circle). When T is perturbed a resonant Normal Form, which in complex notation looks like

$$z \mapsto R_{2\pi\frac{m}{n}}(e^{2\pi i \gamma(|z|)} z + c\bar{z}^{n-1} + \hat{\mathcal{R}}_n(z, \bar{z})), \quad \hat{\mathcal{R}}_n = \mathcal{O}(|z|^n)$$

is obtained. The function $\gamma(|z|)$ is given by $\delta + b_1|z|^2 + b_2|z|^4 + \dots$, where b_j are the Birkhoff coefficients. This NF can be approximated by the composition of a rotation of angle $2\pi m/n$ with the flow time ε of a Hamiltonian. The Hamiltonian shows resonant zones as displayed on the left of Figure 6.8. It looks similar to a pendulum but the separatrices which pass close to points p and q in the figure are no longer symmetric. Generically the separatrices split near p (outer splitting) and near q (inner splitting) in a quite different way. More concretely

Theorem 6.4 *In the situation described above and assuming $b_1 \neq 0$ the inner and outer splittings for the m/n resonance have, generically, different σ parameters (see Theorem 6.3). That is, the ratio of the splittings is also exponentially small in ε for $\varepsilon \to 0$ and δ small.*

See Simó–Vieiro (2008c) for proof and details. Concerning methods to effectively compute invariant manifolds in a variety of cases we refer to Simó (1990).

6.3.3 Key Models and Return Maps

There are *key models* which are extremely useful to understand the dynamics of a given system. Most of them are obtained as *return maps*. This is the case for some celebrated results like Smale's horseshoe theorem, the Newhouse theorem and many other results related to passage close to some homoclinic point.

To fix ideas consider a split separatrix or a resonant zone which have invariant curves C_1, C_2 on both sides. One can take a segment Σ which is transversal to both curves and its image (perhaps under some iterate T^k of the map T). The domain $\mathcal{D}$ bounded by $C_1, C_2, \Sigma, T(\Sigma)$ is a *fundamental domain*. Every point in $\mathcal{D}$ returns to it except, perhaps, points lying on stable manifolds of hyperbolic periodic points. The return map $\mathcal{R} : \mathcal{D} \to \mathcal{D}$ describes the return (when it is defined) and we note that different points $z \in \mathcal{D}$ can require a different number of iterates, $k = k(z)$, of T to return to $\mathcal{D}$. The study of $\mathcal{R}$ allows us to know the dynamics of T between C_1 and C_2.

It is clear that when the orbit of a point z approaches a hyperbolic point H the number of iterates $k(z)$ becomes unbounded. Then NF techniques can be used to study analytically the passage near H. This applies to general flows and maps in any dimension when passage near an invariant object of hyperbolic type (perhaps in a weak sense) is produced, including passages near objects at infinity.

Consider the behaviour near a broken separatrix, either single loop (as in Figure 6.8 centre) or with two loops as in a figure eight. A model can be derived by considering passage close to the saddle and gluing maps. A general separatrix map and relevant universal properties has

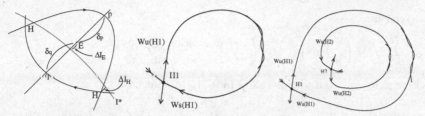

Fig. 6.8. Left: sketch of a resonant zone. The dotted line shows the position of the curve with $\rho = m/n$ before perturbation. E and H are the elliptic and hyperbolic periodic points created, generically, by the perturbation. Centre: A model for the geometry of the separatrix map, associated to a single loop. Right: a model for the geometry of the biseparatrix map.

been studied in Simó–Treschev (2008b). As a simple and typical case we have the Chirikov separatrix map, which scaling actions by the size of the splitting and using symmetry reads as

$$SPM \begin{pmatrix} x \\ y \end{pmatrix} = \begin{pmatrix} \bar{x} \\ \bar{y} \end{pmatrix} = \begin{pmatrix} x + a + b\log(|\bar{y}|) \\ y + \sin(2\pi x) \end{pmatrix},$$

where $a, b \in \mathbb{R}$ and x is taken mod 1.

For b large, SPM looks quite chaotic near $y = 0$. It seems that inside the lobes the dynamics is fully chaotic, but it has been proved that the fraction of ε (in log scale) for which there are stable islands inside the lobe, tends to a positive constant if $\varepsilon \to 0$. See Broer et al. (1998b), Simó–Treschev (2008b), Simó–Vieiro (2008d).

Thanks to contributions of Chirikov, Greene, Mather, MacKay, Rana-de la Llave, Olvera-Simó, based on analysis of the standard map SM it is possible to obtain

Proposition 6.1 *The SPM has only invariant rotational curves (i.e., graphs of $y = g(x)$) if $b/y_0 < \varepsilon_G$, where $\varepsilon_G \approx 0.9716/(2\pi)$, the so-called Greene's critical value.*

For some cases, like in the range $c \in [1.014, 1.015]$ in the Hénon map, see Section 6.5, or in Figure 6.6 for the Michelson system, there are chains of islands, the first and last ones very narrow and becoming larger in the central part. The SPM is not a good model for this. Not only one but two separatrices play a relevant role in the creation of these islands. Similar things occur in the so-called Birkhoff zones. A sketch of the geometry is shown on the right of Figure 6.8. The simplest model is the biseparatrix map, see Simó–Vieiro (2008d), which for $0 < y < d$

is given by

$$BSPM \begin{pmatrix} x \\ y \end{pmatrix} = \begin{pmatrix} \bar{x} \\ \bar{y} \end{pmatrix} = \begin{pmatrix} x + a + b_1 \log(\bar{y}) - b_2 \log(d - \bar{y}) \\ y + \sin(2\pi x) \end{pmatrix},$$

where $d, b_1, b_2 > 0$ and x is taken mod 1. In that case

Proposition 6.2 *The condition for the existence of* r.i.c. *for the BSPM can be written as* $\dfrac{b_1}{y_0} + \dfrac{b_2}{d - y_0} < \varepsilon_G$. *In particular, there are no such* r.i.c. *if* $(\sqrt{b_1} + \sqrt{b_2})^2/d > \varepsilon_G$.

6.4 Computational Tools

We turn to presenting some ideas about relevant computational tools for the study of global dynamics.

6.4.1 Long-Time Integrations: Taylor Methods

Consider an IVP with a vector field f analytic in a neighbourhood of $(t_0, x_0) \in \Omega \subset \mathbb{R} \times \mathbb{R}^n$

$$\dot{x} = f(t, x), \qquad x(t_0) = x_0.$$

It is possible to produce in an easy way the Taylor expansion $x(t_0 + h)$ to high order for suitable h and use it as a one-step method. It is easy to implement, accurate and fast using automatic differentiation methods and with truncation errors τ largely below roundoff error. Except for the effect of roundoff and its propagation along the orbit the method can be considered as "exact".

Theorem 6.5 *For a very large class of analytic, non-stiff, ODE the Taylor method has the following properties:*
1) Asymptotically, for small enough τ, the optimal step size (concerning efficiency) is almost independent of the number of digits and equal to $\rho(t)/\exp(2)$, where $\rho(t)$ is the local radius of convergence.
2) The optimal order is approximately linear in the number of digits.
3) For a given equation and fixed t_0, t_f, the global computational cost, is $O(D^4)$, where D is the number of digits.

For further information and many examples we can refer to Simó (2007) and for software implementing the method to Jorba–Zou (2005).

As an example we consider the energy change in a Sun-Jupiter-asteroid model. The asteroid is moving in a large domain around L_5 (see Section 6.2). Of a total of 175 test particles 43% subsisted for 10^9 Jupiter revolutions (more than twice the age of Solar System). The variations of energy with respect to the initial value are displayed in Figure 6.9 for the worst case. The initial value of the energy is of the order of units.

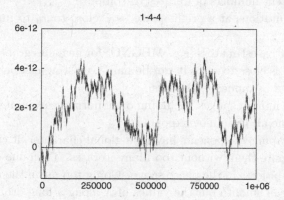

Fig. 6.9. Variation of the energy in a 3D RTBP describing the motion of an asteroid in the Sun-Jupiter system. The initial position is far away from the S-J plane ($z = 0.8$). The variations of the energy are given in dimensionless units and time is given in the horizontal axis using 10^3 Jupiter revolutions as unit.

It is worth mentioning additional checks for the Sun-Jupiter-Saturn-Uranus-Neptune system. For integrations during a time interval of 4.5 Gyr the maximal variations of the energy satisfy $\Delta H/H < 10^{-11}$. The CPU computing time on a Xeon processor at 3.2 GHz is 3 days.

Assuming *iid* random roundoff errors the effect of the roundoff behaves as $O(t^{1/2})$ in the first integrals and, for integrable systems, as $O(t^{3/2})$ in the angles. These results coincide with the expected behaviour for a random walk.

6.4.2 Dynamics Indicators: Lyapunov Exponents

Lyapunov exponents measure rates of exponential divergence of nearby orbits. To compute the maximal Lyapunov exponent for a map at a point x_0, let $v_0, |v_0| = 1$ be an initial vector on the tangent space at x_0 to the phase space. Initialize the "Lyapunov sum" $S_0 = 0$ and compute

$$x_{k+1} = T(x_k), \qquad w_{k+1} = DT(x_k)(v_k),$$

$$v_{k+1} = w_{k+1}/|w_{k+1}|, \qquad S_{k+1} = S_k + \log|w_{k+1}| \quad \text{for} \quad k \geq 0.$$

Then the limit slope of S_k as a function of k gives $\Lambda_{\max}$. The limit exists for almost all x_0, v_0. Some comments on the procedure are:

- Use smoothing, fitting, extrapolation and self-consistency of estimates using different numbers of iterates to compute $\Lambda_{\max}$.
- Stop computations at a "right place" x_N close to x_0 to improve the estimates.
- Use alternative estimators (e.g. MEGNO) for near-integrable systems, when $\Lambda_{\max}$ is close to zero. It can be hard to distinguish between zero and positive Lyapunov exponents.
- With some minor changes, including orthonormalization, it is possible to determine all Lyapunov exponents.
- Despite Lyapunov exponents having a global character, it can be useful to compute them without too many iterates. Then one can locate hyperbolic objects in the phase space. Going too far in the number of iterations can smooth out the values of Λ along a long orbit.
- On the other hand an initial point on a chaotic zone which is quite close to an invariant torus can require a large number of iterates to check that the orbit is really chaotic.

Summarizing, in conservative systems any computation of Lyapunov exponents requires a careful examination of results to avoid misinterpretations. See Broer–Simó (1998a), Cincotta et al. (2003), Ledrappier et al. (2003) for different strategies to extract correct information.

6.4.3 Dynamics Indicators: Frequency Analysis

Let $f(t)$ be an observable of a system, i.e., one of the coordinates of a solution $x(t)$ of a conservative system or a function of them. Assume we have some reason to believe that the dynamics is quasi-periodic. Given a sample $\{f(jT/N)\}_{j=0}^{N-1}$ of $f(t)$ on $[0, T]$, the problem is to determine a trigonometric polynomial,

$$Q_f(t) = A_0^c + \sum_{l=1}^{N_f} \left(A_l^c \cos(2\pi\nu_l t/T) + A_l^s \sin(2\pi\nu_l t/T) \right)$$

whose frequencies, $\{\nu_l\}_{l=1}^{N_f}$, and amplitudes, $\{A_l^c\}_{l=0}^{N_f}$, $\{A_l^s\}_{l=1}^{N_f}$, are a good approximation of the corresponding ones of $f(t)$. The size of the sample N_f can be fixed by the available experimental data or, if there is some freedom (e.g., in case data are obtained by numerical integration) can be determined by the procedure itself (in terms of some input parameters). The goal is to use N_f as small as possible, keeping high accuracy in the computed $\nu_l, A_l^{c,s}$, see Gómez et al. (2009).

Key ideas of the procedure are: a collocation method to compute $\nu_l, A_l^{c,s}$ for $|A_l^{c,s}|$ larger than a given threshold, solving the condition equations by a convergent Newton method, followed by a selection of a decreasing set of thresholds until the agreement between the sample and the trigonometric polynomial is good enough (or one decides that the results are against the assumption of quasi-periodicity). The theory supporting the method is based on

Theorem 6.6 *If f is analytic and quasi-periodic with Diophantine frequencies, then there exist explicit formulas for the errors in $\nu_l, A_l^{c,s}$, depending on the Cauchy estimates for f, the Diophantine constants and the values of T and N. (Gómez-Mondelo-Simó).*

6.4.4 Computing Quasi-Periodic Invariant Curves

According to Theorem 6.1 perturbed twist maps have many invariant curves for small perturbations. For a general APM T the behaviour is similar close (or even not so close) to an elliptic fixed point of T^k for some k. One can consider then the distance to the fixed point as a measure of the perturbation. It is interesting to compute quasi-periodic invariant curves for a given map (e.g. a Poincaré map). One can consider the case of a given rotation number or just check if an invariant curve passes through a given initial point. Several approaches can be used.

Working with a Fourier representation:

Assume $x(t) = \sum_{k \in K} c_k \exp(kit)$, $t \in \mathbb{S}^1$ is a representation of the curve for some set of indices K. Then

a) Look for invariance: Take a grid $\{x(2\pi j/N)\}$, compute images $\{T(x(2\pi j/N))\}$, analyse them and require them to have the same c_k, or

b) Look for conjugation: Search for a transformation C which conjugates T to a rotation.

In both cases one has to use some normalization, because of the arbitrariness of the origin of angles.

Working in phase space:

Select a line $\mathcal{L}$ transversal to the curve and an initial point p on it.

Compute iterates T^k and take some, with $k = k_1, k_2, \ldots$, which return close to p. Interpolate them to find a point q in $\mathcal{L}$. Impose $q = p$ and solve with respect to p.

Curve-fitting:

This a method suitable to answer the question: Given a point p, is it on an invariant curve for T? A procedure can be:

1) Compute iterates of p and keep the ones close to p.
2) Fit a curve to the iterates in some local coordinates. It is convenient to use orthogonal polynomials with respect to the set of abscissas.
3) Find residuals: size and distribution as a function of the iteration.
4) Use some test of acceptance, e.g., based on standard deviation.

As an example of curve-fitting we present the case of an outer cometary problem. Consider the planar circular RTBP with primaries Sun-Jupiter as described in Section 6.2 with $\mu \approx 1/1047.3486$. A relevant question is: Given a value of the Jacobi constant C and considering motions external to Jupiter's orbit, up to which distance one can find *r.i.c.* which prevent escape? How is that distance evolving with C? What about dependence in μ? Some theoretical results in this problem are studied by Kaloshin (2007).

A suitable Poincaré map can be obtained using pericentre passages: we compute passages of the comet through a pericentre and record the radius r and the angle β. Some results are shown in the top of Figure 6.10 for $C = 4$. Several initial conditions have been used and for each one 1000 Poincaré iterates have been computed. The curve-fitting method has been used (with the necessary number of iterates) to detect if one can accept that the initial points are in a *r.i.c.* In blue in the same figure: approximate location where a *r.i.c.* has been found for the first time. In suitable domains the step in the initial conditions has been decreased to refine the estimate.

If we consider the osculating semimajor axis a_{osc} and eccentricity e_{osc}, that is, the values of the Keplerian elements if μ is set to zero, the following relation holds

$$C = 1/a_{\mathrm{osc}} + 2a_{\mathrm{osc}}^{1/2}(1 - e_{\mathrm{osc}}^2)^{1/2}.$$

The curve in blue corresponds to $a_{\mathrm{osc}} \approx 9.55858$, $e_{\mathrm{osc}} \approx 0.77661$, $q_{\mathrm{osc}} \approx 2.13524$, $T_{\mathrm{osc}} \approx 185.682$, where q and T denote the pericentre distance and the period of the osculating orbit. We recall that in the dimensionless units the period of Jupiter is 2π.

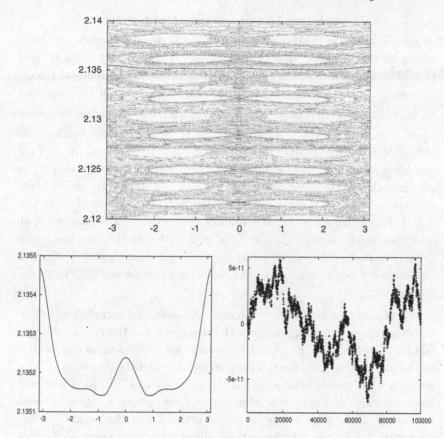

Fig. 6.10. The outer cometary problem. Top: Poincaré iterates using the pericentre passage as surface of section. Orbits of several points are shown and the curve in blue represents the first appearance of a *r.i.c.* No such curves have been found below. Bottom: on the left the above *r.i.c.* with more detail and on the right the errors of the curve fitting method as a function of the iterates on a small interval. See text for details.

To apply the curve-fitting method we have selected $[-0.1, 0.1]$ as the angle β interval and iterates of the Poincaré map have been computed until 10^4 of them fall in this interval (in practice this number of iterates can be decreased considerably). After curve fitting the errors have a normal distribution with standard deviation $\approx 2.5 \times 10^{-11}$. At the bottom of Figure 6.10 (left side) we show the selected *r.i.c.* using a small vertical scale and the evolution of the fitting errors as a function of the number of iterates falling in the β interval (right side). They are due to the propagation of roundoff errors along the integration.

6.5 Some Details of the Hénon map

A general overview of the abundance of non-escaping points has been shown in the top right of Figure 6.7. In Figure 6.11 we show several magnifications which reveal clearly the self-similarity at different scales.

Now we consider the near-integrable case. For c small the map can be approximated by a flow. Let $d = \sqrt{c/\sqrt{2}}$. The change of variables $(x, y) \to (u + dv, u - dv)/\sqrt{2}$ shows that H_c is $\mathcal{O}(d^2)$-close to the time d flow of Hamiltonian $K(u, v) = v^2 - u + \frac{1}{6}u^3$. The level $K = \frac{2}{3}\sqrt{2}$ containing the hyperbolic point $u = -\sqrt{2}, v = 0$ corresponds to a separatrix, enclosing the elliptic point $u = \sqrt{2}, v = 0$. According to Theorem 6.3 the splitting for the manifolds of the map is exponentially small in d and, hence, from the SPM it follows that *r.i.c.* of the map exist at an exponentially small distance of the manifolds. As the area inside the separatrix of K is finite, undoing the scaling it follows that the set of non-escaping points is $\mathcal{O}(\sqrt{c})$ for c small, in agreement with Figure 6.7 top right.

Let us pass to a low-order resonance. As an example consider the parameter value $c = 1.015$, just after the destruction of the *r.i.c.* confining islands of period 4. The global behaviour for this resonance is similar to the one shown in Figure 6.6 bottom for the Michelson system, with a rotation of $\pi/4$ and different symmetries. For $c = 1.015$, as it happens for $\varepsilon = 0.02607$ in Figure 6.6 (blue points), one can see a region without subsisting points close to the invariant manifolds of the period-4 orbit, except in the domain which contains the elliptic fixed point E. In fact this domain has quite sharp boundaries, see the bottom plot in Figure 6.12. It looks as if there exist branches of the period-4 hyperbolic orbits which connect these points and bound the stable domain around E.

A computation of the splitting angles at suitable points in the outer and inner invariant manifolds gives the values $s_o \approx -0.951063 \times 10^{-2}$, $s_i \approx 0.294215 \times 10^{-58}$. The top of Figure 6.12 shows a part of the stable and unstable period-4 manifolds. They seem to coincide because even the outer splitting s_o is hard to detect. In the middle and bottom plots the non-escaping points appear in red, while the escaping ones appear in white. In the middle plot one can see that, despite s_o being moderate it produces escape on a relatively large zone. It is clear that one has to consider separately the tiny islands around an island of period 4 and the ones which are at the external part of the four islands. The tiny islands around an island of period 4 can be seen in detail in the bottom of Figure 6.12.

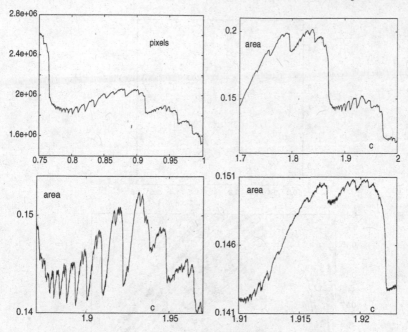

Fig. 6.11. Abundance of non-escaping points for the Hénon map as a function of the parameter c. Top left: a detail after the jump corresponding to the break down of *r.i.c.* around the period-5 islands (measured in pixels, next plots are measure in area in the (x, y)-plane). Top right: the results close to $c = 2$ when the period-doubling occurs. Bottom: successive magnifications close to $c = 1.92$. Compare both right plots.

From the values of s_o and s_i and the dominant eigenvalue at the period-4 hyperbolic periodic orbit one can derive a "composed separatrix" map. It describes the dynamics of the orbits which pass close to the outer separatrix and then close to the inner one. From this, one can predict the distance from the manifolds at which stable islands and invariant curves appear. These values agree with the results obtained by direct simulation. However, one should be careful about how this depends on the place where we look. This is due to the shape of the level curves of a Hamiltonian which gives an approximation of the dynamics of H_c^4.

For the outer islands and concerning the existence or not of *r.i.c.* surrounding the four islands, one requires the use of the $BSPM$. Indeed, the two periodic hyperbolic orbits which play a role are the period-4 orbit

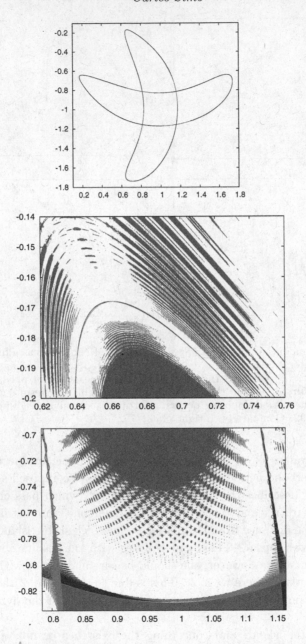

Fig. 6.12. Details on the 4:1 resonance, immediately after *r.i.c.* surrounding the period-4 islands are destroyed. Top: the relevant manifolds. Middle: a detail of the upper part of the manifolds and the non-escaping points around the top period-4 island. Bottom: a detail of the lower part of the manifold and the chains of islands surrounding the period-4 island.

and the hyperbolic fixed point of H_c. Locally, in what concerns nearby islands, one can again approximate by the SPM.

6.6 Summary and Open Problems

Careful numerical methods combined with basic theoretical results are of great help to have a global idea of the dynamics in low-dimensional systems. Several examples, theoretical results and numerical methods have been presented. An interesting feature is the appearance of common patterns and mechanisms in quite different problems. As a conclusion, one can "understand" reasonably well the main features in the phase space and, hence, make predictions on the behaviour, derive auxiliary models which can better studied analytically, etc.

It is clear that for larger dimensions the difficulties increase, both concerning the theoretical and the computational aspects. Let us present some of the aspects of the dynamics and problems to be faced.

a) Many theoretical results assure the validity of a dynamical property if some parameter (say, a perturbation) is sufficiently small. But in many cases the predictions are valid in surprisingly large domains (e.g. a prediction based on a NF at high degree, a computation done by the Lindstedt–Poincaré method, an exponentially small splitting, ...). An effort has to be made to produce realistic estimates for the range of applicability of the theoretical results, with good bounds on the errors.

b) Converse results are interesting. They prove, say, that some property is not satisfied if a parameter is larger than some value. It would be nice to decrease the size of the gap between positive and negative results and to understand the mechanisms which explain the differences in the dynamics. And also to check the relevance of the results: do they play a key role in the dynamics or are they confined to a tiny domain in phase or parameter space?

c) For objects which are far from any "simple object of an unperturbed system" proving rigorous facts should rely on CAP (Computer Assisted Proofs). An effort must be done to convert careful non-rigorous computations into CAP in a systematic way.

d) While in APM there are clear boundaries, i.e., invariant curves, the situation is worse in higher dimensions. Cantor sets of invariant tori are close to some kind of "quasi-boundary" or "fuzzy boundary". Do the frequencies on them satisfy some special DC? Which are the frequencies which play the role of the noble numbers? Certainly there are codimension-1 manifolds which can play a key role, see Simó (1998), but they do not enclose a volume, i.e., have some splitting. Why do they act as "practical boundaries"?

e) Diffusion through the invariant tori can be extremely slow in some regions, but fast enough (and, hence, statistically relevant) in others where the "abundance of tori" decreases. How to identify the different regions? How to model the dynamics as a diffusion process, highly heterogeneous and highly anisotropic? Do invariant Cantor sets play a role on this?

Acknowledgments The author is indebted to a large number of collaborators. Working and discussing with them is a real pleasure. This research has been supported by grant MTM2006-5849/Consolider (Spain).

References

V. I. Arnol'd and A. Avez (1967), *Problèmes Ergodiques de la Mécanique Classique*, Gauthier-Villars, Paris.

H. Broer, I. Hoveijn, M. van Noort, C. Simó and G. Vegter (2004), 'The parametrically forced pendulum: A case study in $1\frac{1}{2}$ degree of freedom', *J. of Dynamics and Differential Equations* **16**, 897–947.

H. Broer, R. Roussarie and C. Simó (1996), 'Invariant circles in the Bogdanov-Takens bifurcation for diffeomorphisms', *Ergod. Th. & Dynam. Sys.* **16**, 1147–1172.

H. Broer and C. Simó (1998a), 'Hill's equation with quasi–periodic forcing: resonance tongues, instability pockets and global phenomena', *Bul. Soc. Bras. Mat.* **29**, 253–293.

H. Broer, C. Simó and J. C. Tatjer (1998b), 'Towards global models near homoclinic tangencies of dissipative diffeomorphisms', *Nonlinearity* **11**, 667–770.

P. M. Cincotta, C. M. Giordano and C. Simó (2003), 'Phase space structure of multidimensional systems by means of the Mean Exponential Growth factor of Nearby Orbits (MEGNO)', *Physica D* **182**, 151–178.

F. Dumortier, S. Ibáñez, H. Kokubu and C. Simó (2009), 'About the unfolding of a Hopf-zero singularity', work in progress.

E. Fontich and C. Simó (1990), 'The splitting of separatrices for analytic diffeomorphisms', *Erg. Th. & Dyn. Systems* **10**, 295–318.

V. Gelfreich (1999), 'A proof of the exponentially small transversality of the separatrices for the standard map', *Comm. Math. Phys.* **201**, 155–216.

V. Gelfreich and V. Lazutkin (2001), 'Splitting of separatrices: perturbation theory and exponential smallness', *Russian Mathematical Surveys* **56**, 499–558.

V. Gelfreich and C. Simó (2008), 'High-precision computations of divergent asymptotic series and homoclinic phenomena', *Discrete and Continuous Dynamical Systems B* **10**, 511–536.

A. Giorgilli, A. Delshams, E. Fontich, L. Galgani and C. Simó (1989), 'Effective stability for a Hamiltonian system near an elliptic equilibrium point, with an application to the restricted three body problem', *J. Diff. Eq.* **77**, 167–198.

G. Gómez, J. M. Mondelo and C. Simó (2009), 'A collocation method for the numerical Fourier analysis of quasi-periodic functions. I: Numerical tests and examples; II: Analytical error estimates', work in progress.

M. Hénon (1969), 'Numerical study of quadratic maps', *Quarterly of App. Math* **XXVII**, 291–311.

À. Jorba and M. Zou (2005), 'A software package for the numerical integration of ODE by means of high-order Taylor methods', *Experimental Mathematics* **14**, 99–117.

V. Kaloshin (2007), private communication.

F. Ledrappier, M. Shub, C. Simó and A. Wilkinson (2003), 'Random versus deterministic exponents in a rich family of diffeomorphisms', *J. Stat Phys.* **113**, 85–149.

J. J. Morales, J. P. Ramis and C. Simó (2007), 'Integrability of Hamiltonian systems and differential Galois groups of higher variational equations', *Annales Sci. de l'ENS* 4e série, **40**, 845–884.

J. Moser (1973), *Stable and Random Motions in Dynamical Systems*, Annals of Mathematics Studies, Princeton Univ. Press.

A. I. Neishtadt (1984), 'The separation of motions in systems with rapidly rotating phase', *Prikladnaja Matematika i Mekhanika* **48**, 133–139.

D. Puigjaner, J. Herrero, F. Giralt and C. Simó (2008), 'Bifurcation analysis of steady Rayleigh-Bénard convection in a cubical cavity with conducting sidewalls', *J. of Fluid Mechanics* **598**, 393–427.

C. Simó (1990), 'Analytical and numerical computation of invariant manifolds', in *Modern Methods in Celestial Mechanics*, (D. Benest, C. Froeschlé, eds), Editions Frontières, Paris, pp. 285–330. Also available at http://www.maia.ub.es/dsg/2004.

C. Simó (1994), 'Averaging under fast quasiperiodic forcing', in *Integrable and Chaotic Behaviour in Hamiltonian Systems* (I. Seimenis, ed.), Plenum Pub. Co., New York, 13–34.

C. Simó (1997), 'An overview on some problems in Celestial Mechanics', lecture notes, http://www-ma1.upc.es/escorial/ .

C. Simó (1998), 'Effective computations in celestial mechanics and astrodynamics', in *Modern Methods of Analytical Mechanics and their Applications*, (V.V. Rumyantsev, A.V. Karapetyan, eds), CISM Courses and Lectures, **387**, Springer, pp. 55–102.

C. Simó (2007), 'Taylor method for the integration of ODE', lecture notes, http://www.maia.ub.es/dsg/2007.

C. Simó, D. Puigjaner, J. Herrero and F. Giralt (2008a), 'Dynamics of particle trajectories in a Rayleigh-Bénard problem', *Commun. Nonlinear Sci Numer. Simulat.*, DOI:10.1016/j.cnsns.2008.07.012

C. Simó and D. Treschev (2008b), 'Stability islands in the vicinity of separatrices of near-integrable symplectic maps', *Discrete and Continuous Dynamical Systems B* **10**, 681–698.

C. Simó and A. Vieiro (2008c), 'Resonant zones, inner and outer splittings in generic and low order resonances of APM', submitted.

C. Simó and A. Vieiro (2008d), 'Dynamics in chaotic zones: close to separatrix and Birkhoff zones', preprint.

C. Simó and A. Vieiro (2009), 'Global study of area preserving maps', work in progress.

7

A Panoramic View of Asymptotics

R. Wong

Department of Mathematics
City University of Hong Kong
Tat Chee Avenue, Kowloon, Hong Kong
e-mail: mawong@cityu.edu.hk

Abstract

Asymptotic methods include asymptotic evaluation of integrals, asymptotic expansion of solutions to differential equations, singular perturbation techniques, discrete asymptotics, etc. In this survey, we present some of the most significant developments in these areas in the second half of the 20th century. Also mentioned will be a new method known as the Riemann-Hilbert approach, which has had a significant impact in the field in recent years.

7.1 Introduction

What is asymptotics? It is the branch of analysis that deals with problems concerning the determination of the behavior of a function as one of its parameters tends to a specific value, or a sequence as its index tends to infinity. Thus, it includes, for example, Stirling's formula, asymptotic expansion of the Lebesgue constant in Fourier series, and even the prime number theorem. But, in general, it refers to just the two main areas: (i) asymptotic evaluation of integrals, and (ii) asymptotic solutions to differential equations. The second area sometimes also includes the subject of singular perturbation theory. But the results in this subarea are mostly formal (i.e., not mathematically rigorous). Although occasionally one may also include the methods of asymptotic enumeration in the general area of asymptotics, the development of this area is far behind those in the two areas mentioned above. For instance, a turning point theory for difference equations was not introduced until just around the turn of this century, while the corresponding theory for differential equations was developed in the 1930's.

In this survey, we will present some of the most important results in these areas in the second half of the 20th century. We begin with

190

asymptotic evaluation of integrals in the next section. Here we mention only two ideas, namely, (i) the cubic transformation introduced by Chester, Friedman & Ursell (1957) and (ii) the distributional method of McClure & Wong (1978, 1979). In Section 7.3, we mention also only two ideas; they are (i) the double asymptotic nature of the Liouville–Green (WKB) approximation and (ii) error bounds associated with asymptotic solutions to second-order differential equations; see Olver (1974). A more recent topic known as "exponential asymptotics" is discussed in Section 7.4, where we explain the difference between Stokes' phenomenon and Berry's transition (1989), and the significance of what Berry & Howls (1991) have called "adjacent saddle" and "adjacent contour". In Section 7.5, we use specific examples to show the deficiency of the method of matched asymptotics when exponentially small terms are required. These examples involve turning-points and nested layers (triple decks). We also discuss how the shooting method can be used to give rigorous analysis in the study of two nonlinear boundary-value problems of Carrier & Pearson (1968). An asymptotic theory for difference equations is described in Section 7.6, where one can also find Airy-type and Bessel-type expansions. The final section is devoted to a brief illustration of a new asymptotic method introduced by Deift and Zhou in 1993. This method was originally designed to study the asymptotic behavior of solutions to nonlinear differential equations; see, e.g., Deift & Zhou (1993, 1994, 1995). But very quickly it was realized that the method is also applicable to finding asymptotic behavior of orthogonal polynomials; see, e.g., Deift (1999), Deift et al. (1999a, 1999b), Bleher & Its (1999), Kuijlaars & McLaughlin (2001), Kuijlaars et al. (2004), Wong & Zhang (2006) and Dai & Wong (2008). Here we will illustrate the method with orthogonal polynomials associated with the exponential weight $w(x) = \exp(-Q(x))$, where $Q(x)$ is a polynomial of even degree and with positive leading coefficient. We show, in particular, by modifying the steepest-descent method of Deift–Zhou, that we can obtain asymptotic expansions for the (scaled) polynomials which hold uniformly in much wider regions.

7.2 Integral Methods

Classical methods for asymptotic evaluation of integrals include Watson's lemma, Laplace's approximation, Kelvin's principle of stationary phase, and Debye's method of steepest descent, etc. Information on these methods can be found in Copson (1965), Bleistein & Handelsman

(1975) and Wong (1989). The last reference also contains methods that were discovered more recently, such as the Mellin transform technique, the summability method and the distributional approach.

Here we shall briefly mention only an extension of the steepest descent method given by Chester, Friedman & Ursell (1957), and the distributional approach introduced by McClure & Wong (1978, 1979). We begin with the paper of Chester, Friedman & Ursell (1957), in which they considered contour integrals of the form

$$I(\lambda; \alpha) = \int_C g(z) e^{-\lambda f(z;\alpha)} \, dz, \tag{7.1}$$

where λ is a large positive number tending to $+\infty$, f and g are analytic functions of the complex variable z, and f is also an analytic function of the parameter α. If α is fixed, an asymptotic expansion of $I(\lambda; \alpha)$ for large λ can usually be found by the classical method of steepest descent, which shows that the major contribution to the integral comes from the saddle points, i.e., the points at which $\partial f/\partial z$ vanishes. If α is allowed to vary, then the position of the saddle points may vary and even coalesce. For simplicity, suppose that there exists a critical value of α, say α_0, such that for $\alpha \neq \alpha_0$ there are two distinct saddle points $z_+ = z_+(\alpha)$ and $z_- = z_-(\alpha)$ of multiplicity 1, but at $\alpha = \alpha_0$ these two points coincide and give a single saddle point of multiplicity 2. Since the simplest function which exhibits two coalescing saddle points is a cubic polynomial, Chester, Friedman & Ursell (1957) introduced the change of variable $z \to u$ defined by

$$f(z; \alpha) = \frac{1}{3} u^3 - \zeta u + \eta, \tag{7.2}$$

where the coefficients $\zeta = \zeta(\alpha)$ and $\eta = \eta(\alpha)$ are determined so that (7.2) results in a single-valued analytic function $z = z(u)$ with neither dz/du nor du/dz vanishing in the relevant regions. Substituting (7.2) into (7.1) gives

$$I(\lambda; \alpha) = e^{-\lambda \eta} \int_{C^*} \varphi_0(u) e^{-\lambda(u^3/3 - \zeta u)} \, du, \tag{7.3}$$

where C^* is the image of C and $\varphi_0(u) = g(z)(dz/du)$.

To derive an asymptotic expansion for the integral $I(\lambda; \alpha)$ which holds uniformly for α in a neighborhood of α_0, Chester, Friedman & Ursell (1957) used the two-point expansion

$$\varphi_0(u) = \sum_{m=0}^{\infty} p_m(\alpha)(u^2 - \zeta)^m + \sum_{m=0}^{\infty} q_m(\alpha) u (u^2 - \zeta)^m, \tag{7.4}$$

where the coefficients can be found by repeated differentiation and use of the correspondences $z_+ \leftrightarrow \zeta^{\frac{1}{2}}, z_- \leftrightarrow -\zeta^{\frac{1}{2}}$. The contour C^* in (7.3) can be any one of the three curves γ_0, γ_1 or γ_2, where γ_0 runs from $\infty e^{-\frac{2}{3}\pi i}$ to $\infty e^{\frac{2}{3}\pi i}, \gamma_1$ runs from ∞e^{i0-} to $\infty e^{-\frac{2}{3}\pi i}$, and γ_2 runs from $\infty e^{\frac{2}{3}\pi i}$ to ∞e^{i0+}. Just for the purpose of illustration, let us take C^* to be the curve γ_0. Inserting (7.4) into (7.3) and carrying out integration term-by-term, one obtains a uniform asymptotic expansion of the form

$$
e^{\lambda \eta} I(\lambda; \alpha) = \frac{\mathrm{Ai}\,(\lambda^{2/3}\zeta)}{\lambda^{1/3}} \left[\sum_{s=0}^{n-1} \frac{a_s(\alpha)}{\lambda^s} + O\left(\frac{1}{\lambda^n}\right) \right]
$$
$$
+ \frac{\mathrm{Ai}'\,(\lambda^{2/3}\zeta)}{\lambda^{2/3}} \left[\sum_{s=0}^{n-1} \frac{b_s(\alpha)}{\lambda^s} + O\left(\frac{1}{\lambda^n}\right) \right]
\tag{7.5}
$$

as $\lambda \to \infty$, where Ai is the Airy function and $a_s(\alpha), b_s(\alpha)$ can be determined by a finite number of the coefficients $p_0, p_1, p_2, \ldots$ and $q_0, q_1, q_2, \ldots$ in (7.4). The two O-terms are independent of α.

A disadvantage of the expansion (7.4) is that it is valid only when α is near α_0 and $|u|$ is small. As a consequence, the region of validity in α is not as large as desired. An alternative derivation of (7.5) was later provided by Bleistein (1966, 1967), which is based on a repeated application of an integration-by-parts technique. An additional advantage of Bleistein's method is that it gives an explicit expression for the remainder term associated with the expansion (7.5), from which error bounds may be obtained; see Wong (1980, §12). A similar result could be obtained by using a two-point expansion with remainder given in López & Temme (2002).

The method of Chester, Friedman & Ursell (1957) has had a tremendous impact in this area of research, and subsequently has been modified, improved and extended by several authors, including Copson (1965), Bleistein (1966, 1967), Frenzen & Wong (1988), and Wong & Zhao (2003). Also, it has been successfully applied by Ursell (1960) to ship wave pattern, Berry (1969, 2005) to diffraction theory and Tsunami waves, Cartwright & Oughstun (2007) to pulse propagation, and by Bo & Wong (1994) and Jin & Wong (1998) to orthogonal polynomials.

We conclude our discussion of uniform asymptotic expansion with a very brief mention of double integrals of the form

$$
J(\lambda; \alpha) = \iint_D g(x, y; \alpha) e^{i\lambda f(x,y;\alpha)}\, dx\, dy,
\tag{7.6}
$$

where λ is again a large positive number and α is an auxiliary parameter.

Such integrals appear frequently in asymptotic solutions of wave equations; see, e.g., Poston & Stewart (1978, p. 260). In (7.6), D is a bounded domain, and f and g are C^∞-functions of (x, y) in D. Furthermore, we assume that the phase function $f_\alpha(x, y) := f(x, y; \alpha)$ has two stationary points $(x_\pm(\alpha), y_\pm(\alpha))$ which coalesce at a point (x_0, y_0) in the interior of D when α tends to a critical value, say α_0. Under these conditions, it is readily seen that (x_0, y_0) is a degenerate stationary point of f_{α_0}. If, in addition, the Hessian matrix of f_α at (x_0, y_0) has rank 1 and the third-order derivatives of f_α at (x_0, y_0) satisfy a certain non-vanishing condition, then by using the *splitting lemma* from catastrophe theory (see Poston & Stewart (1978, p. 95)) and a C^∞-version of the cubic transformation given in (7.2) (see Hörmander (1990, p. 204)) it can be shown that there exists a C^∞ and one-to-one mapping $(x, y) \to (u, v)$ given by $u = u_\alpha(x, y)$ and $v = v_\alpha(x, y)$ such that

$$f_\alpha(x, y) = \varepsilon u^2 + \frac{1}{3}v^3 + \xi(\alpha)v + \eta(\alpha), \qquad (7.7)$$

where $\xi(\alpha)$ and $\eta(\alpha)$ are C^∞-functions with $\xi(\alpha_0) = \eta(\alpha_0) = 0$ and, moreover, $u_{\alpha_0}(x_0, y_0) = v_{\alpha_0}(x_0, y_0) = 0$. In (7.7), ε denotes the sign of $\partial^2 f_{\alpha_0}/\partial x^2$ at (x_0, y_0) if the value of this derivative is not zero; otherwise, we take ε to be the sign of $\partial^2 f_{\alpha_0}/\partial y^2$ at (x_0, y_0). Note that these two derivatives can not both vanish at this point, since the Hessian matrix of f_{α_0} at (x_0, y_0) is of rank 1.

Substituting (7.7) into (7.6) gives

$$I(\lambda; \alpha) = e^{i\lambda\eta(\alpha)} \iint_{D'} \varphi_0(u, v; \alpha) e^{i\lambda(\varepsilon u^2 + \frac{1}{3}v^3 + \xi(\alpha)v)} \, du \, dv, \qquad (7.8)$$

where D' is the image of D under the transformation $(x, y) \to (u, v)$,

$$\varphi_0(u, v; \alpha) = g(x, y; \alpha) \left| \frac{\partial(x, y)}{\partial(u, v)} \right|$$

and $\partial(x, y)/\partial(u, v)$ is the Jacobian of the transformation. An Airy-type expansion can then be derived again by repeated application of an integration-by-parts technique.

The above problem arose in a study on propeller acoustics in the 1990's; see Chapman (1992) and Prentice (1994). The material presented here is taken from Qiu & Wong (2000), where detailed analysis is given. For more discussions (although not all rigorous) on uniform asymptotic expansions of double integrals of the form (7.6), we mention the book by Borovikov (1994).

We now turn our attention to the distributional approach introduced by McClure & Wong (1978, 1979). Consider the Stieltjes transform given by

$$S_f(z) = \int_0^\infty \frac{f(t)}{t+z}\, dt, \tag{7.9}$$

where $f(t)$ is a locally integrable function on $[0, \infty)$ and z is a complex variable in the cut plane $|\arg z| < \pi$. To obtain the large z-behavior of $S_f(z)$, we assume that $f(t)$ possesses an asymptotic expansion of the form

$$f(t) = \sum_{s=0}^{n-1} a_s t^{-s-\alpha} + f_n(t) \tag{7.10}$$

for each $n \geq 1$, where $0 < \alpha < 1$ and $f_n(t) = O(t^{-n-\alpha})$ as $t \to \infty$. To each function in (7.10), we can assign a distribution on the space $\mathcal{S}$ of rapidly decreasing functions. For instance, f defines a regular distribution on $\mathcal{S}$ by $\langle f, \varphi \rangle = \int_0^\infty f(t)\varphi(t)\, dt$ for all $\varphi \in \mathcal{S}$. Similarly, $t^{-\alpha}$ defines a regular distribution by $\langle t^{-\alpha}, \varphi \rangle = \int_0^\infty t^{-\alpha}\varphi(t)\, dt$. Through the remaining portion of this section, we shall only consider distributions which are supported on $[0, \infty)$. Note that $t^{-s-\alpha}$ is not locally integrable on $[0, \infty)$, when $s \geq 1$. Hence, we can not use the same approach to define distributions associated with the functions $t^{-s-\alpha}, s = 1, 2, \ldots$. However, since the sth derivative of $t^{-\alpha}$ is $(-1)^s (\alpha)_s t^{-s-\alpha}$, one can consider $t^{-s-\alpha}$ as the sth distributional derivative of $(-1)^s t^{-\alpha}/(\alpha)_s$; that is, we can assign $\langle t^{-s-\alpha}, \varphi \rangle = \langle t^{-\alpha}, \varphi^{(s)} \rangle / (\alpha)_s$ for all $\varphi \in \mathcal{S}$. For the function $f_n(t)$ in (7.10), we first denote by $f_{n,n}(t)$ the nth iterated integral of f_n, i.e.,

$$f_{n,n}(t) = \frac{(-1)^n}{(n-1)!} \int_t^\infty (\tau - t)^{n-1} f_n(\tau)\, d\tau.$$

Note that for each $n \geq 1, f_{n,n}(t)$ is locally integrable on $[0, \infty)$ and $O(t^{-\alpha})$ as $t \to \infty$. Furthermore, f_n is the nth derivative of $f_{n,n}$. Hence, we may define the distribution associated with $f_n(t)$ by $\langle f_n, \varphi \rangle = (-1)^n \langle f_{n,n}, \varphi^{(n)} \rangle$ for all $\varphi \in \mathcal{S}$. An interesting observation is that when each of the functions in (7.10) is interpreted as a distribution, this equation no longer holds. Instead, we have

$$\langle f, \varphi \rangle = \sum_{s=0}^{n-1} a_s \langle t^{-s-\alpha}, \varphi \rangle - \sum_{s=1}^{n} c_s \langle \delta^{(s-1)}, \varphi \rangle + \langle f_n, \varphi \rangle \tag{7.11}$$

for all $\varphi \in \mathcal{S}$, where δ is the Dirac delta function and $c_s = f_{s,s}(0)$. In order to deduce an asymptotic expansion from (7.11), we choose $\varphi(t) =$

$e^{-\varepsilon t}/(t+z)$ for $t > 0$, and take the limit on both sides of (7.11) as $\varepsilon \to 0$. The result is

$$S_f(z) = \frac{\pi}{\sin \alpha \pi} \sum_{s=0}^{n-1} (-1)^s \frac{a_s}{z^{s+\alpha}} - \sum_{s=1}^{n} (s-1)! \frac{c_s}{z^s} + \varepsilon_n(z), \qquad (7.12)$$

for $0 < \alpha < 1$, where

$$\varepsilon_n(z) = n! \int_0^\infty \frac{f_{n,n}(t)}{(t+z)^{n+1}} \, dt.$$

This result becomes even easier to apply, when one realizes that the coefficients c_s can be expressed in terms of the Mellin transform of $f(t)$, denoted by $M[f; z]$. Indeed, we have $c_s = \frac{(-1)^s}{(s-1)!} M[f; s]$. Moreover, by integration by parts, the remainder $\varepsilon_n(z)$ can be written as

$$\varepsilon_n(z) = \frac{(-1)^n}{z^n} \int_0^\infty \frac{\tau^n f_n(\tau)}{\tau + z} \, d\tau. \qquad (7.13)$$

Note that the integral in (7.13) is just the Stieltjes transform of $t^n f_n(t)$. Therefore, it is no more complicated than the original integral in (7.9), except that it is now multiplied by a decreasing factor $1/z^n$. When $\alpha = 1$ in (7.10), a corresponding result can be derived with the leading-order term $z^{-\alpha}$ in (7.12) replaced by $(\log z)/z$.

The expansion in (7.12) can be used to give a similar result for the one-sided Hilbert transform

$$H_f^+(x) = \fint_0^\infty \frac{f(t)}{t - x} \, dt, \qquad\qquad x > 0, \qquad (7.14)$$

where the bar on the integral sign indicates that the integral is the Cauchy principal value at $t = x$. This is easily seen from the well-known formula of Plemelj (see Bremermann (1965))

$$H_f^+(x) = \frac{1}{2} \lim_{\varepsilon \to 0} \int_0^\infty \left[\frac{1}{t - x + i\varepsilon} - \frac{1}{t - x - i\varepsilon} \right] f(t) \, dt,$$

and the final expansion takes the form

$$H_f^+(x) = \pi \cot \alpha \pi \sum_{s=0}^{n-1} \frac{a_s}{x^{s+\alpha}} - \sum_{s=1}^{n} (s-1)! \frac{(-1)^s c_s}{x^s} + \varepsilon_n^*(x), \qquad (7.15)$$

where

$$\varepsilon^*(x) = \frac{(-1)^n}{x^n} \fint_0^\infty \frac{t^n f_n(t)}{t - x} \, dt.$$

An advantage of the distributional approach is that it gives explicit expressions for the remainders in (7.12) and (7.15), from which it would be easier to derive numerical bounds for the error terms. Although there are now other methods to obtain similar results (see Wong (1979), Ursell (1983) and López (2007)), it is this approach that first showed what the remainder terms should look like.

The distributional method can also be extended to construct asymptotic expansions of other integrals, such as Fourier and Laplace transforms near the origin, and convolution integrals of the form

$$I(x) = \int_0^\infty f(t)h(xt)\,dt$$

and

$$J(x) = \int_0^x f(t)g(x-t)\,dt$$

for large value of x. For references, we mention Wong & McClure (1984), Li & Wong (1994) and López (2000).

7.3 Differential Equation Theory

The two best known results in asymptotic theory of differential equations are the treatment of irregular singular points at infinity and the Liouville–Green (WKB) approximation. Since the first result dates back to Poincaré (1886), here we just concentrate on the second result concerning the differential equation

$$y''(x) = \{\lambda^2 a(x) + b(x)\}y(x), \tag{7.16}$$

where λ is a positive parameter. We first consider the simplest case, in which $a(x)$ is a real, positive and twice continuously differentiable function in a given finite or infinite interval (a_1, a_2). We also assume that $b(x)$ is a continuous real- or complex-valued function. Let

$$\xi = \int a^{1/2}(x)\,dx, \qquad w = a^{1/4}(x)y(x). \tag{7.17}$$

It is readily verified that under this transformation, equation (7.16) becomes

$$\frac{d^2 w}{d\xi^2} = \{\lambda^2 + \psi(\xi)\}w, \tag{7.18}$$

where

$$\psi(\xi) = -\frac{5}{16}\frac{a'^2(x)}{a^3(x)} + \frac{1}{4}\frac{a''(x)}{a^2(x)} + \frac{b(x)}{a(x)}.$$

The change of variable from (x, y) to (ξ, w) is known as the *Liouville transformation*. If we discard ψ in (7.18), then we obtain two linearly independent solutions $e^{\pm\lambda\xi}$. In terms of the original variables, we have

$$
\begin{aligned}
y(x) \sim \; & Aa^{-1/4}(x) \exp\left\{ \lambda \int a^{1/2}(x)\, dx \right\} \\
& + Ba^{-1/4}(x) \exp\left\{ -\lambda \int a^{1/2}(x)\, dx \right\},
\end{aligned}
\tag{7.19}
$$

where A and B are arbitrary constants. Equation (7.19) is known as the *Liouville–Green approximation*, whereas physicists refer to (7.19) as the *WKB approximation*. This is a very old result, and has been extended in a variety of ways. Here we only mention two important observations made by Olver in the 1960's; see Olver (1961) and Olver (1974, Ch. 6). Let $F(x)$ denote the *control function*

$$
F(x) = \int \left\{ \frac{1}{a^{1/4}} \frac{d^2}{dx^2}\left(\frac{1}{a^{1/4}} \right) + \frac{b}{a^{1/4}} \right\} dx,
\tag{7.20}
$$

and use the notation

$$
\mathcal{V}_{d,x}(F) = \int_d^x |F'(t)|\, dt
$$

for the total variation of F over an interval (d, x). Olver's first observation was that the linearly independent solutions of (7.16) can be expressed as

$$
y_1(x) = a^{-1/4}(x) \exp\left\{ \lambda \int a^{1/2}(x)\, dx \right\} [1 + \varepsilon_1(\lambda, x)]
\tag{7.21}
$$

and

$$
y_2(x) = a^{-1/4}(x) \exp\left\{ -\lambda \int a^{1/2}(x)\, dx \right\} [1 + \varepsilon_2(\lambda, x)]
\tag{7.22}
$$

with

$$
|\varepsilon_j(\lambda, x)| \le \exp\left\{ \frac{1}{2\lambda} \mathcal{V}_{a_j, x}(F) \right\} - 1, \qquad j = 1, 2.
\tag{7.23}
$$

For fixed x and large λ, the right-hand side of (7.23) is $O(\lambda^{-1})$. Hence, a general solution to (7.16) has the asymptotic behavior given in (7.19).

Note that the inequalities in (7.23) provide tractable and realistic bounds for the error terms $\varepsilon_1(\lambda, x)$ and $\varepsilon_2(\lambda, x)$. Olver's second observation was that these bounds can also be used to give asymptotic properties of the approximations (7.21) and (7.22) in the neighborhood of a singularity of the differential equation. Because of this *double* asymptotic feature, the Liouville–Green approximation is a remarkably powerful tool for approximating solutions of linear second-order differential equations. To illustrate this point, we let $\lambda = 1$ in (7.16), and assume that $\mathcal{V}_{a_1,a_2}(F) < \infty$ and also

$$\int a^{1/2}(x)\, dx \to \infty \qquad \text{as } x \to a_2^-.$$

Under these conditions, it can be proved that the error term $\varepsilon_1(x) := \varepsilon_1(1, x)$ in (7.21) satisfies

$$\varepsilon_1(x) \to \text{a constant} \qquad \text{as } x \to a_2^-;$$

see Olver (1974, pp. 197–199). This, together with (7.21), shows that there is a solution $y_3(x)$ such that

$$y_3(x) \sim a^{-1/4}(x)\exp\left\{\int a^{1/2}(x)\, dx\right\} \qquad \text{as } x \to a_2^-. \qquad (7.24)$$

Coupling (7.22) and (7.24), we obtain two linearly independent asymptotic solutions as $x \to a_2^-$.

As an illustration, we consider the equation

$$y''(x) = x^x y(x)$$

as $x \to \infty$. In view of the rapid growth of the coefficient function x^x as $x \to \infty$, it is really not easy to guess the large x-behavior of the solution $y(x)$. With $\lambda = 1$, $a(x) = x^x$ and $b(x) = 0$ in (7.16), the control function $F(x)$ in (7.20) is given by

$$F(x) = \frac{1}{16}\int \left[(1+\ln x)^2 x^{-x/2} - 4x^{-(1+x/2)}\right] dx.$$

Clearly, $\mathcal{V}_{x,\infty}(F) \to 0$ as $x \to \infty$. Hence, (7.22) gives the recessive solution

$$y_2(x) \sim x^{-x/4}\exp\left(-\int x^{x/2}\, dx\right), \qquad x \to \infty.$$

A dominated solution is provided by (7.24), namely,

$$y_3(x) \sim x^{-x/4}\exp\left(\int x^{x/2}\, dx\right), \qquad x \to \infty.$$

An asymptotic expansion of the integral $\int x^{x/2}\, dx$ can be obtained by integration by parts. This problem has been treated by Rosenlicht (1983) in a paper on Hardy fields.

The problem of finding asymptotic solutions to the differential equation (7.16) becomes much more complicated when the coefficient function $a(x)$ has a zero, say at $x = x_0$, in the interval (a_1, a_2). Such a point is known as a *turning point* of the differential equation. In this case, there is an ambiguity in taking the square root of the function $a(x)$, and hence the Liouville transform (7.17) is not well-defined.

For definiteness, we assume that $a(x)$ has the same sign as $x - x_0$; i.e.,

$$a(x)(x - x_0) > 0 \qquad \text{for all } x \neq x_0.$$

Instead of (7.17), we now make the change of variables

$$\begin{cases} \frac{2}{3}\zeta^{3/2} = \displaystyle\int_{x_0}^{x} a(t)^{1/2}\, dt, & x_0 \leq x, \\[2mm] \frac{2}{3}(-\zeta)^{3/2} = \displaystyle\int_{x}^{x_0} [-a(t)]^{1/2}\, dt, & x < x_0, \end{cases} \qquad (7.25)$$

and

$$w(\zeta) = \left(\frac{d\zeta}{dx}\right)^{1/2} y. \qquad (7.26)$$

It is easily verified that

$$\left(\frac{d\zeta}{dx}\right)^2 = \frac{a(x)}{\zeta}. \qquad (7.27)$$

The transformation $(x, y) \to (\zeta, w)$ was first introduced by Langer (1931, 1932) in the 1930's, under which equation (7.16) becomes

$$\frac{d^2 w}{d\zeta^2} = \{\lambda^2 \zeta + \psi(\zeta)\} w, \qquad (7.28)$$

where

$$\psi(\zeta) = \frac{5}{16}\zeta^{-2} + \{4a(x)a''(x) - 5[a'(x)]^2\}\frac{\zeta}{16a^3(x)} + \frac{\zeta b(x)}{a(x)}.$$

If ψ in (7.28) is neglected, then we have the Airy equation

$$\frac{d^2 w}{d\zeta^2} = \lambda^2 \zeta w.$$

Its two linearly independent solutions are the Airy functions $\mathrm{Ai}\,(\lambda^{2/3}\zeta)$ and $\mathrm{Bi}\,(\lambda^{2/3}\zeta)$. From this, it is reasonable to expect that equation (7.16) has twice continuously differentiable solutions expressible in the forms

$$y_1(x) = \widehat{a}^{\,-1/4}(x)[\mathrm{Ai}\,(\lambda^{2/3}\zeta) + \varepsilon_1(\lambda, x)], \qquad (7.29)$$

$$y_2(x) = \widehat{a}^{\,-1/4}(x)[\mathrm{Bi}\,(\lambda^{2/3}\zeta) + \varepsilon_2(\lambda, x)], \qquad (7.30)$$

where $\widehat{a}(x) = a(x)/\zeta$; see (7.26) and (7.27). To give estimates for the error terms $\varepsilon_1(\lambda, x)$ and $\varepsilon_2(\lambda, x)$, we let $c = -0.36604\ldots$ be the real root of the equation

$$\mathrm{Ai}\,(x) = \mathrm{Bi}\,(x)$$

of smallest absolute value, and define the *envelopes* of $\mathrm{Ai}\,(x)$ and $\mathrm{Bi}\,(x)$ by

$$\mathrm{env}\,\mathrm{Ai}\,(x) = \mathrm{env}\,\mathrm{Bi}\,(x) = [\mathrm{Ai}^2(x) + \mathrm{Bi}^2(x)]^{1/2}, \qquad -\infty < x \le c,$$

$$\mathrm{env}\,\mathrm{Ai}\,(x) = \sqrt{2}\mathrm{Ai}\,(x), \qquad \mathrm{env}\,\mathrm{Bi}\,(x) = \sqrt{2}\mathrm{Bi}\,(x), \qquad c \le x < \infty.$$

These envelopes are continuous functions of x. Furthermore, we assume that ζ defined in (7.25) ranges over a finite or infinite interval (α_1, α_2). If the total variation of the control function

$$\Psi(\zeta) = \int_0^\zeta \psi(v) v^{-1/2}\, dv$$

on (α_1, α_2) is finite (i.e., $\mathcal{V}_{\alpha_1, \alpha_2}(\Psi) < \infty$), then the error terms in (7.29) and (7.30) satisfy

$$\varepsilon_1(\lambda, x) = \mathrm{env}\,\mathrm{Ai}\,(\lambda^{2/3}\zeta) O\!\left(\frac{1}{\lambda}\right), \qquad (7.31)$$

$$\varepsilon_2(\lambda, x) = \mathrm{env}\,\mathrm{Bi}\,(\lambda^{2/3}\zeta) O\!\left(\frac{1}{\lambda}\right) \qquad (7.32)$$

uniformly with respect to $x \in (a_1, a_2)$.

For bounded values of ζ, the asymptotic nature of the approximations (7.29) and (7.30) was established by Langer (1931, 1932, 1949). For unrestricted ζ, these results were proved by Olver (1954, 1958). The present form of the error estimates in (7.31) and (7.32) is given in Olver & Wong (2009); see also Olver (1974, Ch. 11). Explicit bounds for $\varepsilon_i(\lambda, x), i = 1, 2$, can be found in Olver (1963, 1964).

Returning to equation (7.16), we now assume that $a(x)$ has a simple pole (say) at x_0 and $(x - x_0)^2 b(x)$ is analytic. For simplicity, we also assume that $a(x)$ has the same sign as $x - x_0$. In this case, it is again Langer (1935) who introduced the transformation

$$\begin{cases} \zeta^{1/2} = \displaystyle\int_{x_0}^x a^{1/2}(t)\, dt, & x \ge x_0, \\[2mm] (-\zeta)^{1/2} = \displaystyle\int_x^{x_0} [-a(t)]^{1/2}\, dt, & x \le x_0, \end{cases}$$

and

$$w = \left(\frac{d\zeta}{dx}\right)^{1/2} y,$$

which transforms (7.16) into the new equation

$$\frac{d^2 w}{d\zeta^2} = \left\{\frac{\lambda^2}{4\zeta} + \widehat{\psi}(\zeta)\right\} w, \qquad (7.33)$$

where

$$\widehat{\psi}(\zeta) = \frac{b(x)}{\widehat{a}(x)} + \frac{1}{\widehat{a}^{1/4}(x)} \frac{d^2}{d\zeta^2}\{\widehat{a}^{1/4}(x)\}$$

and $\widehat{a}(x) = (d\zeta/dx)^2 = 4\zeta a(x)$. If $b(x)$ has a simple or double pole at x_0, then $\widehat{\psi}(\zeta)$ has the same kind of singularity at $\zeta = 0$. Denote the value of $\zeta^2\widehat{\psi}(\zeta)$ at $\zeta = 0$ by $\frac{1}{4}(\nu^2 - 1)$, and write (7.33) in the form

$$\frac{d^2 w}{d\zeta^2} = \left\{\frac{\lambda^2}{4\zeta} + \frac{\nu^2 - 1}{4\zeta^2} + \frac{\psi(\zeta)}{\zeta}\right\} w \qquad (7.34)$$

with $\psi(\zeta) = \zeta\widehat{\psi}(\zeta) - \frac{1}{4}(\nu^2 - 1)\zeta^{-1}$. Note that $\psi(\zeta)$ is analytic at $\zeta = 0$, and that equation (7.34) has a regular singularity there. We suppose that the range of ζ is a real interval (α_1, α_2) which contains $\zeta = 0$ and may be unbounded, and consider separately the intervals $[0, \alpha_2)$ and $(\alpha_1, 0]$.

If the term $\psi(\zeta)/\zeta$ is neglected, then (7.34) becomes

$$\frac{d^2 w}{d\zeta^2} = \left\{\frac{\lambda^2}{4\zeta} + \frac{\nu^2 - 1}{4\zeta^2}\right\} w. \qquad (7.35)$$

If ζ is positive, two linearly independent solutions of (7.35) using modified Bessel functions are $\zeta^{1/2} I_\nu(\lambda\zeta^{1/2})$ and $\zeta^{1/2} K_\nu(\lambda\zeta^{1/2})$. If $\zeta^{-1/2}\psi(\zeta)$ is absolutely integrable on $[0, \alpha_2)$, then equation (7.34) has two linearly independent solutions $w_1(\lambda, \zeta)$ and $w_2(\lambda, \zeta)$ such that

$$w_1(\lambda, \zeta) = \zeta^{1/2} I_\nu(\lambda\zeta^{1/2})[1 + O(\lambda^{-1})],$$

$$w_2(\lambda, \zeta) = \zeta^{1/2} K_\nu(\lambda\zeta^{1/2})[1 + O(\lambda^{-1})],$$

as $|\lambda| \to \infty$, where the O−terms hold uniformly with respect to $\zeta \in [0, \alpha_2)$.

When ζ is negative, two linearly independent solutions of (7.35) using Bessel functions are $|\zeta|^{1/2} J_\nu(\lambda|\zeta|^{1/2})$ and $|\zeta|^{1/2} Y_\nu(\lambda|\zeta|^{1/2})$, and if

$|\zeta|^{-1/2}\psi(\zeta)$ is absolutely integrable on $(\alpha_1, 0]$ then equation (7.34) has two solutions, namely,

$$w_1(\lambda, \zeta) = |\zeta|^{1/2} J_\nu(\lambda|\zeta|^{1/2})[1 + O(\lambda^{-1})],$$

$$w_2(\lambda, \zeta) = |\zeta|^{1/2} Y_\nu(\lambda|\zeta|^{1/2})[1 + O(\lambda^{-1})],$$

as $|\lambda| \to \infty$, which hold uniformly for $\zeta \in (\alpha_1, 0]$.

Langer (1935) was the first to derive asymptotic approximations for solutions in terms of Bessel functions, but the validity of the result was confined to a shrinking neighborhood of the singular point. Asymptotic approximations valid uniformly in fixed intervals were established by Swanson (1956) and Olver (1956, 1958).

For extensions to purely imaginary λ in (7.18) or complex ζ in (7.28) and (7.34), more delicate error estimates, and infinite asymptotic expansions, see Olver (1974, Chapters 10, 11 & 12). There are also problems in differential equations that correspond to the coalescence of saddle points discussed in Section 7.2. For two coalescing turning points, we mention Olver (1975, 1976) and Dunster (1996). The approximants in this case are parabolic cylinder functions. For the coalescence of a turning point and a simple pole, see Dunster (1994).

7.4 Exponential Asymptotics

Due to the needs in physical applications, there was a sudden surge of interest in the 1980's in finding ways to pick up exponentially small terms or to derive asymptotic expansions whose error terms are exponentially small; see, e.g., Meyer (1980) and Boyd (1999). Amongst the many papers in this subarea of asymptotic analysis, there are two that have drawn the widest attention. The first was by Kruskal and Segur (1991) with the title "Asymptotics beyond all orders in a model of crystal growth", and the second paper was by Berry (1989) with the title "Uniform asymptotic smoothing of Stokes' discontinuity". The former eventually led to an international conference in San Diego (see Segur et al. (1991)), and the latter generated enough activities to form a half-year program in Cambridge (1995). Since the first paper is more related to singular perturbation theory, we will defer the problem in that paper to Section 7.5 when we discuss that topic. In the current section, we shall be concerned with only the second paper.

Consider the Airy function $\mathrm{Ai}\,(z)$ defined by

$$\mathrm{Ai}\,(z) = \frac{1}{2\pi i} \int_L \exp\left(\frac{1}{3}t^3 - zt\right)\, dt, \qquad (7.36)$$

where L is any contour which begins at infinity in the sector $-\frac{1}{2}\pi < \arg z < -\frac{1}{6}\pi$ and ends at infinity in the sector $\frac{1}{6}\pi < \arg z < \frac{1}{2}\pi$. It has been long known that the asymptotic behavior of $\mathrm{Ai}\,(z)$ is given by

$$
\begin{aligned}
\mathrm{Ai}\,(z) \sim{}& \frac{1}{2\pi z^{1/4}} \exp\left(-\frac{2}{3}z^{3/2}\right) \sum_{m=0}^{\infty} \frac{(-1)^m \Gamma(3m + \frac{1}{2})}{(2m)!\,9^m} z^{-3m/2} \\
&+ \frac{C}{2\pi z^{1/4}} \exp\left(\frac{2}{3}z^{3/2}\right) \sum_{m=0}^{\infty} \frac{\Gamma(3m + \frac{1}{2})}{(2m)!\,9^m} z^{-3m/2},
\end{aligned}
\qquad (7.37)
$$

where C is a constant which is 0 for $\arg z \in (-\frac{1}{3}\pi, \frac{1}{3}\pi)$ and i for $\arg z \in (\frac{1}{3}\pi, \frac{5}{3}\pi)$; see Wong (1989, pp. 93–94). The coefficient C is called a *Stokes multiplier*, and is domain dependent. The discontinuous change of the coefficient C, when the argument of z changes in a continuous manner, is known as *Stokes' phenomenon*.

Since we are dealing with a continuous (in fact, analytic) function, it is rather unsatisfactory to have a discontinuous coefficient in the asymptotic expansion (7.37). In 1989, Berry wrote an innovative paper, in which he adopted a different interpretation of the Stokes phenomenon. In his view, the coefficient of the second series in (7.37) should be a continuous function of $\arg z$, instead of a discontinuous constant; see also Paris & Wood (1995). He gave a beautiful, although not mathematically rigorous, demonstration of this theory with the well-known asymptotic expansion associated with the WKB-approximation for second-order differential equations. Berry's theory has since become known variously as "exponential asymptotics" or "superasymptotics" (see Boyd (1999) and Olde Daalhuis (2003)), and has been successfully applied to several known asymptotic expansions in a mathematically rigorous manner (see, e.g., Boyd (1990), Olver (1991a, 1991b, 1993), Olde Daalhuis & Olver (1994, 1995), Wong & Zhao (1999, 2002)).

In this section, we shall illustrate Berry's theory with the simple Airy function given in (7.36). Our approach is based on another brilliant idea of Berry & Howls (1991), which is a modification of the classical method of steepest descent. However, in view of space limitation, our presentation will be very sketchy. In (7.36), we make the change of variable $t = z^{1/2} u$. The integral in (7.36) then becomes

$$\mathrm{Ai}\,(z) = \frac{z^{1/2}}{2\pi i} \int_L e^{z^{3/2}(\frac{1}{3}u^3 - u)}\, du.$$

Put $\xi := \frac{2}{3} z^{3/2}$ and $f(u) = \frac{1}{2}(u^3 - 3u)$ so that

$$\mathrm{Ai}\,(z) = \frac{1}{2\pi i} \left(\frac{3}{2}\xi\right)^{1/3} \int_L e^{\xi f(u)}\, du.$$

The saddle points of $f(u)$ are located at $u = \pm 1$. Clearly, $f(\pm 1) = \mp 1$. Let $\theta := \arg \xi$ and consider the steepest descent curves

$$\Gamma_{\pm 1}(\theta) : \arg\{\xi[f(\pm 1) - f(u)]\} = \arg\{e^{i\theta}[\mp 1 - f(u)]\} = 0. \qquad (7.38)$$

From Wong (1989, p. 92), we know that only the saddle point at $u = 1$ is relevant. Deform L into $\Gamma_1(\theta)$, and write

$$\mathrm{Ai}\,(z) = \frac{1}{2\pi i} \left(\frac{3}{2}\right)^{1/3} \xi^{-1/6} e^{-\xi} I^{(1)}(\xi), \qquad (7.39)$$

where

$$I^{(\pm 1)}(\xi) = \xi^{1/2} \int_{\Gamma_1(\theta)} e^{\xi[f(u) \pm 1]}\, du. \qquad (7.40)$$

In (7.40), we make the change of variable $\tau = \xi[-1 - f(u)]$. For $u \in \Gamma_1(\theta)$, τ is real and positive; see (7.38). Expanding $f(u)$ into a Taylor series at $u = 1$, we have

$$-\tau = \frac{3}{2}\xi[(u - 1)^2 + \frac{1}{3}(u - 1)^3],$$

from which it follows that

$$\pm i \left(\frac{2}{3}\frac{\tau}{\xi}\right)^{1/2} = (u - 1)\left[1 + \frac{1}{3}(u - 1)\right]^{1/2},$$

where the branch of the square root is chosen so that it reduces to 1 at $u = 1$. By Lagrange's inversion formula,

$$u^{\pm} = 1 + \sum_{n=1}^{\infty} \alpha_n \left(\pm i \sqrt{\frac{2\tau}{3\xi}}\right)^n.$$

Breaking the integration path $\Gamma_1(\theta)$ in (7.40) at $u = 1$, we can rewrite the integral $I^{(1)}(\xi)$ as

$$I^{(1)}(\xi) = \xi^{1/2} \int_0^{\infty} \left[\frac{du^+}{d\tau} - \frac{du^-}{d\tau}\right] e^{-\tau}\, d\tau. \qquad (7.41)$$

Instead of (7.41), Berry & Howls (1991) used the equivalent representation

$$I^{(1)}(\xi) = \xi^{-1/2} \int_0^\infty \left[\frac{1}{f'(u^-(\tau))} - \frac{1}{f'(u^+(\tau))} \right] e^{-\tau} \, d\tau. \tag{7.42}$$

Furthermore, they observed that the quantity inside the square brackets in (7.42) can be expressed as a residue, and showed that it is equal to

$$\frac{1}{2\pi i} \frac{\xi^{3/2}}{\tau^{1/2}} \int_{C_1(\theta)} \frac{[-1 - f(u)]^{1/2}}{\xi[-1 - f(u)] - \tau} \, du, \tag{7.43}$$

where $C_1(\theta)$ is a positively oriented curve surrounding the steepest-descent path $\Gamma_1(\theta)$; for details, see Wong (2001). (Since $\Gamma_1(\theta)$ is an infinite contour, $C_1(\theta)$ actually consists of two infinite curves embracing $\Gamma_1(\theta)$.) Replacing the integrand in (7.42) by the quantity in (7.43) gives a double integral, and upon interchanging the order of integration one obtains

$$I^{(1)}(\xi) = \frac{1}{2\pi i} \int_{C_1(\theta)} [-1 - f(u)]^{-1/2} \int_0^\infty \frac{e^{-\tau}\tau^{-\frac{1}{2}}}{1 - \{\tau/\xi[-1 - f(u)]\}} \, d\tau \, du.$$

The quantity inside the inner integral can be expanded into a geometric series with remainder, and the result is

$$I^{(1)}(\xi) = \sum_{s=0}^{N} c_s \xi^{-s} + R_N(\xi), \tag{7.44}$$

where the coefficients c_s can be evaluated exactly and the remainder $R_N(\xi)$ is given by

$$R_N(\xi) = \frac{\xi^{-N}}{2\pi i} \int_0^\infty e^{-\tau} \tau^{N-\frac{1}{2}}$$
$$\times \int_{C_1(\theta)} [-1 - f(u)]^{-N-\frac{1}{2}} \frac{1}{1 - \{\tau/\xi[-1 - f(u)]\}} \, du \, d\tau. \tag{7.45}$$

Now consider all steepest descent paths $\Gamma_1(\theta)$ passing through $u = 1$ for different values of θ; see Figure 7.1. Since $f(1) - f(-1) = -2$, the path

$$\Gamma_1(\pi): \qquad \arg\{e^{i\pi}[f(1) - f(u)]\} = 0$$

runs into the saddle point $u = -1$. Berry and Howls call $u = -1$ an *adjacent saddle* of $u = 1$, and the steepest-descent path

$$\Gamma_{-1}(\pi): \qquad \arg\{e^{i\pi}[f(-1) - f(u)]\} = 0$$

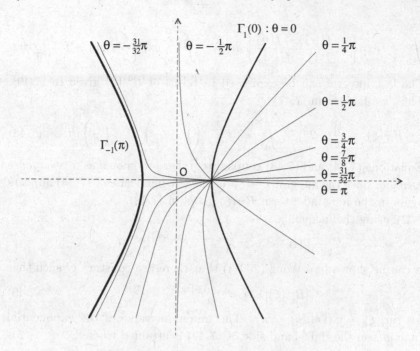

$\Gamma_1(0) : \theta = 0$

$\theta = -\frac{31}{32}\pi$ $\theta = -\frac{1}{2}\pi$ $\theta = \frac{1}{4}\pi$

$\theta = \frac{1}{2}\pi$

$\Gamma_{-1}(\pi)$ $\theta = \frac{3}{4}\pi$
$\theta = \frac{7}{8}\pi$
$\theta = \frac{31}{32}\pi$
$\theta = \pi$

O

Fig. 7.1. Contour $\Gamma_1(\theta)$, $-\pi < \theta < \pi$.

an *adjacent contour*; see also Boyd (1993). The next step is to deform $C_1(\theta)$ into $\Gamma_{-1}(\pi)$, and make the change of variable

$$\tau = t\,\frac{f(1) - f(u)}{f(1) - f(-1)};\qquad (7.46)$$

recall that $f(1) = -1$. Dingle (1973) called the quantity $f(1) - f(-1)$ a *singulant*; see also Berry & Howls (1991). Substituting (7.46) in (7.45), we obtain

$$R_N(\xi) = \frac{\xi^{-N}}{2\pi i}(-2)^{-N-\frac{1}{2}}\int_0^\infty e^{-t}t^{N-\frac{1}{2}}\left(1 + \frac{t}{2\xi}\right)^{-1}$$

$$\times \int_{\Gamma_{-1}(\pi)} \exp\left\{-t\,\frac{f(-1) - f(u)}{f(1) - f(-1)}\right\}\,du\,dt.$$

$$(7.47)$$

In the inner integral, we write $u = -w$. Since $f(u)$ is an odd function and $f(1) - f(-1) = -2$, it follows that

$$\int_{\Gamma_{-1}(\pi)} \exp\left\{-t\frac{f(-1)-f(u)}{f(1)-f(-1)}\right\} du = -\int_{\Gamma_1(0)} \exp\left\{\frac{t}{2}[f(w)-f(1)]\right\} dw.$$

The last integral can be expressed in terms of $I^{(1)}(\xi)$ given in (7.40). Thus, we have from (7.47)

$$R_N(\xi) = \frac{1}{2\pi}(-2\xi)^{-N}\int_0^\infty e^{-t}t^{N-1}\left(1+\frac{t}{2\xi}\right)^{-1}I^{(1)}\left(\frac{t}{2}\right)dt. \quad (7.48)$$

Equations (7.44) and (7.48) coupled together are known as a *resurgence formula*, since the integral $I^{(1)}(\xi)$ on the left-hand side of (7.44) appears again in the remainder term $R_N(\xi)$ given in (7.48).

By using the inequality

$$\Gamma(N) < \sqrt{2\pi}N^{N-\frac{1}{2}}e^{-N+(1/12N)},$$

it can be shown (see Wong (2001)) that there is a constant C such that

$$|R_N(\xi)| < Ce^{N(-1+\log N-\log|2\xi|)} \quad (7.49)$$

for $|\arg\xi| \le \pi$ (i.e., $|\theta| \le \pi$). The minimum value of the exponential function on the right-hand side of (7.49) is attained when

$$\frac{d}{dN}(-N + N\log N - N\log|2\xi|) = \log N - \log(2|\xi|) = 0.$$

Therefore, optimal truncation occurs near $N = N^* = 2|\xi|$. With N given by this value, we obtain from (7.49)

$$|R_N(\xi)| \le Ce^{-2|\xi|}, \quad (7.50)$$

as $|\xi| \to \infty$, for $|\theta| \le \pi$. Olver (1991a) called the expansion (7.44) with error given by (7.50) a *uniform, exponentially improved*, asymptotic expansion in the sector $|\theta| \le \pi$. Optimally truncated asymptotic expansions have now been also called *superasymptotic expansions* by Berry & Howls (1990).

Returning to (7.48), we now replace the function $I^{(1)}(t/2)$ by its asymptotic expansion (7.44). Termwise integration gives a series of integrals which can be expressed in terms of Dingle's *terminant function*

$$\int_0^\infty \frac{t^{k-1}e^{-t}}{1+t/\zeta}dt := 2\pi i(-\zeta)^k e^\zeta T_k(\zeta);$$

see Olver (1991a). More precisely, we have

$$R_N(\xi) = ie^{2\xi}\sum_{r=0}^{N'-1}(-1)^r\frac{c_r}{\xi^r}T_{N-r}(2\xi) + R_{N,N'}(\xi), \quad (7.51)$$

where

$$R_{N,N'}(\xi) = \frac{1}{2\pi}(-2|\xi|)^{-N} \int_0^\infty e^{-t} t^{N-1}\left(1 + \frac{t}{2\xi}\right)^{-1} R_{N'}\left(\frac{t}{2}\right) dt. \quad (7.52)$$

The idea of re-expanding the remainder term in optimally truncated asymptotic series was introduced by Berry & Howls (1990). They called this theory *hyperasymptotics*; see also Olver (1993), Olde Daalhuis & Olver (1994) and Boyd (1999). Another way to derive hyperasymptotic expansions is to use Hadamard's expansion, as has been done by Paris (2001a, 2001b).

With $\arg \zeta = \phi$ and $k = |\zeta| + \alpha, \alpha$ being a bounded positive number, one can show by using an existing theory that the terminant function $T_k(\zeta)$ has the uniform asymptotic expansion

$$T_k(\zeta) \sim \frac{1}{2}\text{erfc}\,(Z) - \frac{i}{\sqrt{2\pi|\zeta|}}e^{-Z^2}\sum_{s=0}^\infty \left(\frac{1}{2}\right)_s g_{2s}(\phi,\alpha)\left(\frac{2}{|\zeta|}\right)^s, \quad (7.53)$$

where $Z := c(\phi)\sqrt{|\zeta|/2}$ and

$$\frac{1}{2}[c(\phi)]^2 := -e^{i(\phi-\pi)} + i(\phi-\pi) + 1.$$

The coefficients $g_{2s}(\phi,\alpha)$ can be given explicitly; see Olver (1991a).

Coupling (7.44) and (7.51) gives

$$I^{(1)}(\xi) = \sum_{s=0}^{N-1} c_s \xi^{-s} + ie^{2\xi}\sum_{n=0}^{N'-1}(-1)^r c_r \xi^{-r} T_{N-r}(2\xi) + R_{N,N'}(\xi). \quad (7.54)$$

The remainder $R_{N,N'}(\xi)$ is given in (7.52), and can be estimated as before. Of course, it is expected to be of lower order of magnitude, and hence can be neglected. Inserting (7.53) into (7.54), we obtain from (7.39)

$$\text{Ai}\,(z) \sim \frac{(3/2)^{1/2}}{2\pi i} z^{-1/4}\left[e^{-\xi}\sum_{s=0}^{2|\xi|-1} c_s \xi^{-s}\right.$$
$$\left. + \frac{i}{2}\,\text{erfc}\,\{c(\theta)|\xi|^{1/2}\}e^\xi \sum_{r=0}^{N'-1}(-1)^r c_r \xi^{-r}\right], \quad (7.55)$$

where $\xi = \frac{2}{3}z^{3/2}$ and $\theta = \arg \xi$. Note that in (7.55), we have truncated the first series at an optimal place. When θ is near π, erfc $\{c(\theta)|\xi|^{1/2}\}$ will have an abrupt but smooth change. In Berry's terminology, this function is called a *Stokes multiplier*. A similar result holds for θ near

$-\pi$. We suggest that the abrupt but smooth change be called *Berry's transition*. The material in this section is taken from Wong (2001).

7.5 Singular Perturbation Theory

This is a subarea of asymptotics which in my view has not attracted the attention it deserves from mathematicians, since there is still a vast amount of results in the area that need mathematically rigorous justifications. In the 50's, 60's or even 70's of the last century, there were at least a few notable figures working in the area, trying to build a mathematical foundation for the singular perturbation theory; e.g., Wasow (1956, 1965), Erdélyi (1962, 1968), Howes (1978) and Smith (1985). Nowadays, it is difficult even to find enough researchers to participate in a workshop on this topic. Nevertheless, some progress is being made in boundary layer problems, a major subject in this theory.

First, we consider the problem in the paper of Kruskal and Segur (1991), which has already been briefly mentioned in Section 7.4. The problem is to study the third order, nonlinear, differential equation

$$\varepsilon^2 \theta''' + \theta' = \cos\theta, \qquad\qquad -\infty < s < \infty.$$

An odd monotonic solution satisfying

$$\theta(s,\varepsilon) \to \pm\pi/2 \qquad\qquad \text{as } s \to \pm\infty,$$

or, equivalently, a monotonic solution satisfying

$$\theta(0,\varepsilon) = 0, \qquad\qquad \theta(+\infty,\varepsilon) = \frac{1}{2}\pi$$

and

$$\theta''(0,\varepsilon) = 0,$$

is called a *needle crystal*. What Kruskal and Segur had shown was that no needle crystal exists for small ε; in fact, they proved that as $\varepsilon \to 0$,

$$\theta''(0,\varepsilon) \sim 2\Gamma\varepsilon^{-5/2}e^{-\pi/2\varepsilon}, \tag{7.56}$$

where Γ is a fixed, nonzero constant. In view of the fact that the term in (7.56) is exponentially small, they titled their paper *Asymptotics beyond all orders*, meaning that one may have to go beyond all orders in an asymptotic expansion in order to determine whether a solution even exists. Since the publication of their paper, people began to point out that exponentially small terms appear in many other problems in applied mathematics, except that it is often very difficult to spot them if one stays within the framework of Poincaré's definition of an asymptotic expansion.

To see just how exponentially small terms can be easily missed out in some of the well-known asymptotic approximations, let us look at the familiar two-point boundary value problem

$$\varepsilon y''(x) + a(x)y'(x) + b(x)y(x) = 0, \tag{7.57a}$$

$$y(-1) = A, \qquad y(1) = B. \tag{7.57b}$$

It is now widely known that if $a(x)$ is positive, then the asymptotic solution to (1.57) which holds uniformly in the interval $[-1, 1]$ is given by

$$
\begin{aligned}
y_{\text{unif}}(x) = {} & B \exp\left(\int_x^1 \frac{b(t)}{a(t)}\, dt\right) \\
& + \left\{A - B \exp\left(\int_{-1}^1 \frac{b(t)}{a(t)}\, dt\right)\right\} e^{-a(-1)(1+x)/\varepsilon},
\end{aligned} \tag{7.58}
$$

meaning that

$$y(x) = y_{\text{unif}}(x) + O(\varepsilon) \qquad \text{as } \varepsilon \to 0, \tag{7.59}$$

where the $O-$term is uniform with respect to $x \in [-1, 1]$. The formula in (7.58) is given in at least eight standard texts; see, e.g., Bender & Orszag (1978, p. 425), Kevorkian & Cole (1981, pp. 53 & 58), Lagerstrom (1988, p. 59), Logan (1987, p. 68), Murdock (1991, p. 421), Nayfeh (1981, p. 289), O'Malley (1991, p. 94) and Simmonds & Mann (1986, p. 109). Despite its usefulness, equation (7.58) is not entirely correct. For instance, if the boundary value B in (7.57b) is zero, then (7.58) becomes

$$y_{\text{unif}}(x) = A e^{-a(-1)(1+x)/\varepsilon},$$

which is exponentially small for $x > -1$, and asymptotically zero with respect to the order estimate in (7.59). The more accurate formula is

$$
\begin{aligned}
y(x) = {} & B \exp\left(\int_x^1 \frac{b(t)}{a(t)}\, dt\right)[1 + O(\varepsilon)] \\
& + \frac{a(0)}{a(x)}\left\{A - B \exp\left(\int_{-1}^1 \frac{b(t)}{a(t)}\, dt\right)\right\} \exp\left(\int_{-1}^x \frac{b(t)}{a(t)}\, dt\right) \\
& \times \exp\left(-\frac{1}{\varepsilon}\int_{-1}^x a(t)\, dt\right)[1 + O(\varepsilon)].
\end{aligned} \tag{7.60}
$$

One can establish this result by using the WKB approximation given in

Section 7.3; it can also be found in O'Malley (1968). To illustrate our point, we consider the simple example

$$\varepsilon y'' + (3 + x)y' + y = 0, \qquad y(-1) = 1, \qquad y(1) = 0.$$

Equations (7.58) gives

$$y_{\text{unif}}(x) = e^{-2(1+x)/\varepsilon}. \tag{7.61}$$

In particular, we have

$$y_{\text{unif}}(0) = e^{-2/\varepsilon}. \tag{7.62}$$

But, from (7.60) it follows that

$$y(x) = \frac{3}{2} \exp\left\{ -\frac{1}{\varepsilon}\left(\frac{1}{2}x^2 + 3x + \frac{7}{2} \right) \right\}[1 + O(\varepsilon)]. \tag{7.63}$$

Both approximations (7.61) and (7.63) are exponentially small when x is near 0. However, (7.59) and (7.62) together give only

$$y(x) = O(\varepsilon),$$

whereas from (7.63) we have

$$y(0) \sim \frac{3}{2} e^{-7/2\varepsilon}.$$

The same kind of problem arises in the case when the coefficient function $a(x)$ in equation (7.57a) has a zero. More specifically, we assume that

$$a(x) \sim \alpha x \qquad \text{and} \qquad b(x) \sim \beta \qquad \text{as } x \to 0,$$

where $\alpha \neq 0$ and β are constants. If $\alpha < 0$ and $\beta/\alpha \neq 0, -1, -2, \ldots$, Bender & Orszag (1978, p. 460) gave the leading order uniform asymptotic solution

$$y_{\text{unif}}(x) = Ae^{-a(-1)(x+1)/\varepsilon} + Be^{a(1)(1-x)/\varepsilon} \tag{7.64}$$

to the boundary value problem (7.57a)–(7.57b), without any justification. While this solution may appear to behave like the true solution near the boundary layers, it is incorrect in the middle of the interval as pointed out in Wong & Yang (2002), where a rigorous derivation of an asymptotic solution is given which is uniformly valid in the interval $[-1, 1]$. Unlike (7.64), the correct asymptotic formula involves parabolic cylinder functions and the values of the coefficient functions in the entire interval $[-1, 1]$. As a comparison, let us look at the example

$$\varepsilon y'' - 2xy' + (1 + x^2)y = 0, \qquad y(-1) = 2, \qquad y(1) = 1. \tag{7.65}$$

Bender & Orszag (1978, p. 460) gave the asymptotic solution

$$y_{\text{unif}}(x) = 2e^{-2(x+1)/\varepsilon} + e^{-2(1-x)/\varepsilon},$$

which, in particular, yields

$$y_{\text{unif}}(0) = 3e^{-2/\varepsilon},$$

whereas, according to Wong & Yang (2002), the correct solution of (7.65) behaves like

$$y(0) \sim -\frac{6\sqrt{2\pi}}{\Gamma(\frac{1}{4})} e^{1/4} \varepsilon^{-3/4} e^{-1/\varepsilon}. \tag{7.66}$$

The minus sign in (7.66) also explains why the graph shown in Bender & Orszag (1978, p. 460 , Fig. 9.17) lies below the x-axis near the origin.

Next, we show that nested boundary layer problems also exhibit a similar kind of phenomena. Nested boundary layers mean that there is one boundary layer which lies inside another one. To illustrate, we consider the specific equation

$$\varepsilon^3 xy''(x) + x^2 y'(x) - (x^3 + \varepsilon)y(x) = 0, \qquad 0 < x < 1, \tag{7.67}$$

with the boundary conditions

$$y(0) = 1, \qquad y(1) = \sqrt{e}. \tag{7.68}$$

By using the method of matched asymptotics, Bender & Orszag (1978, p. 453) constructed the asymptotic solution

$$y_{\text{unif}}(x) = \frac{2\sqrt{x}}{\varepsilon} K_1\left(\frac{2\sqrt{x}}{\varepsilon}\right) + e^{-\varepsilon/x} + e^{x^2/2} - 1, \tag{7.69}$$

where $K_1(x)$ is a modified Bessel function and the first three terms on the right represent the leading terms in the expansions of the *inner-inner solution*, *inner solution* and *outer solution*, respectively, and the fourth term results from the matching of the inner solution and outer solution. However, the solution given in (7.69) does not reveal the fact that the true solution is exponentially small in the interval $O(\varepsilon^2) \ll x \ll O(\varepsilon)$. The correct asymptotic solution can be found in Liang & Wong (to appear), where equation (7.67) is transformed into the canonical form

$$\frac{d^2 U}{d\zeta^2} = \left\{\frac{1}{4\zeta} + \frac{\phi(\zeta)}{4\zeta}\right\} U \tag{7.70}$$

by a Liouville-like transformation $(x, y) \mapsto (\zeta, U)$ defined by

$$\zeta^{\frac{1}{2}}(x) = \frac{1}{\varepsilon} \int_0^x \sqrt{\frac{1}{t} + \frac{t^2}{4\varepsilon^4} + \frac{1}{2\varepsilon}(1 + 2t^2)} \, dt$$

and

$$y(x) = (\zeta'(x))^{-\frac{1}{2}} \exp\left\{-\frac{x^2}{4\varepsilon^2}\right\} U(x).$$

The function $\phi(x)$ in (7.70) satisfies

$$\phi(x) = O\left(\frac{x^2}{\varepsilon^2}\right) \qquad \text{for } 0 \le x \le \rho\varepsilon^{4/3},$$

$$\phi(x) = O\left(\frac{\varepsilon^6}{x^4}\right) \qquad \text{for } \rho\varepsilon^{4/3} \le x \le 1,$$

where ρ is a positive constant. From (7.70) it was further shown that for small ε, equation (7.67) has two linearly independent solutions

$$y_1(x) = \sqrt{\frac{\varepsilon}{2}}\left[\frac{1}{x} + \frac{x^2}{4\varepsilon^4} + \frac{1}{2\varepsilon}(1 + 2x^2)\right]^{-\frac{1}{4}}$$
$$\times \exp\left\{-\frac{x^2}{4\varepsilon^3}\right\}\zeta^{\frac{1}{4}}(x)K_1(\sqrt{\zeta})(1 + O(\varepsilon^{\frac{1}{3}})),$$

$$y_2(x) = \sqrt{\frac{\varepsilon}{2}}\left[\frac{1}{x} + \frac{x^2}{4\varepsilon^4} + \frac{1}{2\varepsilon}(1 + 2x^2)\right]^{-\frac{1}{4}}$$
$$\times \exp\left\{-\frac{x^2}{4\varepsilon^3}\right\}\zeta^{\frac{1}{4}}(x)I_1(\sqrt{\zeta})(1 + O(\varepsilon^{\frac{1}{3}})),$$

where $K_1(z)$ and $I_1(z)$ are the modified Bessel functions of order 1. Moreover, we have

$$\lim_{x \to 0} y_1(x) = \frac{\varepsilon}{2}(1 + O(\varepsilon^{\frac{1}{3}})), \qquad \lim_{x \to 0} y_2(x) = 0,$$

$$y_1(1) = \frac{\sqrt{\varepsilon\pi}}{2}\left(\frac{1}{4\varepsilon^4}\right)^{-\frac{1}{4}}\exp\left\{-\frac{1}{4\varepsilon^3}\right\}\exp\left\{-\sqrt{\zeta(1)}\right\}(1 + O(\varepsilon^{\frac{1}{3}})).$$

The last equation indicates that $y_1(1)$ is exponentially small as $\varepsilon \to 0$. The unique solution to (7.67) – (7.68) is given by

$$Y(x) = \frac{2}{\varepsilon}y_1(x) + \frac{\sqrt{e}}{y_2(1)}y_2(x). \tag{7.71}$$

By using the asymptotic formulas of $K_1(z)$ and $I_1(z)$, it can be verified that $Y(x)$ is exponentially small for $O(\varepsilon^2) \ll x \ll O(\varepsilon)$; for details, see Liang & Wong (to appear). In particular, Bender and Orszag's solution (7.69) gives

$$y_{\text{unif}}(\varepsilon^{7/6}) = \frac{2\sqrt{\varepsilon^{7/6}}}{\varepsilon}K_1\left(\frac{2\sqrt{\varepsilon^{7/6}}}{\varepsilon}\right) + e^{-\varepsilon/\varepsilon^{7/6}} + e^{\frac{1}{2}(\varepsilon^{7/6})^2} - 1 \sim \frac{1}{2}\varepsilon^{7/3},$$

whereas our solution (7.71) yields

$$Y(\varepsilon^{7/6}) \sim \exp\left\{\frac{(\varepsilon^{7/6})^2}{2} - \frac{\varepsilon}{\varepsilon^{7/6}}(1 + o(1))\right\} \sim \exp\left\{-\frac{1}{\varepsilon^{1/6}}\right\}.$$

Finally, let us look at two nonlinear problems of Carrier & Pearson (1968) and we begin with the simpler one

$$\varepsilon u'' + u^2 = 1, \qquad -1 < x < 1, \qquad (7.72a)$$

$$u(-1) = u(1) = 0. \qquad (7.72b)$$

By using the matched asymptotic method, it has been shown that this boundary value problem has at least four approximate solutions

$$\begin{aligned}
u_{\text{unif}}(x) = &-1 + 3 \operatorname{sech}^2\left(\pm\frac{1-x}{\sqrt{2\varepsilon}} + \ln(\sqrt{3} + \sqrt{2})\right) \\
&+ 3 \operatorname{sech}^2\left(\pm\frac{1+x}{\sqrt{2\varepsilon}} + \ln(\sqrt{3} + \sqrt{2})\right),
\end{aligned} \qquad (7.73)$$

depending on the two choices of plus and minus signs, and these solutions are uniformly valid in the entire interval $[-1, 1]$. Note that all four approximate solutions in (7.73) have two boundary layers in the interval $[-1, 1]$, one at each endpoint of the interval. Furthermore, let $x_0 \in (-1, 1), 1 - |x_0| \gg \sqrt{\varepsilon}$, and define

$$\tilde{u} = -1 + 3 \operatorname{sech}^2\left(\frac{x - x_0}{\sqrt{2\varepsilon}}\right) \qquad (7.74)$$

for x near x_0. The second quantity on the right-hand side takes the value 3 at $x = x_0$, and decays to zero with exponential rapidity as $|x - x_0|/\sqrt{\varepsilon} \to \infty$; thus it behaves like a *spike* near x_0 for sufficiently small ε. Matching (7.74) with the outer solution and the two inner solutions near ± 1, we get a composite formula

$$\begin{aligned}
u_{\text{unif}}(x) = &-1 + 3 \operatorname{sech}^2\left(\pm\frac{1-x}{\sqrt{2\varepsilon}} + \ln(\sqrt{3} + \sqrt{2})\right) \\
&+ 3 \operatorname{sech}^2\left(\pm\frac{1+x}{\sqrt{2\varepsilon}} + \ln(\sqrt{3} + \sqrt{2})\right) \\
&+ 3 \operatorname{sech}^2\left(\frac{x - x_0}{\sqrt{2\varepsilon}}\right).
\end{aligned} \qquad (7.75)$$

Formula (7.75) *appears* to be a valid approximation for x in the entire interval. But, by using phase plane analysis, Carrier & Pearson (1968, p. 204) showed that (7.75) can approximate an exact solution only if

$$x_0 = 0.$$

Thus, for most values of x_0, the solutions given in (7.75) cannot be valid, and they are called *spurious solutions*; see also Lange (1983) and MacGillivray (1997).

In Ou & Wong (2003), the problem in (1.72) has been investigated from a rigorous point of view. By using a "shooting method", the authors proved that the formal solutions in (7.75) obtained from the method of matched asymptotics approximate the true solutions with exponentially small errors. The so-called spurious solutions turn out to be approximations of true solutions, when the locations of their "spikes" are properly assigned. They also gave an estimate for the maximum number of spikes that these solutions can have.

As a continuation of their earlier work, Ou & Wong (2004) extended their method to include the singularly perturbed two-point problem

$$\varepsilon u'' + Q(u) = 0, \qquad -1 < x < 1, \qquad (7.76a)$$

with boundary conditions

$$u(-1) = u(1) = 0 \qquad (7.76b)$$

or

$$u'(-1) = u'(1) = 0. \qquad (7.76c)$$

The nonlinear term $Q(u)$ vanishes at s_-, 0, s_+ and nowhere else in $[s_-, s_+]$, with $s_- < 0 < s_+$. Furthermore, they assumed that $Q'(s_\pm) < 0$, $Q'(0) > 0$ and

$$\int_{s_-}^{s_+} Q(s)\, ds = 0.$$

Simple examples of functions satisfying these conditions are $Q(u) = u(1 - u^2)$ and $Q(u) = \sin \pi u$ for $u \in [-1, 1]$.

Equation (7.76a) can be considered as the equation of motion of a nonlinear spring with spring constant large compared to the mass. It is also the steady state version of many partial differential equations arising in physics and biochemistry. Unlike the case $Q(u) = u^2 - 1$, now the solutions exhibit a new phenomenon, known as the *shock* layer, i.e., solutions vary rapidly from one value to another in a very short interval.

The second problem of Carrier and Pearson consists of the non-autonomous nonlinear equation

$$\varepsilon u'' + 2(1 - x^2)u + u^2 = 1 \qquad (7.77a)$$

and the boundary conditions

$$u(-1) = u(1) = 0. \tag{7.77b}$$

So far the shooting method introduced in Ou & Wong (2003, 2004) can be used only to construct an asymptotic approximation for the maximum number of spikes that a solution of (1.77) can have, which in turn gives an estimate for the number of solutions to this problem; see Wong & Zhao (2008). No rigorous analysis has yet been found to establish the asymptotic nature of the approximate solutions obtained by using the matched asymptotic method.

7.6 Difference Equations

After Poincaré in 1886 introduced the notion of an asymptotic expansion in his theory of irregular singular points at infinity for ordinary differential equations, Birkhoff (1911, 1930), Adams (1928) and Birkhoff & Trjitzinsky (1932) began to develop a corresponding theory for linear difference equations. However, their analysis is very complicated and not easily understood even by specialists in asymptotics. For this reason, the development of this theory is far behind that of the asymptotic theory for linear differential equations. In Frank Olver's words (2003), "the work of B & T set back all research into the asymptotic solution of difference equations for most of the 20*th* century".

We first summarize some of the results in Wong & Li (1992a, 1992b), which were attempts to make Birkhoff and Trjitzinsky's theory more accessible. Then we present some more recent results in this area such as turning point theory and Bessel-type expansions; see Wang & Wong (2002, 2003, 2005a). We begin with the simplest and most familiar difference equation

$$y(n+2) + a(n)y(n+1) + b(n)y(n) = 0, \tag{7.78}$$

where $a(n)$ and $b(n)$ have asymptotic expansions of the form

$$a(n) \sim \sum_{s=0}^{\infty} \frac{a_s}{n^s} \quad \text{and} \quad b(n) \sim \sum_{s=0}^{\infty} \frac{b_s}{n^s} \tag{7.79}$$

for large values of n, and $b_0 \neq 0$.

Asymptotic solutions of (7.78) are classified by the roots of the *characteristic equation*

$$\rho^2 + a_0\rho + b_0 = 0. \tag{7.80}$$

Two possible values of ρ are

$$\rho_1, \rho_2 = -\frac{1}{2}a_0 \pm \left(\frac{1}{4}a_0^2 - b_0\right)^{1/2}.$$

If $\rho_1 \neq \rho_2$, i.e., $a_0^2 \neq 4b_0$, then Birkhoff (1911) showed that (7.78) has two linearly independent solutions $y_j(n), j = 1, 2$, such that

$$y_j(n) \sim \rho_j^n n^{\alpha_j} \sum_{s=0}^{\infty} \frac{c_{s,j}}{n^s}, \qquad n \to \infty, \qquad (7.81)$$

where

$$\alpha_j = -\frac{a_1\rho_j + b_1}{2\rho_j^2 + \rho_j a_0} = \frac{a_1\rho_j + b_1}{a_0\rho_j + 2b_0}$$

and $c_{0,j} = 1$. The series in (7.81) are known as *normal series* or *normal solutions*.

If $\rho_1 = \rho_2$ but their common value $\rho = -\frac{1}{2}a_0$ is not a root of the auxiliary equation

$$a_1\rho + b_1 = 0, \qquad (7.82)$$

i.e., $2b_1 \neq a_0a_1$, then Adams (1928) showed that (7.78) has two linearly independent solutions $y_j(n), j = 1, 2$, such that

$$y_j(n) \sim \rho^n \exp((-1)^j \gamma\sqrt{n})n^\alpha \sum_{s=0}^{\infty} (-1)^{js}\frac{c_s}{n^{s/2}}, \qquad (7.83)$$

where

$$\gamma = 2\sqrt{\frac{a_0a_1 - 2b_1}{2b_0}},$$

$$\alpha = \frac{1}{4} + \frac{b_1}{2b_0},$$

and $c_0 = 1$. Series of the form (7.83) are called *subnormal series* or *subnormal solutions*. Recursive formulas can be derived for higher coefficients in (7.81) and (7.83) by formal substitution.

When the double root of the characteristic equation (7.80) satisfies the auxiliary equation (7.82), i.e., when $2b_1 = a_0a_1$, we have three (exceptional) cases to consider, depending on the values of the zeros α_1, α_2 (Re $\alpha_1 \geq$ Re α_2) of the *indicial polynomial*

$$q(\alpha) = \alpha(\alpha - 1)\rho^2 + (a_1\alpha + a_2)\rho + b_2$$
$$= b_0\alpha^2 - \left(b_0 + \frac{1}{2}a_0a_1\right)\alpha - \frac{1}{2}a_0a_2 + b_2. \qquad (7.84)$$

Case (i): $\alpha_2 - \alpha_1 \neq 0, 1, 2, \ldots$. In this case, (7.78) has independent solutions $y_j(n), j = 1, 2$, of the form

$$y_j(n) \sim \rho^n n^{\alpha_j} \sum_{s=0}^{\infty} \frac{c_{s,j}}{n^s}, \qquad n \to \infty, \qquad (7.85)$$

with $c_{0,j} = 1$.

Case (ii): $\alpha_2 - \alpha_1 = 1, 2, \ldots$. Here, (7.85) applies only in the case of $j = 1$. A second independent solution is given by

$$y_2(n) \sim \rho^n n^{\alpha_2} \sum_{s=0}^{\infty}{}' \frac{d_s}{n^s} + c(\log n)y_1(n),$$

where the prime on $\sum$ denotes that the term for $s = \alpha_2 - \alpha_1$ is absent. The coefficients c and d_s can be determined by formal substitution, beginning with $d_0 = 1$.

Case (iii): $\alpha_2 = \alpha_1$. As in case (ii), (7.85) again gives only one solution $y_1(n)$. The second solution is given by

$$y_2(n) \sim \rho^n n^{\alpha_2} \sum_{s=1}^{\infty} \frac{d_s}{n^s} + (\log n)y_1(n).$$

For proofs and examples of the above results, see Wong & Li (1992a). Very often, the difference equations arising from applications are not of the form (7.78), but of the more general form

$$y(n+2) + n^p a(n)y(n+1) + n^q b(n)y(n) = 0, \qquad (7.86)$$

where p and q are integers, and $a(n)$ and $b(n)$ are as given in (7.79). For example, the recurrence relation for the Charlier polynomial is

$$C_{n+1}^{(a)}(x) + (n + a - x)C_n^{(a)}(x) + anC_{n-1}^{(a)}(x) = 0, \qquad a \neq 0,$$

and the number $T(n)$ of idempotent elements in the symmetric group of order n satisfies

$$T_n = T_{n-1} + (n-1)T_{n-2}, \qquad T_0 = T_1 = 1.$$

A discussion of the more general equation (7.86) is given in Wong & Li (1992b).

In equations (7.78) and (7.86), the coefficient functions $a(n)$ and $b(n)$ are free from auxiliary parameters. If they depend on a parameter, then the asymptotic solutions constructed above may no longer be valid since the roots of the characteristic equation (7.80) may coalesce, and the difference $\alpha_2 - \alpha_1$ of the two roots of the indicial polynomial (7.84) may tend to a nonnegative integer, as the parameter varies. Let us consider just the simplest case

$$p_{n+1}(x) = (a_n x + b_n)p_n(x) - c_n p_{n-1}(x), \qquad n = 1, 2, \ldots, \quad (7.87)$$

where a_n, b_n and c_n are constants. This is the frequently encountered three-term recurrence relation in the study of orthogonal polynomials; see Szegő (1975, p. 42). It is also satisfied by many other special functions of mathematical physics such as the Bessel and the Legendre functions; see Olver (1974) . If x is a fixed number, then (7.87) is equivalent to the second-order difference equation (7.78). By introducing the sequence $\{K_n\}$ defined recursively by $K_{n+1}/K_{n-1} = c_n$, with K_0 and K_1 being arbitrary, (7.87) can be transformed into the canonical form

$$P_{n+1}(x) - (A_n x + B_n)P_n(x) + P_{n-1}(x) = 0. \qquad (7.88)$$

As in (7.78), we assume that the coefficients A_n and B_n have asymptotic expansions of the form

$$A_n \sim n^{-\theta} \sum_{s=0}^{\infty} \frac{\alpha_s}{n^s} \qquad \text{and} \qquad B_n \sim \sum_{s=0}^{\infty} \frac{\beta_s}{n^s}, \qquad (7.89)$$

where θ is a real number and $\alpha_0 \neq 0$.

Motivated by the uniform asymptotic expansions of the Hermite polynomial (see Olver (1974, p. 403)) and the Laguerre polynomial (see Frenzen & Wong (1988)), we let τ_0 be a constant and put $\nu = n + \tau_0$. Clearly, the expansions in (7.89) can be recast in the form

$$A_n \sim \nu^{-\theta} \sum_{s=0}^{\infty} \frac{\alpha'_s}{\nu^s} \qquad \text{and} \qquad B_n \sim \sum_{s=0}^{\infty} \frac{\beta'_s}{\nu^s}. \qquad (7.90)$$

In (7.88), we now let $x = \nu^\theta t$ and $P_n = \lambda^n$. Substituting (7.90) into (7.88) and letting $n \to \infty$ (and hence $\nu \to \infty$), we obtain the characteristic equation

$$\lambda^2 - (\alpha'_0 t + \beta'_0)\lambda + 1 = 0. \qquad (7.91)$$

The roots of this equation are given by

$$\lambda = \frac{1}{2}\left[(\alpha'_0 t + \beta'_0) \pm \sqrt{(\alpha'_0 t + \beta'_0)^2 - 4}\right],$$

and they coincide when $t = t_\pm$, where

$$\alpha'_0 t_\pm + \beta'_0 = \pm 2.$$

The values $t_\pm$ play an important role in the asymptotic theory of the three-term recurrence relation (7.88), and they correspond to the transition points (i.e., turning point and poles) occurring in differential equations; see Olver (1974, p. 362). For this reason, we shall also call them *transition points*. Since t_+ and t_- have different values, we may restrict ourselves to just the case $t = t_+$. In terms of the exponent θ in (7.89) and the transition point t_+, we have three cases to consider; namely, (i) $\theta \neq 0$ and $t_+ \neq 0$; (ii) $\theta \neq 0$ and $t_+ = 0$; and (iii) $\theta = 0$. Here we present only the results in cases (i) and (iii). (The investigation of case (ii) has not yet been completed.)

We begin with case (i), and assume for simplicity that $\theta > 0$ and $|\beta_0'| < 2$ (i.e., t_+ and t_- are of opposite signs). The analysis for the case $\theta < 0$ is very similar; for an important example with $\theta = -1$, the interested reader is referred to Wang & Wong (2002). Furthermore, we choose

$$\tau_0 = -\frac{(\alpha_1 t_+ + \beta_1)}{(2 - \beta_0)\theta}$$

so that

$$\alpha_1' t_+ + \beta_1' = 0, \tag{7.92}$$

and define the function $\zeta(t)$ by

$$\frac{2}{3}[\zeta(t)]^{3/2} := \alpha_0' t^{1/\theta} \int_{t_+}^{t} \frac{s^{-1/\theta}}{\sqrt{(\alpha_0' s + \beta_0')^2 - 4}} \, ds$$
$$- \log \frac{\alpha_0' t + \beta_0' + \sqrt{(\alpha_0' t + \beta_0')^2 - 4}}{2}, \qquad t \geq t_+, \tag{7.93a}$$

and

$$\frac{2}{3}[-\zeta(t)]^{3/2} := \cos^{-1} \frac{\alpha_0' t + \beta_0'}{2}$$
$$- \alpha_0' t^{1/\theta} \int_{t}^{t_+} \frac{s^{-1/\theta}}{\sqrt{4 - (\alpha_0' s + \beta_0')^2}} \, ds, \qquad t < t_+. \tag{7.93b}$$

Moreover, we put

$$H_0(\zeta) := -\sqrt{\frac{(\alpha_0' t + \beta_0')^2 - 4}{4\zeta}}$$

and

$$\Phi(t) := -\frac{1}{\zeta^{1/2}} \int_{t_+}^{t} \frac{\alpha_1' \tau + \beta_1'}{2\theta \tau \zeta^{\frac{1}{2}} H_0} \, d\tau. \tag{7.94}$$

Now we are ready to state our first result on uniform asymptotic expansions for difference equations.

Theorem 7.1 *Assume that the coefficients A_n and B_n in the recurrence relation (7.88) have asymptotic expansions of the form given in (7.89) with $\theta \neq 0$ and $|\beta_0| < 2$. Let ζ and Φ be given as in (1.93) and (7.94), respectively. Then equation (7.88) has, for each value of ν and each nonnegative integer p, a pair of solutions $P_n(x)$ and $Q_n(x)$, given by*

$$
P_n(\nu^\theta t) = \left(\frac{4\zeta}{(\alpha_0' t + \beta_0')^2 - 4} \right)^{\frac{1}{4}} \left[\mathrm{Ai} \left(\nu^{\frac{2}{3}} \zeta + \frac{\Phi}{\nu^{\frac{1}{3}}} \right) \sum_{s=0}^{p} \frac{\widetilde{A}_s(\zeta)}{\nu^{s - \frac{1}{6}}} \right.
$$
$$
\left. + \mathrm{Ai}' \left(\nu^{\frac{2}{3}} \zeta + \frac{\Phi}{\nu^{\frac{1}{3}}} \right) \sum_{s=0}^{p} \frac{\widetilde{B}_s(\zeta)}{\nu^{s + \frac{1}{6}}} + \varepsilon_p(\nu, t) \right]
$$

and

$$
Q_n(\nu^\theta t) = \left(\frac{4\zeta}{(\alpha_0' t + \beta_0')^2 - 4} \right)^{\frac{1}{4}} \left[\mathrm{Bi} \left(\nu^{\frac{2}{3}} \zeta + \frac{\Phi}{\nu^{\frac{1}{3}}} \right) \sum_{s=0}^{p} \frac{\widetilde{A}_s(\zeta)}{\nu^{s - \frac{1}{6}}} \right.
$$
$$
\left. + \mathrm{Bi}' \left(\nu^{\frac{2}{3}} \zeta + \frac{\Phi}{\nu^{\frac{1}{3}}} \right) \sum_{s=0}^{p} \frac{\widetilde{B}_s(\zeta)}{\nu^{s + \frac{1}{6}}} + \delta_p(\nu, t) \right],
$$

where

$$
|\varepsilon_p(\nu, t)| \leq \frac{M_p}{\nu^{p + \frac{5}{6}}} \widetilde{\mathrm{Ai}} \left(\nu^{\frac{2}{3}} \zeta + \frac{\Phi}{\nu^{\frac{1}{3}}} \right) \tag{7.95}
$$

and

$$
|\delta_p(\nu, t)| \leq \frac{M_p}{\nu^{p + \frac{5}{6}}} \widetilde{\mathrm{Bi}} \left(\nu^{\frac{2}{3}} \zeta + \frac{\Phi}{\nu^{\frac{1}{3}}} \right) \tag{7.96}
$$

for $\delta \leq t < \infty$, $0 < \delta < t_+$ and M_p being a constant.

The coefficient functions $\widetilde{A}_s(\zeta)$ and $\widetilde{B}_s(\zeta)$ can be determined successively from their predecessors $\widetilde{A}_0(\zeta), \widetilde{B}_0(\zeta), \ldots, \widetilde{A}_{s-1}(\zeta)$ and $\widetilde{B}_{s-1}(\zeta)$, with

$$
\widetilde{A}_0(\zeta) = 1, \qquad \zeta^{\frac{1}{2}} \widetilde{B}_0(\zeta) = 0
$$

for $t \geq t_+$, and

$$
\widetilde{A}_0(\zeta) = \cos \left(\int_{t_+}^{t} (-\zeta)^{\frac{1}{2}} \varphi \, d\tau \right),
$$
$$
(-\zeta)^{\frac{1}{2}} \widetilde{B}_0(\zeta) = \sin \left(\int_{t_+}^{t} (-\zeta)^{\frac{1}{2}} \varphi \, d\tau \right)
$$

for $0 < \delta \le t < t_+$. The upper bounds $\widetilde{\mathrm{Ai}}(z)$ and $\widetilde{\mathrm{Bi}}(z)$ in (7.95) and (7.96) are the modulus functions frequently used in uniform asymptotic expansions (see Wong (1989, p. 381)), and are defined by

$$\widetilde{\mathrm{Ai}}(z) = \begin{cases} \mathrm{Ai}(z) & \text{if } z \ge 0, \\ [\mathrm{Ai}^2(z) + \mathrm{Bi}^2(z)]^{\frac{1}{2}} & \text{if } z < 0, \end{cases}$$

and

$$\widetilde{\mathrm{Bi}}(z) = \begin{cases} \mathrm{Bi}(z) & \text{if } z \ge 0, \\ [\mathrm{Ai}^2(z) + \mathrm{Bi}^2(z)]^{\frac{1}{2}} & \text{if } z < 0. \end{cases}$$

Now we consider case (iii), i.e., $\theta = 0$ in (7.89). The characteristic equation (7.91) is obtained in the same manner as before, except that t is replaced by x. The characteristic roots again coincide when $x = x_\pm$, where

$$\alpha_0 x_\pm + \beta_0 = \pm 2. \tag{7.97}$$

We assume, throughout the remaining part of this section, that

$$\alpha_1 = \beta_1 = 0$$

so that

$$\alpha_1 x_+ + \beta_1 = 0,$$

which was used in case (i); see (7.92). It is interesting to note that in a paper on WKB methods for difference equations, Dingle & Morgan (1967a, 1967b) also assumed this condition. In fact, they assumed the stronger condition that all coefficients α_s and β_s in (7.89) with odd indices vanish. Since $\widetilde{P}_n(x) := (-1)^n P_n(x)$ satisfies the recurrence relation $\widetilde{P}_{n+1}(x) + (A_n x + B_n)\widetilde{P}_n(x) + \widetilde{P}_{n-1}(x) = 0$, we may, without loss of generality, assume $\alpha_0 > 0$.

Let $\tau_0 := -(\alpha_3 x_+ + \beta_3)/2(\alpha_2 x_+ + \beta_2)$ and $N := n + \tau_0$. Define

$$\nu := \pm\left(\alpha_2 x_+ + \beta_2 + \frac{1}{4}\right)^{\frac{1}{2}} \quad \text{and} \quad \zeta^{\frac{1}{2}} = \cosh^{-1}\left(\frac{\alpha_0 x + \beta_0}{2}\right).$$

We are now ready to state our second result on uniform asymptotic expansions for difference equations.

Theorem 7.2 *Assume that the coefficients A_n and B_n in the recurrence relation (7.88) are real, and have asymptotic expansions given in (7.89) with $\theta = 0$. Let $x_\pm$ be the transition points defined in (7.97). Then,* *for each nonnegative integer p, equation (7.88) has a pair of linearly independent solutions*

$$P_n(x) = \left(\frac{4\zeta}{(\alpha_0 x + \beta_0)^2 - 4}\right)^{\frac{1}{4}} \left[N^{\frac{1}{2}} I_\nu(N\zeta^{\frac{1}{2}}) \sum_{s=0}^{p} \frac{A_s(\zeta)}{N^s}\right.$$
$$\left. + N^{\frac{1}{2}} \zeta^{\frac{1}{2}} I_{\nu-1}(N\zeta^{\frac{1}{2}}) \sum_{s=0}^{p} \frac{B_s(\zeta)}{N^s} + \varepsilon_p(N, x)\right]$$

and

$$Q_n(x) = \left(\frac{4\zeta}{(\alpha_0 x + \beta_0)^2 - 4}\right)^{\frac{1}{4}} \left[N^{\frac{1}{2}} K_\nu(N\zeta^{\frac{1}{2}}) \sum_{s=0}^{p} \frac{A_s(\zeta)}{N^s}\right.$$
$$\left. - N^{\frac{1}{2}} \zeta^{\frac{1}{2}} K_{\nu-1}(N\zeta^{\frac{1}{2}}) \sum_{s=0}^{p} \frac{B_s(\zeta)}{N^s} + \delta_p(N, x)\right],$$

where I_ν and K_ν are the modified Bessel functions. The error terms satisfy

$$|\varepsilon_p(N, x)| \le \frac{M_p}{N^{p+\frac{1}{2}}} \left[|I_\nu(N\zeta^{\frac{1}{2}})| + |I_{\nu-1}(N\zeta^{\frac{1}{2}})|\right]$$

and

$$|\delta_p(N, x)| \le \frac{M_p}{N^{p+\frac{1}{2}}} \left[|K_\nu(N\zeta^{\frac{1}{2}})| + |K_{\nu-1}(N\zeta^{\frac{1}{2}})|\right]$$

for $x_- + \delta \le x < \infty$, where M_p is a positive constant. The coefficients $A_s(\zeta)$ and $B_s(\zeta)$ can be determined successively for any given $A_0(\zeta)$ and $B_0(\zeta)$.

7.7 Riemann–Hilbert Approach

A significant development in asymptotic analysis took place in the 1990's, when Deift, Zhou and their associates introduced the steepest-descent method for Riemann–Hilbert problems (RHP); see Deift (1999), Deift et al. (1999a, 1999b), and Bleher & Its (1999). The important feature of their method is that it can be used to find asymptotic formulas for the solutions to nonlinear differential equations, such as the modified KdV, the nonlinear Schrödinger, and the Painlevé equations; see Deift & Zhou (1993, 1994, 1995). In this section, we shall illustrate this method with an application to orthogonal polynomials. Even just in this restricted area, the impact of this discovery has been significant.

Let Γ be a simple and smooth curve in the plane, and Γ^0 denote the interior of Γ (i.e., Γ with endpoints removed). In the case of classical orthogonal polynomials, it can be either the real line $\mathbb{R}$ (Hermite polynomials), half of the real line (Laguerre polynomials), a finite interval (Jacobi polynomials), or a circle (Bessel polynomials). Let $w(z)$ be a weight function defined on Γ, which is continuous. If Γ has an endpoint a, then we assume that $w(z) = O(|z-a|^\alpha)$ as $z \to a$ for some $\alpha > -1$. If Γ is an unbounded curve, then we assume that $w(z)$ decays sufficiently fast as $z \to \infty$ along Γ so that all moments

$$\mu_n = \int_\Gamma \zeta^n w(\zeta)\, d\zeta, \qquad n = 0, 1, 2, \ldots$$

exist. Let $p_n(z)$ denote the polynomials orthonormal with respect to $w(z)$; that is,

$$\int_\Gamma p_n(\zeta) p_m(\zeta) w(\zeta)\, d\zeta = \delta_{n,m}, \qquad n, m = 0, 1, 2, \ldots, \tag{7.98}$$

where $\delta_{n,m}$ is the Kronecker delta. Write $p_n(z) = \gamma_n z^n + \cdots$, and put $\pi_n(z) := p_n(z)/\gamma_n = z^n + \cdots$. So, $\pi_n(z)$ is the monic orthogonal polynomial associated with $w(z)$. The fundamental connection between orthogonal polynomials and the steepest descent method of Deift and Zhou is provided by the following Riemann–Hilbert problem of Fokas, Its & Kitaev (1992): Find a 2×2 matrix-valued function Y satisfying

(Y_1) $Y(z)$ is analytic in $\mathbb{C} \setminus \Gamma$;

(Y_2) $Y_+(\zeta) = Y_-(\zeta) \begin{pmatrix} 1 & w(\zeta) \\ 0 & 1 \end{pmatrix}$ for $\zeta \in \Gamma$;

(Y_3) $Y(z) = \left[I + O\left(\frac{1}{z}\right) \right] \begin{pmatrix} z^n & 0 \\ 0 & z^{-n} \end{pmatrix}$ as $z \to \infty$ in $\mathbb{C} \setminus \Gamma$;

(Y_4) if Γ has an endpoint a, then $\lim_{z \to a}(z - a)Y(z) = 0$.

Here $Y_\pm(\zeta)$ denote the limits of $Y(z)$ as $z \to \zeta \in \Gamma$ from the positive (respectively negative) side of Γ. If Γ has no endpoint, there is no need to impose condition (Y_4).

Theorem 7.3 (Fokas, Its & Kitaev) *The above Riemann–Hilbert problem for Y has the unique solution given by*

$$Y(z) = \begin{pmatrix} \pi_n(z) & C(\pi_n w)(z) \\ c_n \pi_{n-1}(z) & c_n C(\pi_{n-1} w)(z) \end{pmatrix},$$

where π_n and π_{n-1} are the monic orthogonal polynomials of degree n

and $n - 1$, respectively, $C(\pi_j w)(z)$ denotes the Cauchy transform

$$C(\pi_j w)(z) = \frac{1}{2\pi i} \int_\Gamma \frac{\pi_j(\zeta) w(\zeta)}{\zeta - z} \, d\zeta, \qquad j = n - 1, n,$$

and $c_n = -2\pi i \gamma_{n-1}^2, \gamma_n$ being the leading coefficient of the polynomial $p_n(z)$.

As an illustration, we shall take $p_n(z)$ to be the polynomials orthogonal with respect to the exponential weight

$$w(x) = e^{-Q(x)}, \tag{7.99}$$

where $Q(x) = q_{2m} x^{2m} + q_{2m-1} x^{2m-1} + \cdots$ is a polynomial of degree $2m$ with $q_{2m} > 0$, and the curve Γ is the whole real line $\mathbb{R}$.

The Deift–Zhou method of steepest descent consists of a sequence of transformations

$$Y \to U \to T \to S \to R.$$

The first transformation $Y \to U$ is simply a rescaling, and in many applications this step is not necessary. The second transformation $U \to T$ is most crucial. It is a normalization process; i.e., it takes Y to T so that $T(z) \sim I$ as $z \to \infty$. Since condition (Y_3) can be written as $Y(z) = [I + O(1/z)]z^{n\sigma_3}$, where σ_3 is the Pauli matrix $\begin{pmatrix} 1 & 0 \\ 0 & -1 \end{pmatrix}$, normalization can be achieved by setting $T(z) = Y(z)z^{-n\sigma_3}$. But, this simple-minded transformation brings the problem at infinity to the origin. What Deift and his collaborators (1999a, 1999b) have shown is that normalization can be accomplished with the aid of a so-called g-function defined by

$$g(z) = \int \log(z - x) \, d\mu(x), \tag{7.100}$$

where $\mu(x)$ is the equilibrium measure which can be regarded as the limit of the zero distribution of $\pi_n(z)$ as $n \to \infty$. In our case, the support of the measure is the interval $[-1, 1]$ after a rescaling. The transformation T is given by

$$T(z) = e^{\frac{1}{2} n l_n \sigma_3} U(x) e^{-(ng(z) + \frac{1}{2} n l_n)\sigma_3},$$

where l_n is a constant known as the Lagrange constant.

The matrix T now satisfies a Riemann–Hilbert problem on $\mathbb{R}$ with the normalization condition $T(z) \sim I$ as $z \to \infty$, but the jump matrix J_T associated with T is even more complicated, and the Riemann–Hilbert problem for T can not be solved explicitly. However, the jump matrix on $[-1, 1]$ can be factored, and the factorization suggests that through contour deformation, the Riemann–Hilbert problem for T can be transformed into a Riemann–Hilbert problem for S on a system of curves consisting of not only the real line $\mathbb{R}$, but also two arcs connecting -1 and 1, one in the upper half-plane and the other in the lower half-plane. On these two arcs, the jump matrix for S tends to the identity matrix I as $n \to \infty$. On the interval $[-1, 1]$, the jump matrix is a constant. It is this contour deformation that Deift & Zhou (1993) referred to as the steepest descent method for Riemann–Hilbert problems. The limiting Riemann–Hilbert problem can now be solved explicitly. Suppose that its solution is denoted by $N(z)$. Then, formally, through the use of inverse transformations, Y has the asymptotic behavior

$$Y(z) \sim e^{-\frac{1}{2}nl_n \sigma_3} N(z) e^{(ng(z) + \frac{1}{2}nl_n)\sigma_3} \qquad \text{as } n \to \infty. \qquad (7.101)$$

Since the jump matrices for S and N are not uniformly close to each other near the endpoints of the equilibrium measure and other special points such as singularities, the next step in the Deift–Zhou method is to do some local analysis and to construct parametrices near these points. In our example, the endpoints of the equilibrium measure are at $z = \pm 1$, and there is no singularity. In view of the property $p_n(\overline{z}) = \overline{p_n(z)}$ in our case, we may restrict z to the closed upper half-plane $\{z \in \mathbb{C} : \operatorname{Im} z \geq 0\}$. Thus, we need be concerned with only the points $z = -1$ and $z = 1$. If P and $\widetilde{P}$ denote the parametrices of S in neighborhoods of 1 and -1, respectively, then the final transformation $S \to R$ is defined by $R := SP^{-1}$ near $z = 1$ and $R := S\widetilde{P}^{-1}$ near $z = -1$. Outside the neighborhoods of -1 and 1, R is defined by $R = SN^{-1}$. The matrix-valued function $R(z)$ is now analytic in the complex plane, except for the curves on which S has a jump matrix, but the jump matrices of R are close to the identity matrix for large values of n.

A disadvantage of the final step $S \to R$ is that the validity of the results is often established in neighborhoods of special points. For instance, in Deift et al. (1999b) six asymptotic approximations are constructed in six different regions. Here, we wish to indicate that by making a modification of the Deift–Zhou method, much more global results can be obtained; see Wang & Wong (2005b). We continue to use our present example as an illustration. Our method is to guess an asymptotic approximation $\widetilde{U}(z)$, based directly on the asymptotic formula (7.101). The matrix $\widetilde{U}(z)$ has the same jump as $U(z)$ on the real line $\mathbb{R}$, and has the same behavior as $U(z)$ as $z \to \infty$, but it is not analytic on some specified curves denoted by Σ (see Figure 7.2), which divide the whole complex plane into four regions $\Omega_i, i = -1, 0, 1, 2$. We then consider the transformation $U \to R$ defined by

$$R(z) = e^{-\frac{1}{2}nl_n\sigma_3} U(z)\tilde{U}(z)^{-1} e^{\frac{1}{2}nl_n\sigma_3}, \qquad (7.102)$$

and show that (i) it has no jump on $\mathbb{R}$, (ii) it is analytic for $z \in \mathbb{C} \setminus \Sigma$, (iii) it is normalized at infinity, and (iv) its jump matrix J on Σ has a uniform asymptotic expansion of the form

$$J(z) \sim I + \sum_{s=2m}^{\infty} \frac{\Delta_s(z)}{n^{s/2m}} \qquad \text{as } n \to \infty,$$

where

$$\Delta_s(z) = O\left(\frac{1}{z^{2m}}\right), \qquad s = 2m, 2m+1, \ldots, \qquad \text{as } z \to \infty.$$

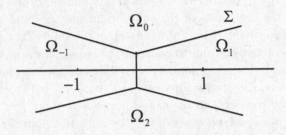

Fig. 7.2. Curve Σ and regions $\Omega_i, i = -1, 0, 1, 2$.

To prove that $R(z)$ has a similar asymptotic expansion, we appeal to a theorem given in Qiu & Wong (2008). Let $\Gamma_1, \ldots, \Gamma_m$ be simple, smooth, and oriented curves in the complex plane $\mathbb{C}$, and let $\Gamma = \Gamma_1 + \cdots + \Gamma_m$ be the union of these curves. Assume that Γ has no endpoint, denote by Γ^0 the interior of Γ (i.e., Γ with points of self-intersection removed), and consider the following matrix RHP for $R : \mathbb{C} \setminus \Gamma \to \mathbb{C}^2 \times \mathbb{C}^2$ with parameter N:

(R_a) $R(z) = R(z, N)$ is analytic in $\mathbb{C} \setminus \Gamma$;
(R_b) $R_+(\zeta) = R_-(\zeta)V(\zeta)$ for $\zeta \in \Gamma^0$;
(R_c) $R(z) = I + o(1)$, as $z \to \infty$, for $z \in \mathbb{C} \setminus \Gamma$ and N fixed;
(R_d) at every point of self-intersection $\zeta_0 \in \Gamma \setminus \Gamma^0$, $\lim_{z \to \zeta_0} (z - \zeta_0)R(z) = 0$ for fixed N.

Suppose that the jump matrix $V(\zeta) = V(\zeta, N)$ satisfies the following conditions:

(V_1) If V^j is the restriction of V to Γ_j^0, i.e., $V^j = V|_{\Gamma_j^0}$, then there exists a $\delta > 0$ such that every V^j has an analytic continuation to $\Gamma_j(\delta)$ for every $j = 1, 2, \ldots, m$, and every $V^j(z)$ is invertible for $z \in \Gamma_j(\delta)$, where $\Gamma_j(\delta) = \{z \in \mathbb{C} : d(z, \Gamma_j) < \delta\}$ is a δ-neighborhood of Γ_j.

(V_2) There exists an $\alpha > 0$ such that if Γ_j is unbounded for some j, then for $z \in \Gamma_j(\delta)$

$$V^j(z) = I + O(1/|z|^\alpha) \qquad \text{as} \quad z \to \infty$$

uniformly with respect to N.

(V_3) Suppose ζ_0 is a point of self-intersection of Γ. Let $\Gamma_{j_1}, \Gamma_{j_2}, \ldots, \Gamma_{j_q}$, $1 \leq j_1, j_2, \ldots, j_q \leq m$, be the branches of Γ which meet at ζ_0. We order them in the counterclockwise direction around ζ_0. All V^{j_1}, V^{j_2}, $\ldots, V^{j_q}$ are well defined and analytic in the disk $U(\zeta_0, \delta) = \{z \in \mathbb{C} : |z - \zeta_0| < \delta\}$. Furthermore, suppose that the *cyclic condition*

$$(V^{j_q})^{\varepsilon_q}(z) \cdot (V^{j_{q-1}})^{\varepsilon_{q-1}}(z) \cdots (V^1)^{\varepsilon_1}(z) = I \quad \text{for} \quad |z - \zeta_0| < \delta$$

holds, where $\varepsilon_i = -1$ if Γ_{j_i} is directed away from ζ_0 and $\varepsilon_i = 1$ if Γ_{j_i} is directed towards ζ_0. When $\varepsilon_i = -1$, $(V^{j_i})^{\varepsilon_i}$ means the inverse of V^{j_i}.

Moreover, we assume that for every $j = 1, 2, \ldots, m$, $V^j(z)$ has an asymptotic expansion of the form

$$V^j(z) \sim I + \sum_{k=1}^{\infty} \frac{V_k^j(z)}{N^k} \qquad \text{as} \quad N \to \infty$$

uniformly for $z \in \Gamma_j(\delta)$, where each $V_k^j(z)$ is also analytic in $\Gamma_j(\delta)$, and that for all large $z \in \Gamma_j(\delta)$,

$$\left\| V^j(z) - I - \sum_{k=1}^{p} \frac{V_k^j(z)}{N^k} \right\| \leq \frac{C_p}{N^{p+1}|z|^\alpha},$$

where $\alpha > 0$ and $\| \cdot \|$ denotes the norm defined by the maximum modulus of all entries in a matrix.

Theorem 7.4 *The solution $R(z) = R(z, N)$ to the RHP $(R_a) - (R_d)$ has an asymptotic expansion of the form*

$$R(z) \sim I + \sum_{k=1}^{\infty} \frac{R_k(z)}{N^k} \qquad as \quad N \to \infty$$

uniformly for $z \in \mathbb{C} \setminus \Gamma$, where the coefficients $R_k(z)$ are all analytic in $\mathbb{C} \setminus \Gamma$.

By using (7.102), one can work backwards to find the uniform asymptotic expansion of $\pi_n(z)$ in each region Ω_i, $i = -1, 0, 1, 2$. In particular, for z belonging to Ω_{-1} and Ω_1, the expansions involve the Airy function Ai and its derivative Ai$'$, while for $z \in \Omega_0 \cup \Omega_2$, the expansions involve only elementary functions. Here, we have taken the opportunity to correct an error in the statement of the theorem given in Wang & Wong (2005b).

References

C.R. Adams (1928), On the irregular cases of linear ordinary difference equations, *Trans. Amer. Math. Soc.* **30**, 507–541.

C.M. Bender and S.A. Orszag (1978), *Advanced Mathematical Methods for Scientists and Engineers*, McGraw-Hill, New York.

M.V. Berry (1969), Uniform approximation: a new concept in wave theory, *Science Progress* (Oxford) **57**, 43–64.

M.V. Berry (1989), Uniform asymptotic smoothing of Stokes' discontinuities, *Proc. Roy. Soc. Lond. A* **422**, 7–21.

M.V. Berry (2005), Tsunami asymptotics, *New Journal of Physics* **7** 129, 1–18.

M.V. Berry and C.J. Howls (1990), Hyperasymptotics, *Proc. Roy. Soc. Lond. A* **430**, 653–668.

M.V. Berry and C.J. Howls (1991), Hyperasymptotics for integrals with saddles, *Proc. Roy. Soc. Lond. A* **434**, 657–675.

G.D. Birkhoff (1911), General theory of linear difference equations, *Trans. Amer. Math. Soc.* **12**, 243–284.

G.D. Birkhoff (1930), Formal theory of irregular linear difference equations, *Acta Math.* **54**, 205–246.

G.D. Birkhoff and W.J. Trjitzinsky (1932), Analytic theory of singular difference equations, *Acta Math.* **60**, 1–89.

P. Bleher and A. Its (1999), Semiclassical asymptotics of orthogonal polynomials, Riemann-Hilbert problem, and universality in the matrix model, *Ann. Math.* **150**, 185–266.

N. Bleistein (1966), Uniform asymptotic expansions of integrals with stationary points near algebraic singularity, *Commun. Pure Appl. Math.* **19**, 353–370.

N. Bleistein (1967), Uniform asymptotic expansions of integrals with many nearby stationary points and algebraic singularities, *J. Math. Mech.* **17**, 533–559.

N. Bleistein and R. A. Handelsman (1975), *Asymptotic Expansions of Integrals*, Holt, Rinehart and Winston, New York. (Reprinted by Dover Publications, New York, 1986.)

R. Bo and R. Wong (1994), Uniform asymptotic expansion of Charlier polynomials, *Methods Appl. Anal.* **1**, 294–313.

V. A. Borovikov (1994), *Uniform Stationary Phase Method*, IEE, London.

J.P. Boyd (1999), The Devil's invention: Asymptotics, superasymptotics and hyperasymptotics, *Acta Appl. Math.* **56**, 1–98.

W.G.C. Boyd (1990), Stieltjes transforms and the Stokes' phenomenon, *Proc. Roy. Soc. Lond. A* **429**, 227–246.

W.G.C. Boyd (1993), Error bounds for the method of steepest descent, *Proc. Roy. Soc. Lond. A* **440**, 493–518.

H.J. Bremermann (1965), *Distributions, Complex Variables, and Fourier Transforms*, Addison-Wesley, Reading, Massachusetts.

G. F. Carrier and C. E. Pearson (1968), *Ordinary Differential Equations*, Blaisdell Pub. Co., Waltham, MA. (Reprinted in SIAM Classics in Applied Mathematics series, Vol. 6, SIAM, Philadelphia, 1991.)

N.A. Cartwright and K.E. Oughstun (2007), Uniform asymptotics applied to ultrawideband pulse propagation, *SIAM Rev.* **49**, 628–648.

C.J. Chapman (1992), The asymptotic theory of rapidly rotating sound fields, *Proc. R. Soc. Lond. A* **436**, 511–526.

C. Chester, B. Friedman and F. Ursell (1957), An extension of the method of steepest descent, *Proc. Camb. Phil. Soc.* **53**, 599–611.

E.T. Copson (1965), *Asymptotic Expansions*, Cambridge Tracts in Math. and Math. Phys. No 55, Cambridge University Press, London.

D. Dai and R. Wong (2008), Global asymptotics for Laguerre polynomials with large negative parameters − a Riemann-Hilbert approach, *The Ramanujan Journal* **16**, 181–209.

P. Deift (1999), *Orthogonal Polynomials and Random Matrices: A Riemann-Hilbert Approach*, Courant Lecture in Mathematics 3, New York University, New York. (Reprinted by Amer. Math. Soc., Providence, R. I., 2000.)

P. Deift and X. Zhou (1993), A steepest descent method for oscillatory Riemann-Hilbert problems, Asymptotics for the mKdV equaiton; *Ann. Math.* **137**, 295–370.

P. Deift and X. Zhou (1994), *Long-Time Behavior of the Non-Focusing Nonlinear Schrödinger Equation: A Case Study*, New Series: Lecture in Math. Sciences, Vol. 5, University of Tokyo.

P. Deift and X. Zhou (1995), Asymptotics for the Painlevé II equation, *Comm. Pure Appl. Math.* **48**, 277–337.

P. Deift, T. Kriecherbauer, K.T.-R. McLaughlin, S. Venakides and X. Zhou (1999a), Uniform asymptotics for polynomials orthogonal with respect to varying exponential weights and applications to universality questions in random matrix theory, *Comm. Pure Appl. Math.* **52**, 1335–1425.

P. Deift, T. Kriecherbauer, K.T.-R. McLaughlin, S. Venakides and X. Zhou (1999b), Strong asymptotics of orthogonal polynomials with respect to exponential weights, *Comm. Pure Appl. Math.* **52**, 1491–1552.

R.B. Dingle and G.J. Morgan (1967a), WKB methods for difference equations I, *Appl. Sci. Res.* **18**, 221–237.

R.B. Dingle and G.J. Morgan (1967b), WKB methods for difference equations II, *Appl. Sci. Res.* **18**, 238–245.

R.B. Dingle (1973), *Asymptotic Expansions: Their Derivations and Interpretation*, Academic Press, New York.

T.M. Dunster (1994), Uniform asymptotic solutions of second-order linear differential equations having a simple pole and a coalescing turning point in the complex plane, *SIAM J. Math. Anal.* **25**, 322–353.

T.M. Dunster (1996), Asymptotic solutions of second-order linear differential equations having almost coalescing turning points, with an application to the incomplete gamma function, *Proc. Roy. Soc. London Ser. A* **452**, 1331–1349.

A. Erdélyi (1962), On a nonlinear boundary value problem involving a small parameter, *J. Australian Math. Soc.* **2**, 425–439.

A. Erdélyi (1968), Approximate solutions of a nonlinear boundary value problem, *Arch. Rational Mech. Anal.* **29**, 1–17.

A.S. Fokas, A.R. Its and A.V. Kitaev (1992), The isomonodromy approach to matrix models in 2D quantum gravity, *Comm. Math. Phys.* **147**, 395–430.

C.L. Frenzen and R. Wong (1988), Uniform asymptotic expansions of Laguerre polynomials, *SIAM J. Math. Anal.* **19**, 1232–1248.

L. Hörmander (1990), *The Analysis of Linear Partial Differential Operators*, 2nd ed. Vol. 1, Springer.

F.A. Howes (1978), Boundary-interior layer interactions in nonlinear singular perturbation theory, *Memoirs Amer. Math. Soc.* **203**, 1–108.

X.-S. Jin and R. Wong (1998), Uniform asymptotic expansions for Meixner polynomials, *Constr. Approx.* **14**, 113–150.

J. Kevorkian and J.D. Cole (1981), *Perturbation Methods in Applied Mathematics*, Springer-Verlag, New York.

T. Kriecherbauer and K.T.-R. McLaughlin (1999), Strong asymptotics of polynomials orthogonal with respect to Freud weights, *Internat. Math. Res. Notices*, no. 6, 299–333.

M.D. Kruskal and H. Segur (1991), Asymptotics beyond all orders in a model of crystal growth, *Stud. Appl. Math.* **85**, 129–181.

A.B. Kuijlaars and K.T.-R. McLaughlin (2001), Riemann-Hilbert analysis for Laguerre polynomials with large negative parameter, *Comput. Methods Funct. Theory* **1**, 205–233.

A.B. Kuijlaars, K.T.-R. McLaughlin, W. Van Assche and M. Vanlessan (2004), The Riemann-Hilbert approach to strong asymptotics for orthogonal polynomials on [−1, 1], *Adv. Math.* **188**, 337–398.

P.A. Lagerstrom (1988), *Matched Asymptotic Expansions: Ideas and Techniques*, Springer-Verlag, New York.

C.G. Lange (1983), On spurious solutions of singular perturbation problems, *Stud. Appl. Math.* **68**, 227–257.

R.E. Langer (1931), On the asymptotic solutions of ordinary differential equations, with an application to the Bessel functions of large order, *Trans. Amer. Math. Soc.* **33**, 23–64.

R.E. Langer (1932), On the asymptotic solutions of differential equations, with an application to the Bessel functions of large complex order, *Trans. Amer. Math. Soc.* **34**, 447–480.

R.E. Langer (1935), On the asymptotic solutions of ordinary differential equations, with reference to the Stokes' phenomenon about a singular point, *Trans. Amer. Math. Soc.* **37**, 397–416.

R.E. Langer (1949), On the asymptotic solutions of ordinary differential equations of the second order, with special reference to a turning point, *Trans. Amer. Math. Soc.* **67**, 461–490.

X. Li and R. Wong (1994), Error bounds for asymptotic expansions of Laplace convolutions, *SIAM J. Math. Anal.* **25**, 1537–1553.

X. Liang and R. Wong, On a Nested Boundary-Layer Problem, *Communications on Pure and Applied Analysis*, to appear.

J.D. Logan (1987), *Applied Mathematics*, John Wiley and Sons, New York.

J.L. López (2000), Asymptotic expansions of symmetric standard elliptic integrals, *SIAM J. Math. Anal.* **31**, 754–775.

J.L. López (2007), Asymptotic expansions of Mellin convolutions by means of analytic continuation, *J. Comput. Appl. Math.* **200**, 628–636.

J.L. López and N.M. Temme (2002), Two-point Taylor expansions of analytic functions, *Studies in Appl. Math.* **109**, 297–311.

A.D. MacGillivray (1997), A method for incorporating transcendentally small terms into the method of matched asymptotic expansions, *Stud. Appl. Math.* **99**, 285–310.

J.P. McClure and R. Wong (1978), Explicit error terms for asymptotic expansions of Stieltjes transforms, *J. Inst. Math. Applic.* **22**, 129–145.

J.P. McClure and R. Wong (1979), Exact remainder for asymptotic expansions of fractional integrals, *J. Inst. Math. Applics.* **23**, 139–147.

R. E. Meyer (1980), Exponential asymptotics, *SIAM Review* **22** , 213–224.

J.A. Murdock (1991), *Perturbation: Theory and Methods*, John Wiley & Sons, New York.

A.H. Nayfeh (1981), *Introduction to Perturbation Techniques*, John Wiley & Sons, New York.

A.B. Olde Daalhuis (2003), *Exponential Asymptotics*, Lecture Notes in Math., v. 1817, Springer, Berlin, 211–244.

A.B. Olde Daalhuis and F.W.J. Olver (1994), Exponentially improved asymptotic solutions of ordinary differential equations. II. Irregular singularities of rank one, *Proc. Roy. Soc. Lond. A* **445**, 39–56.

A.B. Olde Daalhuis and F.W.J. Olver (1995), Hyperasymptotic solutions of second-order linear differential equations. I, *Methods Appl. Anal.* **2**, 173–197.

F.W.J. Olver (1954), The asymptotic solution of linear differential equations of the second order for large values of a parameter, *Philos. Trans. Roy. Soc. London Ser. A* **247**, 307–327.

F.W.J. Olver (1956), The asymptotic solution of linear differential equations of the second order in a domain containing one transition point, *Philos. Trans. Roy. Soc. London Ser. A* **249**, 65–97.

F.W.J. Olver (1958), Uniform asymptotic expansions of solutions of linear second-order differential equations for large values of a parameter, *Philos. Trans. Roy. Soc. London Ser. A* **250**, 479–517.

F.W.J. Olver (1961), Error bounds for the Liouville-Green (or WKB) approximation, *Proc. Cambridge Philos. Soc.* **57**, 790–810.

F.W.J. Olver (1963), Error bounds for first approximations in turning-point problems, *J. Soc. Indust. Appl. Math.* **11**, 748–772.

F.W.J. Olver (1964), Error bounds for asymptotic expansions in turning-point problems, *J. Soc. Indust. Appl. Math.* **12**, 200–214.

F.W.J. Olver (1974), *Asymptotic and Special Functions*, Academic Press, New York. (Reprinted by A.K. Peters, Wellesley, 1997.)

F.W.J. Olver (1975), Second-order linear differential equations with two turning points, *Philos. Trans. Roy. Soc. London Ser. A* **278**, 137–174.

F.W.J. Olver (1976), Improved error bounds for second-order differential equations with two turning points, *J. Res. Nat. Bur. Standards Sect. B* **80B**, 437–440.

F.W.J. Olver (1991a), Uniform, exponentially improved, asymptotic expansions for the generalized exponential integrals, *SIAM J. Math. Anal.* **22**, 1460–1474.

F.W.J. Olver (1991b), Uniform, exponentially improved, asymptotic expansions for the confluent hypergeometric function and other integral transforms, *SIAM J. Math. Anal.* **22**, 1475–1489.

F.W.J. Olver (1993), Exponentially improved asymptotic solutions of ordinary differential equations I, the confluent hypergeometric function, *SIAM J. Math. Anal.* **24**, 756–767.

F.W.J. Olver (2003), Private communication.

F.W.J. Olver and R. Wong (2009), *Asymptotic Approximations*, Chapter 2 in NIST Handbook of Mathematical Functions (F.W.J. Olver, D.W. Lozier, C.W. Clark and R.F. Boisvert, eds.), National Institute of Standards and Technology, Gaithersburg, Maryland (available on the web).

R.E. O'Malley Jr. (1968), Topics in singular perturbations, *Adv. Math.* **2**, 365–470.

R.E. O'Malley Jr. (1991), *Singular Perturbation Methods for Ordinary Differential Equations*, Springer-Verlag, New York.

C.H. Ou and R. Wong (2003), On a two-point boundary value problem with spurious solutions, *Stud. Appl. Math.* **111**, 377–408.

C.H. Ou and R. Wong (2004), Shooting method for nonlinear singularly perturbed boundary-value problems, *Stud. Appl. Math.* **112**, 161–200.

R.B. Paris (2001a), On the use of Hadamard expansions in hyperasymptotic evaluation, I: Real variables, *Proc. Roy. Soc. Lond. A* **457**, 2835–2853.

R.B. Paris (2001b), On the use of Hadamard expansions in hyperasymptotic evaluation, II: Complex variables, *Proc. Roy. Soc. Lond. A* **457**, 2855–2869.

R.B. Paris and A.D. Wood (1995), Stokes phenomenon demystified, *Bull. Inst. Math. Appl.* **31** (1-2), 21–28.

H. Poincaré (1886), Sur les intégrales irrégulières des équations linéaires, *Acta Math.* **8** , 295–344.

T. Poston and I. Stewart (1978), *Catastrophe Theory and Its Applications*, Pitman, Boston, MA.

P.R. Prentice (1994), Time-domain asymptotics. I. General theory for double integrals, *Proc. R. Soc. Lond. A* **466**, 341–360.

W.-Y. Qiu and R. Wong (2000), Uniform asymptotic expansions of a double integral: coalescence of two stationary points, *Proc. Roy. Soc. Lond. A* **456**, 407–431.

W.-Y. Qiu and R. Wong (2008), Asymptotic expansions for Riemann-Hilbert problems, *Analysis and Applications* **6**, 269–298.

M. Rosenlicht (1983), Hardy fields, *J. Math. Anal. Appl.* **93**, 297–311.

H. Segur, S. Tanveer and H. Levine, eds. (1991), *Asymptotics Beyond All Orders*, NATO Advanced Science Institute Series B: Physics, Vol. 284, Plenum, New York.

J.G. Simmonds and J.E. Mann (1986), *A First Look at Perturbation Theory*, Robert E. Krieger Publishing Co., Malabar, Florida.

D. Smith (1985), *Singular-Perturbation Theory*, Cambridge University Press, Cambridge.

C.A. Swanson (1956), Differential equations with singular points, Tech. Rep. No. 16, Dept. of Math., California Inst. of Technol., Pasadena, CA.

G. Szegő (1975), *Orthogonal Polynomials*, Fourth edition, Colloquium Publications, Vol. 23, Amer. Math. Soc., Providence, R. I.

F. Ursell (1960), On Kelvin's ship wave pattern, *J. Fluid Mech.* **8**, 418–431.

F. Ursell (1983), Integrals with a large parameter: Hilbert transforms, *Math. Proc. Camb. Phil. Soc.* **93**, 141–149.

Z. Wang and R. Wong (2002), Uniform asymptotic expansion of $J_\nu(\nu a)$ via a difference equation, *Numer. Math.* **91**, 147–193.

Z. Wang and R. Wong (2003), Asymptotic expansions for second-order linear difference equations with a turning point, *Numer. Math.* **94**, 147–194.

Z. Wang and R. Wong (2005a), Linear difference equations with transition points, *Math. Comp.* **74**, 629–653.

Z. Wang and R. Wong (2005b), Uniform asymptotics for orthogonal polynomials with exponential weights – the Riemann–Hilbert approach, *Stud. Appl. Math.* **115**, 139–155.

W. Wasow (1956), Singular perturbations of boundary value problems for nonlinear differential equations of the second order, *Comm. Pure Appl. Math.* **9**, 93–113.

W. Wasow (1965), *Asymptotic Expansions for Ordinary Differential Equations*, Wiley-Interscience, New York.

R. Wong (1979), Explicit error terms for asymptotic expansions of Mellin convolutions, *J. Math. Anal. Appl.* **72**, 740–756.

R. Wong (1980), Error bounds for asymptotic expansions of integrals, *SIAM Review* **22**, 401–435.

R. Wong (1989), *Asymptotic Approximations of Integrals*, Academic Press, Boston, MA. (Reprinted by SIAM, Philadelphia, PA, 2001.)

R. Wong (2001), Exponential asymptotics, in *Special Functions 2000: Current Perspective and Future Directions*, J. Bustoz et al. (eds.), NATO Science Series II, Vol. 30, Kluwer, 505–518.

R. Wong and H. Li (1992a), Asymptotic expansions for second-order linear difference equations, *J. Comput. & Appl. Math.* **41**, 65–94.

R. Wong and H. Li (1992b), Asymptotic expansions for second-order linear difference equations II, *Studies Appl. Math.* **87**, 289–324.

R. Wong and J.P. McClure (1984), Generalized Mellin convolutions and their asymptotic expansions, *Can. J. Math.* **36**, 924–960.

R. Wong and H. Yang (2002), On a boundary-layer problem, *Stud. Appl. Math.* **108**, 369–398.

R. Wong and W.-J. Zhang (2006), Uniform asymptotics for Jacobi polynomials with varying large negative parameters – a Riemann–Hilbert approach, *Trans. Amer. Math. Soc.* **358**, 2663–2694.

R. Wong and Y.-Q. Zhao (1999), Smoothing of Stokes discontinuity for the generalized Bessel function II, *Proc. Roy. Soc. Lond. Ser. A* **455**, 3065–3084.

R. Wong and Y.-Q. Zhao (2002), Exponential asymptotics of the Mittag-Leffler function, *Constr. Approx.* **18**, 355–385.

R. Wong and Y.-Q. Zhao (2003), Estimates for the error term in a uniform asymptotic expansion of the Jacobi polynomials, *Analysis and Applications* **1**, 213–241.

R. Wong and Y. Zhao (2008), On the number of solutions to Carrier's problem, *Stud. Appl. Math.* **120**, 213–245.

8

Tractability of Multivariate Problems

Henryk Woźniakowski

Department of Computer Science
Columbia University
New York, NY 10027, USA, and
Institute of Applied Mathematics
University of Warsaw
ul. Banacha 2, 02-097 Warsaw, Poland
Email: henryk@cs.columbia.edu

Abstract

In this paper I present a history of tractability of continuous problems, which has its beginning in the successful numerical tests for high-dimensional integration of finance problems. Tractability results will be illustrated for two multivariate problems, integration and linear tensor products problems, in the worst case setting. My talk at FoCM'08 in Hong Kong and this paper are based on the book *Tractability of Multivariate Problems*, written jointly with Erich Novak. The first volume of our book has been recently published by the European Mathematical Society.

8.1 Introduction

Many people have recently become interested in studying the tractability of continuous problems. This area of research addresses the computational complexity of multivariate problems defined on spaces of functions of d variables, with d that can be in the hundreds or thousands; in fact, d can even be arbitrarily large. Such problems occur in numerous applications including physics, chemistry, finance, economics, and the computational sciences.

As with all problems arising in information-based complexity, we want to solve multivariate problems to within ε, using algorithms that use finitely many functions values or values of some linear functionals. Let $n(\varepsilon, d)$ be the minimal number of function values or linear functionals that is needed to compute the solution of the d-variate problem to within ε.

For many multivariate problems defined over standard spaces of functions $n(\varepsilon, d)$ is exponentially large in d. This phenomenon is called the *curse of dimensionality*, after Richard Bellman, who coined this phrase in 1957.

What can we do with such intractable problems? Is there some way that we can vanquish the curse of dimensionality? Since $n(\varepsilon, d)$ is already defined for the best algorithm possible, it is impossible to find a cleverer algorithm. The only way is to change our problem, either by shrinking the class of functions, or by switching to a more lenient error setting, in which we redefine what we mean by "to within ε".

The first option means that our functions must satisfy additional "non-standard" properties. What do we mean by this? For many standard classes of functions, the dependence on all variables and groups of variables is the same. In contrast, for many practical computational problems in which d is very large, it is more likely to expect that the dependence of functions on successive variables or groups of variables may vary significantly:

- For many applications in computational finance, the functions depend on successive variables in a diminishing way, due to the discounted value of money, see e.g., Traub, Werschulz (1998) and references cited there.
- In computational physics, the influence of neighboring particles is significant, and functions can be represented or well-approximated by sums of functions of a few variables. In particular, this holds for Coulomb pair potentials, see e.g., Glimm, Jaffe (1987).
- In computational chemistry, there is the need to construct poly-atomic potential energy surfaces that underlie molecular dynamics and spectroscopies. It is observed that functions depending on many variables often can be well approximated by a sum of functions that depend on only few variables, see Ho, Rabitz (2003) and Rabitz, Alis (1999).
- These kinds of functions also appear in optimization; they are called "partially separable", see Griewank, Toint (1982).
- In computational economics, the variables may be moderated by the Cobb-Douglas condition, which guarantees equal partitioning of goods; this condition is used for the Bellman fixed point problem, see Rust, Traub, Woźniakowski (2002).

All these examples suggest to study *weighted* spaces, in which the importance of successive variables and groups of variables is moderated by weights. It is then natural to seek necessary and sufficient conditions on the weights to vanquish the curse of dimensionality. We believe the reason that so many high-dimensional problems are solved efficiently in computational practice is that these problems belong to weighted spaces with weights satisfying the appropriate conditions; for more details see Novak, Woźniakowski (2008).

The second option is to relax the notion of what we mean by "to within ε". Suppose that we have established the curse of dimensionality in the worst case setting, in which we demand that the error is at most ε for *all* functions from a given set. This worst case assurance is very strong. If we are willing to settle for a weaker assurance (given by, e.g., the randomized, probabilistic or average case setting), we might be able to vanquish the curse. To be specific, let us focus on the randomized setting. We now allow randomized algorithms, and we want to guarantee that the *expected* error for all functions from the same set as in the worst case setting is at most ε. The standard example is Monte Carlo for multivariate integration, which may indeed vanquish the curse. Suppose that our class of integrands consists of Lipschitz functions with Lipschitz constant 1. It is known that $n(\varepsilon, d)$ is proportional to ε^{-d} in the worst case setting; however, $n(\varepsilon, d)$ is proportional to at most ε^{-2} in the randomized setting. So the curse of dimensionality that is present in the worst case setting has been vanquished by switching to the randomized setting. We stress that switching to more lenient settings does not always help, and there are problems for which the curse of dimensionality is present in all the settings mentioned above. Characterizing those multivariate problems for which the worst case curse of dimensionality can be vanquished by switching to (e.g.) the randomized setting is a major open problem. The reader is referred to Novak, Woźniakowski (2008) for a more detailed discussion.

This second option of switching to a more lenient setting is beyond the scope of this paper. So, we will only consider here the first option of shrinking the class of functions by studying weighted spaces.

The main goal when studying the tractability of a given multivariate problem is to find classes of functions for which the curse of dimensionality is *not* present. More precisely, for a given multivariate problem, we want to identify for which classes of functions and for which error settings the problem does not suffer from the curse of dimensionality, meaning that the minimal number $n(\varepsilon, d)$ is *not* exponential in ε^{-1} and d. Since there are many different ways of measuring the lack of exponential behavior, we have different types of tractability. In this paper we restrict ourselves to:

- *polynomial tractability*, for which we want to prove that $n(\varepsilon, d)$ can be polynomially bounded in ε^{-1} and d,
- *strong polynomial tractability*, for which we want to prove that $n(\varepsilon, d)$ can be bounded by a polynomial in ε^{-1} independent of d, and
- *weak tractability*, for which we want to prove that $\ln n(\varepsilon, d)$ goes to zero as $\varepsilon^{-1} + d$ approaches infinity.

Today there are many tractability results for continuous problems, and the book *Tractability of Multivariate Problems*, written jointly with Erich Novak, summarizes the state of the art for tractability studies. The first volume is devoted to algorithms using arbitrary linear functionals and has been recently published by the European Mathematical Society. The second volume is devoted to algorithms using function values for approximation of linear functionals. The primary example of such a problem is multivariate integration. Finally, the third volume will be devoted to algorithms using function values. It will mainly deal with the approximation of general linear operators; it will also cover the approximation of certain nonlinear operators. We are now finishing the second volume, and we expect it to be published in late 2009. The third volume is at a more preliminary stage; we hope to complete this volume by the end of 2009, so that it will be published in 2010.

When I was invited to present a plenary talk during FoCM'08 in Hong Kong, it was quite clear to me that I should talk about tractability of continuous problems, since this has been my main research topic for the last 15 years. However, I faced the difficult problem of presenting the vast body of tractability results without boring the listeners of my talk too much. (I now face the same problem regarding the readers of this paper.)

I decided to report on how tractability of continuous problems was born. I do hope that the listeners of my talk were and the readers of my paper will be satisfied with my decision. After all, that is exactly the way we do research. First we ask questions about why something holds or can be true; in time, we gradually discover partial answers; this process usually goes forever. So I decided to describe the tractability process from the very beginning (the early 1990s) and to report gradual progress till today. I want to stress that this process, by all means, is *not* yet finished, and there are still plenty of open tractability problems. In fact, we list 30 open problems in Volume I of our book. I am pleased to report that two of these problems have already been solved by Stefan Heinrich, Heinrich (2008a, 2008b). I hope that the readers of this paper will decide to work on open tractability problems, especially on the problem related to financial mathematics that is presented at the end of this paper.

I finished my talk in Hong Kong with a joke by defining a tractability number. I want to repeat it here, since otherwise it will be forgotten. So the point, my dear reader, is to determine your tractability number which is defined as follows.

- If you publish k "good" tractability papers your tractability number is k. (Your k can be zero but, in any case, everyone now has a well-defined tractability number.)
- If you solve p open tractability problems from Volumes I–III of our book, then you can add $2p$ to your tractability number. (Yes, it is unfair to single out our book, but after all, I really want *you* to solve these open problems.)
- If you publish s "good" papers with someone whose tractability number is t, and t is positive, then you add

$$\frac{s}{1+s}\frac{t}{1+t}$$

  to your tractability number. (Yes, you can use this bullet item many times, as long as s is positive and you published with many people having positive t.)

I think the reader would agree that it is not trivial to determine the tractability number. In fact, the problem of determining tractability number is probably *intractable* since the concept of a "good" paper is open for discussion. The reader is asked to compute her/his tractability number under the simplifying assumption that all published papers are good. Needless to say, the larger your tractability number the better. Please let me know if your tractability number is at least 10. Although it seems impossible, you should start worrying if your tractability number is still zero.

8.2 A Little History

In this section I wish to report my (probably) quite subjective story of how tractability of continuous problems was born in the early 1990s. The story is a little unusual, because it started from successful computational experiments that could not have been explained by the current state of the theory. Usually, the situation is the reverse. Typically, we have new algorithms based on a new and promising theory, and we hope to solve many practical problems using these new algorithms. In many cases the results of the new algorithms are mixed and at best they are only partially successful. Our case is different but let me describe it without further ado.

In 1991 I published a paper "Average case complexity of multivariate integration", Woźniakowski (1991), solving a long-standing problem about optimal selection of sample points for multivariate integration

$$I_d(f) = \int_{[0,1]^d} f(t)\, dt$$

for the space of continuous functions equipped with the Wiener sheet measure. The solution was possible by showing that this problem is equivalent to minimizing L_2-discrepancy in the worst case setting. I was lucky because the L_2-discrepancy problem had already been solved by Roth, see Roth (1954, 1980), who provided sharp lower and upper bounds, and by Frolov, see Frolov (1980), who independently provided sharp upper bounds. My paper got some publicity and many people have probably studied it.

I believe that many practitioners realized at that time that multivariate integration can be solved not only by Monte Carlo but also by Quasi-Monte Carlo (QMC) algorithms that were mentioned in my paper. Let me add that a QMC algorithm is a particularly simple algorithm of the form

$$\mathrm{QMC}_n(f) = \frac{1}{n} \sum_{j=1}^{n} f(t_j),$$

with deterministically chosen sample points $t_1, t_2, \ldots, t_n$ from $[0,1]^d$. The name "QMC" was coined to stress the similarity with Monte Carlo (MC), which formally looks exactly the same; however, for MC the sample points $t_1, t_2, \ldots, t_n$ are chosen as independent random points uniformly distributed over $[0,1]^d$. The first QMC algorithms were designed already in the 1960s but the knowledge about their existence was, I believe, quite limited among practitioners up to the early 1990s.

Choosing good sample points for QMC is an art. The choice is studied in the context of a certain Sobolev space, which will be defined later, and the construction of a particular set of sample points usually bears the name of a given author who both proposed the construction and proved an almost optimal worst case error bound. To name a few examples, we currently have Faure, Halton, Hammersley, Niederreiter, Niederreiter-Xing, Sobol, Tezuka, (t, m, s)-points, and shifted lattice points. In my paper I mentioned points used by Roth, which were shifted Hammersley points. All these points are called *low discrepancy* points, to stress the fact that they nearly minimize L_2-discrepancy (at least within a logarithmic factor).

People started to use QMC without bothering to check all the assumptions under which these points should lead to small error bounds. The majority of these people were from banks; they used low discrepancy points for finance applications with large d. In the first examples they used Halton, Sobol and modified Faure and Sobol points. I got a couple of phone calls from people from various banks. Surprisingly enough, they reported very good empirical results. I was initially quite skeptical about these results. I always asked them how come they knew that the results were good. They tested finance problems without knowing the exact value of an integral, and so it was not clear whether the computed result was good. They all gave me the same answer, which finally convinced me that something interesting was really happening. Namely, their answer was that over many years, they had performed many MC tests; they were now comparing QMC and MC results, noticing that the results obtained by QMC with n points were roughly equivalent to MC with n^2 points. Since MC (stochastically) converges like $1/\sqrt{n}$, this was a strong indication that the errors of QMC in these numerical tests were proportional to $1/n$, i.e., that

$$I_d(f) - \mathrm{QMC}_n(f) \approx \frac{1}{n}.$$

Furthermore, the n used in these tests was not particularly large; as we shall see in a moment, this was quite astonishing and seemed to contradict the existing theory. They reported that n was about a thousand or so and that d was 360.

At that time, Joseph Traub and I decided to test these finance problems at Columbia University by ourselves. Whenever two professors are in charge of a problem, it almost always means that the tests will be done by their graduate students. We asked Spassimir Paskov, then a PhD student in Computer Science, to be involved in this project. Paskov got a contact with a major bank in New York, and after a while he was testing financial instruments such as collateralized mortgage obligations. This problem models 30-year mortgages whose values depend, in particular, on future interest rates. Permitting monthly re-financing, we need to evaluate an integral depending on $30 \times 12 = 360$ (months) variables. In other words, $d = 360$ for this problem. It is important to add that one function value of such a problem required some 10^5 floating point operations. Spassimir used Halton and Sobol points and got also very positive results, always beating MC significantly. These empirical results were then published in Paskov, Traub (1995). Further tests using generalized Faure points were reported in Papageorgiou , Traub (1996). QMC once again beat MC considerably. More references can be found in Traub, Werschulz (1998).

A software package called FINDER for computing high-dimensional integrals has been built at Columbia University. FINDER can be obtained free of charge for academic use from Columbia University.

Let me again stress that these surprising tests in the early 1990s were done for $d = 360$; more recent financial tests have been done for $d = 9125$ by Frances Kuo and Ben Waterhouse in Australia. Their problem involves a number of coupon payments each year; the amount of each payment depends on how many stocks still remain above some barrier. The number 9125 comes as the product of 5 stocks considered for 5 years, each with 365 days per year. Details can be found in Giles, Kuo, Sloan, Waterhouse (2008). It is also interesting to note that Christiane Lemieux and Pierre L'Ecuyer, see Lemieux, L'Ecuyer (1999), have run tests with $d = 8000$; in her PhD thesis Christiane Lemieux, see Lemieux (2000), even used tests with $d = 40000$. In both cases they used simple and randomly shifted Korobov rules. All these tests are quite successful showing superiority of QMC over MC.

Given so many positive empirical results, it was quite natural to ask why these numerical tests are so good. The existing theory (which we report in the next section) seemed to indicate that the numerical results should be much worse, and so we were surprised by how good these numerical results were. Thus, the theory faced the challenge of explaining this computational phenomenon. And this was the beginning of tractability studies for continuous problems.

8.3 Challenge to Theory

If we want to understand why we were surprised that the numerical tests were so good, we need to discuss the theoretical background of QMC up to the early 1990s. The typical space of functions for which QMCs are studied is a Sobolev space of real functions defined on $[0,1]^d$. This is a reproducing kernel Hilbert space $H(K_d)$ whose kernel is

$$K_d(x,y) = \prod_{j=1}^{d} \left(1 + \min(x_j, y_j)\right),$$

where $x = [x_1, x_2, \ldots, x_d]$ and $y = [y_1, y_2, \ldots, y_d]$ are from $[0,1]^d$.

This space is a tensor product of univariate spaces of absolutely continuous functions for which the first derivatives are square integrable. The inner product in $H(K_d)$ is given by

$$\langle f, g \rangle_{H(K_d)} = \sum_{\mathfrak{u} \subseteq [d]} \int_{[0,1]^{|\mathfrak{u}|}} \frac{\partial^{|\mathfrak{u}|}}{\partial x_{\mathfrak{u}}} f(x_{\mathfrak{u}}, 0) \frac{\partial^{|\mathfrak{u}|}}{\partial x_{\mathfrak{u}}} g(x_{\mathfrak{u}}, 0) \, dx_{\mathfrak{u}}$$

for all $f, g \in H(K_d)$. Here, $\mathfrak{u}$ is an arbitrary subset of the index set

$$[d] := \{1, 2, \ldots, d\},$$

and by $x_{\mathfrak{u}}$ we mean the vector in $\mathbb{R}^{|\mathfrak{u}|}$ obtained from x by removing the components not in $\mathfrak{u}$. For instance, if $d \geq 7$ and $\mathfrak{u} = \{2, 5, 7\}$ then $x_{\mathfrak{u}} = [x_2, x_5, x_7]$. The vector $(x_{\mathfrak{u}}, 0)$ is a d-dimensional vector with components x_j if $j \in \mathfrak{u}$ and 0 if $j \notin \mathfrak{u}$. For our example with $d = 8$, we have $(x_{\mathfrak{u}}, 0) = [0, x_2, 0, 0, x_5, 0, x_7, 0]$. Clearly, $\partial x_{\mathfrak{u}}$ means that we differentiate once with respect to x_j for each $j \in \mathfrak{u}$. For $\mathfrak{u} = \emptyset$, the integral over an empty set should be interpreted as $f(0)g(0)$.

From Roth (1980) and Frolov (1980), we know that there are QMC algorithms such that for all $f \in H(K_d)$ we have

$$I_d(f) - \text{QMC}_n(f) = \mathcal{O}\left(\frac{[\ln n]^{(d-1)/2}}{n} \|f\|_{H(K_d)}\right), \qquad (8.1)$$

where the factor in the big $\mathcal{O}$ notation is independent of n and f, but it depends on d.

The result (8.1) tells us that for *any* fixed d, the error decreases almost like n^{-1}, asymptotically in n. However, the exponent of $\ln n$ depends linearly on d. As long as d is relatively small, this logarithmic factor does not matter very much. A rule of thumb up to the early 1990s was that as long as (say) $d \leq 12$, the power of $\ln n$ does not matter, and so the QMC enjoys linear convergence, proportional roughly to n^{-1}.

But what happens if d is large? Remember $d = 360$ was used in the numerical tests. Observe that

$$\frac{[\ln n]^{(d-1)/2}}{n}$$

is increasing for $n \leq e^{(d-1)/2}$, and $d = 360$ yields

$$e^{(d-1)/2} \approx 9 \cdot 10^{77}.$$

Hence, for $n \leq 9 \cdot 10^{77}$ the bound (8.1) is completely useless when $d = 360$. It is then natural to hope that maybe (8.1) is *not* sharp. But no, this bound *is* sharp. To show this, we need to explain what we mean by best or minimal error algorithms.

Let us take a step back and ask how the integrals $I_d(f)$ can be computed for f from the space $H(K_d)$. Clearly, the norm of f must affect the error, so let us scale the problem by assuming (say) that $\|f\|_{H(K_d)} \leq 1$. The basic assumption is that we can compute function values at arbitrary sample points from $[0,1]^d$. The sample points t_j may be chosen adaptively, i.e., t_j may depend on the information $f(t_1), f(t_2), \ldots, f(t_{j-1})$ already computed. If we have enough function values, say, $f(t_1), f(t_2), \ldots, f(t_n)$, then we combine them in an arbitrary way $\varphi_n(f(t_1), f(t_2), \ldots, f(t_n))$ and it is our approximation to $I_d(f)$. Here, the mapping $\varphi_n : \mathbb{R}^n \to \mathbb{R}$ can be arbitrary; in particular, it can be nonlinear; the details can be found in Traub, Wasilkowski, Woźniakowski (1988).

As proved in Bahkvalov (1971), adaption does not help for this problem. Furthermore, without loss of generality, we can assume that φ_n is linear, as proved by Smolayk in his PhD thesis in 1965 (see the same paper of Bahkvalov (1971), where Smolyak's result was published for the first time). The proofs of these results, as well as their extensions to linear operators, can be found in Novak, Woźniakowski (2008), and Traub, Wasilkowski, Woźniakowski (1988).

The results of Bahkvalov and Smolyak mean that, without loss of generality, we can consider linear algorithms of the form

$$A_{n,d}(f) = \sum_{j=1}^{n} a_j f(t_j)$$

for some coefficients $a_1, a_2, \ldots, a_n$ and sample points $t_1, t_2, \ldots, t_n$ that can depend on n and d but are independent of f.

We stress that, in general, the choice $a_j = 1/n$ (as in QMC algorithms) is not good. In fact, there are a number of results showing that the choice $a_j = 1/n$ can be quite bad for some spaces other than the space $H(K_d)$ considered here. This subject is beyond the scope of this paper and we refer the reader to Novak, Woźniakowski (2009), and Sloan, Woźniakowski (2001).

We now define the error of $A_{n,d}$ in the worst case setting as

$$e^{\mathrm{wor}}(A_{n,d}) = \sup_{\|f\|_{H(K_d)} \leq 1} |I_d(f) - A_{n,d}(f)|.$$

Obviously the error depends on how well we choose the coefficients and the sample points. So let

$$e^{\mathrm{wor}}(n,d) = \inf_{A_{n,d}} e^{\mathrm{wor}}(A_{n,d})$$

denote the nth minimal worst case error that can be achieved by the best choice of $\{a_j\}$ and $\{t_j\}$, i.e., by the best choice of a linear algorithm $A_{n,d}$.

We are now ready to explain why (8.1) is sharp. Namely, we have

$$e^{\mathrm{wor}}(n,d) = \Theta\left(\frac{[\ln n]^{(d-1)/2}}{n}\right), \tag{8.2}$$

with the factors in the Θ-notation independent of n but depending on d. This last bound is implied by the following results:

- The lower bound was proved in Roth (1954) assuming the QMC case, that is $a_j = 1/n$, and extended in Chen (1985, 1987) to arbitrary a_j.
- The upper bound was proved independently in Roth (1980) and Frolov (1980); the proofs are non-constructive. The first construction of the sample points t_j with $a_j = 1/n$ achieving the upper bound was done in Chen, Skriganov (2002).

Hence, there are some QMC algorithms whose worst case error (asymptotically, as $n \to \infty$) is the best possible for any d. Still, this nice asymptotic behavior only holds if n is exponentially large in d. For large d, such as $d = 360$, there is no way to compute so many function values. Furthermore, the values of n used in the numerical tests were not so large. In many cases, n was at most roughly 1000.

We hope that the reader is starting to share our excitement and surprise as to why the numerical tests were *so* good for such *small* n and such *large* d.

At this point I think it is obvious that we need to know more about the dependence on d. In particular it is compelling to uncover what is hidden in the Θ-notation in (8.2) or in the $\mathcal{O}$-notation in (8.1).

First of all, let me convince the reader that we may have two very different situations depending on the factors in, say, the Θ-notation in (8.2). The first scenario is optimistic. Assume we can show that the factors are bounded from above by $1/d!$, say. Since $x^k/k! \leq e^x$, we would find that the right hand of (8.2) is bounded by

$$\frac{e^{\sqrt{\ln n}}}{n} = o\left(n^{-(1-\delta)}\right) \quad \text{for all} \quad \delta > 0.$$

Hence, we would have an error estimate independent of d and tending to zero almost like n^{-1}. This could indeed explain the success of the tests.

On the other hand, consider the second scenario, which is pessimistic. Assume that we can show that the factors are bounded from below by 1, say. Then we would have to wait exponentially long for good asymptotic behavior, and the success of the tests would still be a mystery.

In any case, we have a new goal in sight: studying the dependence on d in the error bounds. This will make us one step closer to tractability.

8.4 Dependence on d

In the previous section we defined the nth minimal error. When $n = 0$, we do not sample the function f at all, and so the only linear algorithm is $A_{0,d} = 0$. Thus

$$e^{\text{wor}}(0, d) = \sup_{\|f\|_{H(K_d)} \le 1} |I_d(f)| = \|I_d\|$$

is equal to the operator norm of I_d. We call $e^{\text{wor}}(0, d)$ the *initial* error, since it only depends on the formulation of the problem.

Since I_d is a continuous linear functional over $H(K_d)$, by Riesz's theorem it must be of the form

$$I_d(f) = \langle f, h_d \rangle_{H(K_d)} \quad \text{for all} \quad f \in H(K_d)$$

for some $h_d \in H(K_d)$. It is easy to check that

$$h_d(x) = \int_{[0,1]^d} K_d(x, t) \, \mathrm{d}t = \prod_{j=1}^{d} \left(1 + x_j - \tfrac{1}{2} x_j^2\right) \quad \text{for all} \quad x \in [0,1]^d.$$

Furthermore,

$$\|I_d\| = \|h_d\|_{H(K_d)} = \left(\tfrac{4}{3}\right)^{d/2}.$$

This is the first time we see something strange for large d. Namely, the operator norm of multivariate integration is exponentially large in d. Again, as long as d is small, this does not matter. But if $d = 360$ then we want to approximate a linear functional whose norm is

$$\left(\tfrac{4}{3}\right)^{180} \approx 10^{22.488\cdots}.$$

This may be an indication that the problem is not properly scaled. We
will return to this point later but now let us continue, despite the fact
that the norm of I_d is huge.

We wish to decrease the initial error. More precisely, let $\varepsilon \in (0, 1)$
be given. We want to find an algorithm whose worst case error is at
most $\varepsilon \|I_d\|$. Since I_d is so huge for large d, the reader may think that
ε should be appropriately small. But, as we shall see in a moment, we
will get into trouble even if we take any ε smaller than (say) $\frac{1}{2}$ and not
necessarily small. Let

$$n^{\mathrm{wor}}(\varepsilon, d) = \min\{\, n \mid e^{\mathrm{wor}}(n, d) \le \varepsilon \|I_d\| \,\}$$

denote the minimal number of function values that is needed to reduce
the initial error by a factor ε. It is called the *information complexity*. For
multivariate integration, the information complexity multiplied by the
cost of one function value is almost the same as the (total) complexity.
Indeed, to solve the problem we need to compute $n^{\mathrm{wor}}(\varepsilon, d)$ function
values, and since a linear algorithm is optimal, it is then enough to
perform at most $n^{\mathrm{wor}}(\varepsilon, d)$ multiplications and $n^{\mathrm{wor}}(\varepsilon, d) - 1$ additions.
Since the cost of one multiplication and addition is usually much less
than the cost of one function value, we see that the total complexity is
indeed proportional to $n^{\mathrm{wor}}(\varepsilon, d)$.

It is also of interest to study the minimal number of function values
needed to reduce the initial error by QMC algorithms, that is, when the
choice of $a_j = 1/n$ is fixed and we can choose sample points t_j arbitrarily.
This number will be denoted by $n^{\mathrm{wor\text{-}QMC}}(\varepsilon, d)$.

From (8.1) and (8.2), we conclude that $n^{\mathrm{wor\text{-}QMC}}(\varepsilon, d)$ and $n^{\mathrm{wor}}(\varepsilon, d)$
are of the same order and that

$$n^{\mathrm{wor}}(\varepsilon, d) = \Theta\left(\frac{[\ln \varepsilon^{-1}]^{(d-1)/2}}{\varepsilon} \right),$$

with the factors in the Θ notation independent of ε^{-1} but depending
on d. For small d, all looks fine. As before, the question is what happens
for large d. In particular, we ask two basic questions that will lead us
to tractability studies:

- how does $n^{\mathrm{wor}}(\varepsilon, d)$ depend on d?
- do we have the curse of dimensionality, i.e., does $n^{\mathrm{wor}}(\varepsilon, d)$ depend
  exponentially on d?

We now report what is known about specific bounds on $n^{\mathrm{wor\text{-}QMC}}(\varepsilon, d)$

and $n^{\mathrm{wor}}(\varepsilon, d)$. We begin with upper bounds. Using a standard averaging argument for the worst case error, which will be given later, it is easy to show that

$$n^{\mathrm{wor\text{-}QMC}}(\varepsilon, d) \leq \left(\tfrac{9}{8}\right)^d \varepsilon^{-2} = (1.125)^d \, \varepsilon^{-2}.$$

Hence, the right hand depends exponentially on d although for small d the function $\left(\tfrac{9}{8}\right)^d$ behaves innocently. Indeed, even for $d = 50$ we obtain $\left(\tfrac{9}{8}\right)^{50} \approx 361$.

For general linear algorithms we have a slightly better upper bound that follows from Plaskota, Wasilkowski, Zhao (2008). Namely

$$n^{\mathrm{wor}}(\varepsilon, d) \leq \left(3 - \tfrac{4}{3}\sqrt{2}\right)^d \varepsilon^{-2} = (1.1143\dots)^d \, \varepsilon^{-2}.$$

Again we have an exponential dependence in d but for small d it is harmless.

We now switch to lower bounds. For QMC algorithms, it was shown in Sloan, Woźniakowski (1998) that

$$n^{\mathrm{wor\text{-}QMC}}(\varepsilon, d) \geq (1.055)^d (1 - \varepsilon^2).$$

The last estimate is only of interest for large d, since it says that the curse of dimensionality is indeed present. Again, for small d the bound is harmless, but for $d = 360$ we already have

$$(1.055)^d \geq 2 \cdot 10^8.$$

This tells us that QMC algorithms suffer from the curse of dimensionality. Do general algorithms also suffer from this curse? Unfortunately, the answer is *yes*. It was proved in Novak, Woźniakowski (2001), see also Novak, Woźniakowski (2009), that

$$n^{\mathrm{wor}}(\varepsilon, d) \geq (1.0202)^d (1 - \varepsilon^2).$$

These results imply that for large d, there is no way to guarantee good results for small n. We stress that the curse is present even if ε is quite large. For instance, we can take $\varepsilon = \tfrac{1}{2}$. Then d is the troublemaker; we must use exponentially many function values to solve the problem for large d.

From our point of view, these theoretical results do *not* explain the success of the numerical tests; and the mystery is still present.

8.5 Searching for Additional Properties

We can interpret the results from the previous sections as stating that the standard space $H(K_d)$ used for studying QMC algorithms is simply too large. Why? We may give two reasons:

- The norm of multivariate integration is exponentially large in d.
- The curse of dimensionality is present for information complexity and contradicts the numerical evidence.

Let us pause for a moment in order to find a remedy for this puzzling dilemma.

I would like to use an analogy with solving linear systems $Ax = b$ for an $n \times n$ non-singular matrix A. If n is relatively small, say $n \leq 50$, then it is natural to consider solving this problem over the class of *all* non-singular matrices. On the other hand, if n is large then we usually shrink the class of all non-singular matrices by exploiting some additional properties of matrices. For instance, we may switch to a subclass of *sparse* matrices or to a subclass of matrices whose coefficients are generated by a *few* parameters.

Back to our multivariate integration: If d is relatively small then we may consider *all* functions from the unit ball of $H(K_d)$. However, if d is large, then it may be quite natural to restrict the class $H(K_d)$ and switch to an appropriately-chosen subclass of $H(K_d)$. Of course, this subclass should be related to functions occurring in computational practice (in particular, to functions used in the numerical tests for finance).

The space $H(K_d)$ as well as many other standard spaces of functions are *isotropic* in the sense that all variables and groups of variables are equally important. That is, if $f \in H(K_d)$ and we define

$$g(x) = f(x_{j_1}, x_{j_2}, \ldots, x_{j_d})$$

for any permutation of variables $(j_1, j_2, \ldots, j_d)$ then $g \in H(K_d)$ and $\|g\|_{H(K_d)} = \|f\|_{H(K_d)}$.

Is this really the case if d is large for f occurring in computational practice? This condition does *not* hold for functions from finance applications since the initial variables are "more important" than further variables, due to the discounted value of money. As already mentioned in the introduction, this condition does *not* hold for functions that are equal to or well-approximated by sums of functions of a few variables, and such functions occur in many applications,

Hence, there is probably a hidden *structure* of functions for large d. Hopefully, this structure may allow us to vanquish the curse of dimensionality and finally explain the good numerical tests.

This leads us to *weighted* spaces, which will be defined in a moment. Before we do so, we wish to add that the concept of weighted spaces began in Sloan, Woźniakowski (1998) and gradually was refined to consider more general cases of weighted spaces. The reader is referred to Novak, Woźniakowski (2008) for history and particular results for weighted spaces.

To motivate the derivation of weighted spaces, let us return to the reproducing kernel from Section 3, which we rewrite as

$$K_d(x, y) = \sum_{\mathfrak{u} \subseteq [d]} K_\mathfrak{u}(x_\mathfrak{u}, y_\mathfrak{u}),$$

where

$$K_\mathfrak{u}(x_\mathfrak{u}, y_\mathfrak{u}) = \prod_{j \in \mathfrak{u}} \min(x_j, y_j).$$

Note that $K_\mathfrak{u}$ is the reproducing kernel of the space $H(K_\mathfrak{u})$, which is a subset of $H(K_d)$ consisting of functions depending only on the variables present in $\mathfrak{u}$ and that vanish at $x_\mathfrak{u}$ if at least one component of $x_\mathfrak{u}$ is zero.

Functions f from $H(K_d)$ can be uniquely decomposed as

$$f = \sum_{\mathfrak{u} \subseteq [d]} f_\mathfrak{u},$$

where $f_\mathfrak{u} \in H(K_\mathfrak{u})$ and the functions $f_\mathfrak{u}$ are orthogonal, i.e.,

$$\langle f_\mathfrak{u}, f_\mathfrak{v} \rangle_{H(K_d)} = 0 \qquad \text{for} \qquad \mathfrak{u} \neq \mathfrak{v}.$$

We also have

$$\|f\|_{H(K_d)}^2 = \sum_{\mathfrak{u} \subseteq [d]} \|f_\mathfrak{u}\|_{H(K_\mathfrak{u})}^2. \tag{8.3}$$

We again stress that $f_\mathfrak{u}$ depends only on variables in $\mathfrak{u}$. Moreover, $f_\mathfrak{u}$ describes the behavior of f with respect to these variables since $f(x_\mathfrak{u}, 0) = f_\mathfrak{u}(x_\mathfrak{u})$.

This last decomposition is similar in spirit to the ANOVA decomposition, which is often used in statistics and also in the study of QMC algorithms. Much more can be said about such decompositions but it is beyond the scope of this paper. Instead, the reader is referred to Kuo, Sloan, Wasilkowski, Woźniakowski (2008).

The main idea behind (8.3) is that all the components f_u are of equal importance and they equally contribute to the norm of f. But for large d we may know that f depends differently on different functions f_u. For instance, if we know that f can be represented as a sum of functions of, say, ω variables, then we know a priori that $f_u = 0$ for all $|u| > \omega$. If we know that the first variable is more important than the second variable, and the second is more important then the third and so on, then the functions f_u corresponding to the initial indices are more important than the functions corresponding to the remaining indices. In general, we may consider the case when all f_u are of different importance.

How can we model such cases? The answer is through *weighted* spaces, defined as follows. Let

$$\gamma = \{\gamma_{d,u}\}_{d \in \mathbb{N}, \, u \subseteq [d]} \quad \text{with} \quad \gamma_{d,u} \geq 0$$

be a given sequence of non-negative weights. Obviously, $\mathbb{N} = \{1, 2, \dots\}$. We assume that for all d, at least one weight $\gamma_{d,u}$ is positive. Then we define the *weighted* reproducing kernel

$$K_{d,\gamma}(x, y) = \sum_{u \subseteq [d]} \gamma_{d,u} K_u(x_u, y_u)$$

and the *weighted* reproducing kernel Hilbert space $H(K_{d,\gamma})$.

If all weights $\gamma_{d,u}$ are positive, then the space $H(K_{d,\gamma})$ is algebraically the same as the space $H(K_d)$. The norm in the space $H(K_{d,\gamma})$ is given by

$$\|f\|_{H(K_{d,\gamma})}^2 = \sum_{u \subseteq [d]} \frac{1}{\gamma_{d,u}} \int_{[0,1]^{|u|}} \left(\frac{\partial^{|u|}}{\partial x_u} f(x_u, 0) \right)^2 \, dx_u,$$

or, equivalently, by

$$\|f\|_{H(K_{d,\gamma})}^2 = \sum_{u \subseteq [d]} \frac{1}{\gamma_{d,u}} \|f_u\|_{H(K_u)}^2.$$

Clearly, for all $f \in H(K_d)$ we have

$$\frac{1}{\max_{u \subseteq [d]} \gamma_{d,u}^{1/2}} \|f\|_{H(K_d)} \leq \|f\|_{H(K_{d,\gamma})} \leq \frac{1}{\min_{u \subseteq [d]} \gamma_{d,u}^{1/2}} \|f\|_{H(K_d)}.$$

Hence, the norms in $H(K_d)$ and $H(K_{d,\gamma})$ are equivalent. We stress, however, that the equivalence factors can depend arbitrarily on d.

If one of the weights $\gamma_{d,u}$ is zero, then $H(K_{d,\gamma})$ is a proper subspace of $H(K_d)$. This subspace has the same norm as before, but we must now assume that $f_u = 0$ for all u such that $\gamma_{d,u} = 0$, and interpret $0/0$

as 0. For the extreme case if we take $\gamma_{d,\emptyset} = 1$ and all other weights $\gamma_{d,u} = 0$, then $H(K_{d,\gamma}) = \text{span}(1)$ consists of constant functions and $\|f\|_{H(K_{d,\gamma})} = |f(0)|$. On the other hand, if all $\gamma_{d,u} = 1$, then we are back to the unweighted case, and thus $H(K_{d,\gamma}) = H(K_d)$.

Hence, we can model various properties of f by choosing specific weights $\gamma_{d,u}$. There are a number of different types of weights, see again Novak, Woźniakowski (2008) for a survey. We restrict ourselves only to two types:

- **Product weights.** Here, the weights are of the form

$$\gamma_{d,u} = \prod_{j \in u} \gamma_{d,j}$$

for some $\gamma_{d,1} \geq \gamma_{d,2} \geq \cdots \geq \gamma_{d,d} \geq 0$. The weight $\gamma_{d,j}$ moderates the influence of the variable x_j. Product weights that are independent of d (i.e., for which $\gamma_{d,j} = \gamma_j$) were introduced in Sloan, Woźniakowski (1998); product weights depending on d were introduced in Wasilkowski, Woźniakowski (1999). For product weights, the space $H(K_{d,\gamma})$ is the tensor product

$$H(K_{d,\gamma}) = H(K_{1,\gamma_{d,1}}) \otimes H(K_{1,\gamma_{d,2}}) \otimes \cdots \otimes H(K_{1,\gamma_{d,d}}),$$

where $H(K_{1,\gamma_{d,j}})$ is the reproducing kernel Hilbert space of univariate functions defined on $[0,1]$ that are absolutely continuous and whose first derivatives are in $L_2([0,1])$, with the reproducing kernel

$$K_{1,\gamma_{d,j}}(x,y) = 1 + \gamma_{d,j} \min(x,y) \quad \text{for all} \quad x,y \in [0,1].$$

The norm in $H(K_{1,\gamma_{d,j}})$ is given by

$$\|f\|_{H(K_{1,\gamma_{d,j}})}^2 = f^2(0) + \frac{1}{\gamma_{d,j}} \int_0^1 [f'(x)]^2 \, dx,$$

with the convention that for $\gamma_{d,j} = 0$ we assume that $f' \equiv 0$ and $0/0 = 0$. Hence, for $\gamma_{d,j} = 0$ we have $H(K_{1,0}) = \text{span}(1)$.

The reproducing kernel of $H(K_{d,\gamma})$ is now of the product form,

$$K_{d,\gamma}(x,y) = \prod_{j=1}^d (1 + \gamma_{d,j} \min(x_j,y_j)) \quad \text{for all} \quad x,y \in [0,1]^d.$$

- **Finite-order weights.** Here, there exists an $\omega > 0$ such that the weights satisfy

$$\gamma_{d,u} = 0 \quad \text{for all} \quad |u| > \omega.$$

The smallest ω satisfying the condition above is called the *order* of the finite-order weights. Such weights were introduced in Dick, Sloan, Wang, Woźniakowski (2006). For finite-order weights we have

$$f = \sum_{\mathfrak{u} \subseteq [d], \, |\mathfrak{u}| \leq \omega} f_{\mathfrak{u}},$$

i.e., f is a sum of functions depending on at most ω variables. We believe that finite-order weights model at least approximately many multivariate problems with large d.

8.6 Tractability

In the previous section, we defined the weighted space $H(K_{d,\gamma})$. We know that for the unweighted case $\gamma_{d,\mathfrak{u}} \equiv 1$, multivariate integration suffers from the curse of dimensionality in the worst case setting. It is then natural to ask what are necessary and sufficient conditions on the weights $\gamma = \{\gamma_{d,\mathfrak{u}}\}$ for which we can vanquish the curse of dimensionality.

We consider multivariate integration for functions from the weighted space $H(K_{d,\gamma})$. We have

$$I_d(f) = \langle f, h_{d,\gamma} \rangle_{H(K_{d,\gamma})},$$

where $h_{d,\gamma} \in H(K_{d,\gamma})$ is given by

$$h_{d,\gamma}(x) = \int_{[0,1]^d} K_{d,\gamma}(x,t) \, \mathrm{d}t = \sum_{\mathfrak{u} \subseteq [d]} \gamma_{d,\mathfrak{u}} \prod_{j \in \mathfrak{u}} (x_j - \tfrac{1}{2}x_j^2)$$

for all $x \in [0,1]^d$.

Let us denote the norm of I_d over the space $H(K_{d,\gamma})$ by $\|I_d\|_\gamma$. Then

$$\|I_d\|_\gamma = \|h_{d,\gamma}\|_{H(K_{d,\gamma})} = \left(\sum_{\mathfrak{u} \subseteq [d]} \gamma_{d,\mathfrak{u}} \left(\tfrac{1}{3}\right)^{|\mathfrak{u}|} \right)^{1/2}.$$

It seems natural to use weights for which $\|I_d\|_\gamma$ is of order one or at most polynomially dependent on d, and to eliminate weights for which we have an exponential dependence on d. For example, for product weights we have

$$\|I_d\|_\gamma = \prod_{j=1}^{d} \left(1 + \tfrac{1}{3}\gamma_{d,j}\right)^{1/2}.$$

Then $\|I_d\|_\gamma$ is of order one iff

$$\sup_d \sum_{j=1}^d \gamma_{d,j} < \infty.$$

For product weights independent of d, i.e., for $\gamma_{d,j} = \gamma_j$, the last condition simplifies to

$$\sum_{j=1}^\infty \gamma_j < \infty.$$

Remembering that we are still working with product weights, $\|I_d\|_\gamma$ is polynomially dependent on d iff

$$\limsup_{d\to\infty} \frac{\sum_{j=1}^d \gamma_{d,j}}{\ln d} < \infty.$$

Obviously, we *cannot* take $\gamma_{d,j} = 1$ to satisfy the last condition.

For finite-order weights, we have

$$\|I_d\|_\gamma = \left(\sum_{\mathfrak{u}\subseteq[d],\,|\mathfrak{u}|\le\omega} \gamma_{d,\mathfrak{u}} \left(\tfrac{1}{3}\right)^{|\mathfrak{u}|} \right)^{1/2} = \mathcal{O}\left(d^{\omega/2} \max_{\mathfrak{u}\subseteq[d],\,|\mathfrak{u}|\le\omega} \gamma_{d,\mathfrak{u}}^{1/2} \right),$$

with the factor in the $\mathcal{O}$-notation independent of d. Hence, the norm $\|I_d\|_\gamma$ is at most polynomial in d for bounded finite-order weights.

The worst case error of a linear algorithm $A_{n,d}$ over the space $H(K_{d,\gamma})$ is analogously given by

$$e_\gamma^{\mathrm{wor}}(A_{n,d}) = \sup_{\|f\|_{H(K_{d,\gamma})}\le 1} |I_d(f) - A_{n,d}(f)|,$$

and the nth minimal worst case error is given by

$$e_\gamma^{\mathrm{wor}}(n,d) = \inf_{A_{n,d}} e_\gamma^{\mathrm{wor}}(A_{n,d}).$$

For $n = 0$, the initial error is equal to $\|I_d\|_\gamma$. We have already discussed conditions for which the initial error is of order one or polynomially dependent on d.

Finally, we update the definition of the information complexity as being

$$n_\gamma^{\mathrm{wor}}(\varepsilon,d) = \min\left\{ n \mid e_\gamma^{\mathrm{wor}}(n,d) \le \varepsilon \|I_d\|_\gamma \right\}.$$

The main point of determining tractability is to check when $n_\gamma^{\mathrm{wor}}(\varepsilon,d)$ is *not* exponentially dependent on ε^{-1} and d. Since there are various ways of measuring the lack of exponential dependence, we have various types of tractability. In this paper we restrict ourselves to three basic types of tractability, referring the reader to Gnewuch, Woźniakowski (2007) and Novak, Woźniakowski (2008) for more general notions.

We say that $\{I_d\}$ is *polynomially tractable* iff there are non-negative numbers C, q and p such that

$$n_\gamma^{\mathrm{wor}} \leq C\, d^q\, \varepsilon^{-p} \quad \text{for all} \quad \varepsilon \in (0,1) \quad \text{and} \quad d \in \mathbb{N}.$$

If $q = 0$ in the formula above, then we say that $\{I_d\}$ is *strongly polynomially tractable*, and the infimum of those p satisfying this inequality (with $q = 0$) is called the *exponent of strong polynomial tractability*, or shortly the *exponent*.

We say that $\{I_d\}$ is *weakly tractable* iff

$$\lim_{\varepsilon^{-1}+d\to\infty} \frac{\ln n(\varepsilon,d)}{\varepsilon^{-1}+d} = 0,$$

The notion of polynomial tractability is clear. However, strong polynomial tractability looks like a very demanding property, since in this case we can solve the problem using polynomially-many (in ε^{-1}) function values *independently of d*. On the other hand, numerical tests for finance applications seem to indicate that the error is indeed independent of d and proportional to n^{-1}. Assuming that the initial error is of order one, these tests would imply that we have strong polynomial tractability, with exponent one.

The notion of weak tractability, introduced in Gnewuch, Woźniakowski (2007), means that $n_\gamma^{\mathrm{wor}}(\varepsilon,d)$ grows sub-exponentially, since

$$\frac{n_\gamma^{\mathrm{wor}}(\varepsilon,d)}{a^{\varepsilon^{-1}+d}}$$

goes to zero as $\varepsilon^{-1} + d$ tends to infinity, and this holds for any $a > 1$. However, if weak tractability holds, then $n_\gamma^{\mathrm{wor}}(\varepsilon,d)$ may go faster to infinity than any polynomial or any function of the form $e^{(\varepsilon^{-1}+d)^\beta}$ with $\beta < 1$. Weak tractability is a necessary property to eliminate the curse of dimensionality.

We are ready to present a sample of tractability results. For simplicity, we present such results only for product and finite-order weights.

- **For product weights:** $\gamma_{d,\mathfrak{u}} = \prod_{j\in\mathfrak{u}} \gamma_{d,j}$.

  - Strong polynomial tractability holds iff

$$\limsup_{d\to\infty} \sum_{j=1}^{d} \gamma_{d,j} < \infty.$$

If so, then the exponent belongs to $[1, 2]$. If

$$\limsup_{d \to \infty} \sum_{j=1}^{d} \gamma_{d,j}^{1/2} < \infty,$$

then this exponent is one.

– Polynomial tractability holds iff

$$\limsup_{d \to \infty} \frac{\sum_{j=1}^{d} \gamma_{d,j}}{\ln d} < \infty.$$

– Weak tractability holds iff

$$\lim_{d \to \infty} \frac{\sum_{j=1}^{d} \gamma_{d,j}}{d} = 0.$$

• **For finite-order weights:** $\gamma_{d,u} = 0$ for all $|u| > \omega$. We have

$$n_{\gamma}^{\mathrm{wor}}(\varepsilon, d) \leq \frac{\sum_{u \subseteq [d], |u| \leq \omega} \gamma_{d,u} 2^{-|u|}}{\sum_{u \subseteq [d], |u| \leq \omega} \gamma_{d,u} 2^{-|u|}} \frac{1}{\varepsilon^2} \leq \left(\tfrac{3}{2}\right)^{\omega} \frac{1}{\varepsilon^2}.$$

Hence, strong polynomial tractability holds for *all* finite-order weights.

A few comments are in order. For product weights independent of d and for QMC algorithms, the conditions for strong polynomial and polynomial tractability were proved in Sloan, Woźniakowski (1998); for general linear algorithms, these conditions were proved in Novak, Woźniakowski (2001). The case of product weights depending on d can be handled by the same proofs as the case of product weights independent of d. This means that we have the same polynomial tractability conditions for both QMC and general linear algorithms. The conditions on weak tractability were proved in Gnewuch, Woźniakowski (2008).

We now discuss the exponent p of strong polynomial tractability for product weights. The inequality $p \geq 1$ is trivial since for $d = 1$ the nth minimal worst case errors behave like n^{-1}. The inequality $p \leq 2$ follows easily from the averaging argument for QMC algorithms and was presented in Sloan, Woźniakowski (1998). The next step was to show that a stronger condition on weights, $\limsup_d \sum_{j=1}^{d} \gamma_{d,j}^{1/2} < \infty$, implies that $p = 1$; this was proved in Hickernell, Woźniakowski (2000). Whether or not this condition is also necessary for $p = 1$ is an open problem. More results and conjectures on how the exponent of strong polynomial tractability may depend on the summability of various powers of $\gamma_{d,j}$ may be found in Heinrich (2003).

The results reported so far for upper bounds were obtained by non-constructive arguments. The first fully constructive upper bounds were obtained by a component by component (CBC) algorithm for shifted lattice rules designed by the Australian school of Ian Sloan. The major step was made by Kuo, see Kuo (2003), who showed how to achieve $p = 1$ under the assumption $\limsup_d \sum_{j=1}^d \gamma_{d,j}^{1/2} < \infty$. Further improvement was done in Nuyens, Cools (2006), who showed how to implement the CBC algorithm in time proportional to $n\, d \ln n$.

We stress that for product weights, the condition for strong polynomial tractability is exactly the same as the condition for $\|I_d\|_\gamma$ to be of order one. The same holds for polynomial tractability.

The bound presented above for finite-order weights is easy to prove, but non-constructive. We provide this proof as an illustration of a proof technique used to obtain tractability results. Take a QMC algorithm

$$A_{n,d}(f) = \frac{1}{n} \sum_{j=1}^n f(t_j) \quad \text{for} \quad f \in H(K_{d,\gamma})$$

with (as-yet) unspecified sample points $t_1, t_2, \ldots, t_n$. Note that

$$I_d(f) - A_{n,d}(f) = \left\langle f, h_{d,\gamma} - \frac{1}{n} \sum_{j=1}^n K_{d,\gamma}(\cdot, t_j) \right\rangle_{H(K_{d,\gamma})}.$$

This implies that

$$e_\gamma^{\mathrm{wor}}(A_{n,d}) = \left\| h_{d,\gamma} - \frac{1}{n} \sum_{j=1}^n K_{d,\gamma}(\cdot, t_j) \right\|_{H(K_{d,\gamma})}.$$

Let $a = a(t_1, t_2, \ldots, t_d) = e_\gamma^{\mathrm{wor}}(A_{n,d})$: Then

$$a^2 = \|h_{d,\gamma}\|_{H(K_{d,\gamma})}^2 - 2 \sum_{j=1}^n h_{d,\gamma}(t_j) + \frac{1}{n^2} \sum_{i,j=1}^n K_{d,\gamma}(t_i, t_j).$$

We now treat the sample points $t_1, t_2, \ldots, t_n$ as independent random points uniformly distributed over $[0,1]^d$. We compute the average value of $a^2(t)$ for $t = [t_1, t_2, \ldots, t_n] \in \mathbb{R}^{nd}$. Using the fact that

$$\int_{[0,1]^d} h_{d,\gamma}(t)\, \mathrm{d}t = \|h_{d,\gamma}\|_{H(K_{d,\gamma})}^2 = \int_{[0,1]^{2d}} K_{d,\gamma}(x,t)\, \mathrm{d}x\, \mathrm{d}t,$$

we obtain

$$\int_{[0,1]^{dn}} a^2(t)\, \mathrm{d}t = \frac{1}{n} \left(\int_{[0,1]^d} K_{d,\gamma}(t,t)\, \mathrm{d}t - \int_{[0,1]^{2d}} K_{d,\gamma}(x,t)\, \mathrm{d}t\, \mathrm{d}x \right)$$

$$= \frac{1}{n} \sum_{u \subseteq [d], \, |u| \leq \omega} \gamma_{d,u} \left(2^{-|u|} - 3^{-|u|} \right).$$

By the mean value theorem, we conclude that there are sample points $t_1, t_2, \ldots, t_n$ (this makes the proof non-constructive) such that

$$e_\gamma^{\mathrm{wor}}(n, d) \leq e_\gamma^{\mathrm{wor}}(A_{n,d}) \leq \frac{1}{\sqrt{n}} \left(\sum_{u \subseteq [d] \, |u| \leq \omega} \gamma_{d,u} 2^{-|u|} \right)^{1/2}.$$

Therefore

$$\frac{e_\gamma^{\mathrm{wor}}(n, d)}{e_\gamma^{\mathrm{wor}}(0, d)} \leq \frac{1}{\sqrt{n}} \left(\frac{\sum_{u \subseteq [d] \, |u| \leq \omega} \gamma_{d,u} 2^{-|u|}}{\sum_{u \subseteq [d] \, |u| \leq \omega} \gamma_{d,u} 3^{-|u|}} \right)^{1/2}.$$

This yields the bound on the information complexity

$$n_\gamma^{\mathrm{wor}}(\varepsilon, d) \leq \frac{\sum_{u \subseteq [d], \, |u| \leq \omega} \gamma_{d,u} 2^{-|u|}}{\sum_{u \subseteq [d], \, |u| \leq \omega} \gamma_{d,u} 3^{-|u|}} \frac{1}{\varepsilon^2}.$$

Finally note that

$$\frac{\sum_u \gamma_{d,u} 2^{-|u|}}{\sum_u \gamma_{d,u} 3^{-|u|}} = \frac{\sum_u \gamma_{d,u} (3/2)^{|u|} 3^{-|u|}}{\sum_u \gamma_{d,u} 3^{-|u|}} \leq \left(\frac{3}{2} \right)^\omega,$$

which completes the proof.

The essence of this bound is that we always have strong polynomial tractability for finite-order weights although the upper bound on the information complexity depends exponentially on the order of finite-order weights. In fact, it must be so. For large ω we do not control weights for $d \leq \omega$ and we know that for general weights we must have an exponential dependence on d, i.e., on ω. On the other hand, if ω is relatively small, as it is for many applications, the exponential dependence on ω is not so important.

Much more is known about finite-order weights and will be reported in the next sections.

8.7 Semi-Constructive and Constructive Bounds for Finite-Order Weights

The bound reported in the previous section for finite-order weights has two drawbacks: it is non-constructive, and it has a quadratic dependence on ε^{-1}. In this section we will see how to remove at least one of these drawbacks.

We consider a *shifted lattice rule*, which is a special QMC algorithm of the form

$$\mathrm{QMC}_{n,d}(f) = \frac{1}{n} \sum_{j=1}^{n} f\left(\left\{\frac{j-1}{n} z + \Delta\right\}\right),$$

where z is an integer vector from $\{1, 2, \ldots, n-1\}^d$ that can be computed by the CBC (component by component) algorithm with cost $\mathcal{O}(n\, d \ln n)$ as reported before, and the shift vector $\Delta \in [0,1)^d$. For $x = [x_1, x_2, \ldots, x_d]$ we let $\{x\}$ denote the vector of the fractional parts of x_j.

Theorem 7 of Sloan, Wang, Woźniakowski (2004) states that there *exists* a shift vector Δ such that the following estimate of the worst case error of $\mathrm{QMC}_{n,d}$ holds. For any $a \in [1,2)$, there exists a positive number C_a such that

$$\frac{e_\gamma^{\mathrm{wor}}(\mathrm{QMC}_{n,d})}{\|I_d\|_{H(K_{d,\gamma})}} \le C_a\, d^{\omega\,(a-1)/2}\, n^{-a/2}.$$

We stress that C_a is independent of n and d, but depends on a. Using this inequality, we get the estimate

$$n_\gamma^{\mathrm{wor}}(\varepsilon, d) \le C_a^{2/a}\, d^{\omega(1-1/a)}\, \varepsilon^{-2/a} \qquad \forall \varepsilon \in (0,1), d \in \mathbb{N} \qquad (8.4)$$

for the information complexity. The number C_a is known but its form is not important. However, C_a goes to infinity as a approaches 2.

Let us discuss the bound (8.4).

- Let a be close to 2. Then $n_\gamma^{\mathrm{wor}}(\varepsilon, d)$ depends almost linearly on ε^{-1}. This is the best possible dependence on ε^{-1}, since even when $d = 1$ the information complexity is of order ε^{-1}. Furthermore, in this case, we have polynomial tractability, with the exponent of d roughly equal to $\omega/2$.

- Let $a = 1$. Then the dependence on d disappears and we have strong polynomial tractability at the expense of the exponent of ε^{-1} which is now 2.

The bound (8.4) shows a tradeoff between the exponents of ε^{-1} and d, since we can decrease the exponent of ε^{-1} by increasing the exponent of d and vice versa. Clearly, we can choose a to minimize the error bound for any given ε^{-1} and d.

We stress that this result is only *semi*-constructive, since (8.4) holds for the constructed z and for *some* Δ. One possible remedy is to view Δ as a random shift vector; we can then find some a posteriori error estimates, as shown in Sloan, Kuo, Joe (2002). This point is, however, beyond the scope of this paper.

There is one more point we want to stress for the shifted lattice rule. The construction of the vector z and the existence of Δ depend on the set γ of finite-order weights. In other words, for different finite-order weights we should use different z and Δ. It would be much more convenient if we could use the same algorithm for several sets of finite-order weights.

We now present a fully constructive result. Once again, we consider a QMC algorithm

$$\mathrm{QMC}_{n,d}(f) = \frac{1}{n} \sum_{j=1}^{n} f(t_j),$$

using the *Niederreiter sequence* $t_1, t_2, \ldots, t_n$. This is a well known low discrepancy sequence whose definition has nothing whatsoever to do with finite-order weights, see Niederreiter (1992).

Theorem 10 of Sloan, Wang, Woźniakowski (2004) gives us our desired result: for all positive δ, there exists positive C_δ such that

$$n_\gamma^{\mathrm{wor}}(\varepsilon, d) \leq C_\delta \left[d^\omega \ln(d+b) \right]^{1+\delta} \varepsilon^{-(1+\delta)}. \tag{8.5}$$

We stress that the bound (8.5) holds for arbitrary finite-order weights although the algorithm $\mathrm{QMC}_{n,d}$ based on the Niederreiter sequence is *independent* of the weights. To determine the proper n that yields the error of order ε we must know only the order ω of the finite-order weights.

Note that, modulo δ, we have the best possible dependence on ε^{-1}. Although we have lost strong polynomial tractability, we still have polynomial tractability with the exponent roughly ω. As remarked in Sloan, Wang, Woźniakowski (2004), similar bounds hold also for Sobol and Halton sample points.

8.8 General Multivariate Problems

The reader may have a wrong impression that tractability of continuous problems has only been studied for multivariate integration. The purpose of this section is to show the opposite and to introduce tractability for general multivariate problems, see Novak, Woźniakowski (2008) for more details.

Hence, for $d \in \mathbb{N}$ let

$$S_d \colon F_d \;\rightarrow\; G_d$$

be an operator defined over a normed linear space of d-variate functions
with a normed linear space G_d as its target space. In most tractability
papers it is assumed that S_d is linear but there are also papers studying
nonlinear S_d, see e.g., Werschulz, Woźniakowski (2007a, 2007b).

We want to approximate $S_d(f)$ for $f \in F_d$. We assume that f is not
known to us. Instead we can compute finitely many linear functionals
$L_j(f)$ with L_j from a class $\Lambda \subseteq F_d^*$. Typically, two classes of Λ are
studied:

- the class $\Lambda^{\text{all}} = F_d^*$ of *linear* information, in which all continuous
  linear functional are allowed, and
- the class Λ^{std} of *standard* information, in which only function values
  are allowed, i.e., $L_j(f) = f(t_j)$ for some t_j from the common domain
  of all our functions f.

We approximate $S_d(f)$ by algorithms of the form

$$A_{n,d}(f) = \varphi_n(L_1(f), L_2(f), \ldots, L_n(f))$$

with $L_j \in \Lambda$ that can be chosen adaptively and with $\varphi_n \colon \mathbb{R}^n \to G_d$. Note
that we place no restrictions on the mapping φ_n. More details can be
found in Novak, Woźniakowski (2008), Traub, Wasilkowski, Woźniakowski (1988).

The notion of *error* of an algorithm depends on the setting.

- In the *worst case* setting, the error of $A_{n,d}$ is given as

$$e^{\text{wor}}(A_{n,d}) = \sup_{f \in F_d, \|f\|_{F_d} \leq 1} \|S_d(f) - A_{n,d}(f)\|_{G_d}.$$

This is the setting we used when we studied multivariate integration
earlier in this paper.

- In the *average case* setting, we assume that F_d is equipped with a
  probability measure μ_d and define the error of $A_{n,d}$ as

$$e^{\text{avg}}(A_{n,d}) = \left(\int_{F_d} \|S_d(f) - A_{n,d}(f)\|_{G_d}^p \, \mu_d(\mathrm{d}f) \right)^{1/p},$$

where $p \in [1, \infty]$. Typically we choose $p = 2$ to make the average
case analysis easier. Although it looks as if we must also assume that
$\|S_d(\cdot) - A_{n,d}(\cdot)\|_{G_d}$ is measurable, there is a way to get around this
additional assumption.

- In the *randomized* setting we are allowed to choose L_j and φ_n randomly, so that

$$A_{n,d,t} = \varphi_{n,t}(L_{1,t}(f), L_{2,t}(f), \ldots, L_{n,t}(f))$$

with a random parameter $t \in T$ distributed according to a probability measure ϱ. The error of $A_{n,d}$ is given as

$$e^{\mathrm{ran}}(A_{n,d}) = \sup_{f \in F_d, \|f\|_{F_d} \le 1} \left(\int_T \|S_d(f) - A_{n,d,t}(f)\|_{G_d}^p \, \varrho(\mathrm{d}y) \right)^{1/p},$$

where $p \in [1, \infty]$ with $p = 2$ as a popular choice. Although it appears that we need to assume measurability with respect to t, this assumption can once again be omitted.

We will only discuss these three settings; however, there are additional settings, such as the probabilistic and asymptotic settings.

In each setting we want to minimize the error for given n and d. The nth minimal error for the d-variate case in the given setting is given as

$$e^{\mathrm{setting}}(n,d) = \inf_{A_{n,d}} e^{\mathrm{setting}}(A_{n,d}) \quad \text{where} \quad \text{setting} \in \{\text{wor}, \text{avg}, \text{ran}\}.$$

Then the *information complexity* is given as

$$n^{\mathrm{setting}}(\varepsilon, d) = \min\{\, n \mid e^{\mathrm{setting}}(n,d) \le \varepsilon \, \mathrm{CRI}_d \,\},$$

where CRI_d is a chosen *error criterion*:

- For the *absolute* error criterion, we set $\mathrm{CRI}_d = 1$.
- For the *normalized* error criterion, we set $\mathrm{CRI}_d = e^{\mathrm{setting}}(0,d)$, which for a linear S_d is just its norm in a given setting. Note that we used the normalized error criterion in our tractability analysis of multivariate integration given earlier.

Equipped with all these definitions, we are ready to define tractability for the problem $S = \{S_d\}$. The subject of tractability of continuous problems was introduced in Woźniakowski (1994a, 1994b). The main idea behind tractability is to eliminate an exponential dependence of ε^{-1} or d in the information complexity $n^{\mathrm{setting}}(\varepsilon, d)$. Since there are many different ways of measuring the lack of exponential dependence, we have various types of tractability.

In particular, S is *polynomially tractable* iff there are non-negative numbers C, q and p such that

$$n^{\mathrm{setting}}(\varepsilon, d) \le C \, d^q \, \varepsilon^{-p} \quad \text{for all} \quad \varepsilon \in (0,1), \; d \in \mathbb{N}.$$

We say that S is *strongly polynomially tractable* if $q = 0$ in the formula above; the infimum of p in the formula above with $q = 0$ is called the *exponent* of strong polynomial tractability.

Finally, S is *weakly tractable* iff

$$\lim_{\varepsilon^{-1}+d\to\infty} \frac{\ln n^{\text{setting}}(\varepsilon, d)}{\varepsilon^{-1} + d} = 0.$$

If S is *not* weakly tractable then we say that S is *intractable*.

If $n^{\text{setting}}(\varepsilon, d)$ depends exponentially on d then we say that S suffers from the *curse of dimensionality*.

Hence, we have polynomial, strong polynomial, and weak tractability in the worst case, average case, and randomized settings for the absolute and normalized error criteria. So, there are altogether 18 different cases, not to mention other error settings and other error criteria. There are also different types of tractability, such as T-tractability, see Gnewuch, Woźniakowski (2007, 2008), but we do not cover them here and we refer the reader to Novak, Woźniakowski (2008, 2009) for more details and more general cases.

8.9 A Sampling of Tractability Results

We present tractability results for only one setting. To keep things simple, we will only discuss the worst case setting. For the average case setting, we would need to discuss measure theory over infinite-dimensional spaces. For the randomized setting, there are not, frankly speaking, too many general tractability results.

We already discussed the class Λ^{std} for which only function values are allowed. For a change, we now choose the class Λ^{all}, so that *arbitrary* continuous linear functionals can be used. Note that multivariate integration, as well as any $S_d \in F_d^*$, is trivial for the class Λ^{all}, since such a problem can be solved exactly in just one evaluation. Hence, for the class Λ^{all}, we need to assume that S_d is at least a two-dimensional operator.

We add in passing that the class Λ^{all} is usually easier to analyze, but it is not usually available for many practical multivariate problems. Nevertheless, the results for Λ^{all} can be viewed as lower bounds for Λ^{std} and one of the most exciting research problems of information-based complexity is the study of relations between these two classes in various settings. Again this subject is beyond the scope of this paper, and the reader is referred to Hickernell, Wasilkowski, Woźniakowski (2006), Hinrichs, Novak, Vybiral (2008), Kuo, Wasilkowski, Woźniakowski (2008a, 2008b), Wasilkowski, Woźniakowski (2001, 2006).

To further simplify our presentation, we will only discuss a specific class of problems, namely, problems with a tensor product structure.

We first discuss the unweighted case. To do this, we start with the univariate case $d = 1$, letting $S_1 : F_1 \to G_1$ be a linear compact operator between separable Hilbert spaces, with F_1 being a space of real univariate functions $f : D \to \mathbb{R}$ and $D \subseteq \mathbb{R}$. To simplify notation, we assume that $\dim(F_1) = \infty$. Moving to the d-variate case, we let

$$F_d = F_1^{\otimes d} \quad \text{and} \quad G_d = G_1^{\otimes d},$$

i.e., F_d and G_d are d-fold tensor products of F_1 and G_1, respectively. Note that F_d is a space of real d-variate functions $f : D^d \to \mathbb{R}$ with $D^d = D \times D \times \cdots \times D \subseteq \mathbb{R}^d$. Finally, we define

$$S_d = S_1^{\otimes d}$$

as the d-fold tensor product of S_1.

This concludes the definition of the problem $S = \{S_d\}$. We stress that this is an unweighted problem, since all variables and groups of variables play the same role.

Much is known about such tensor product problems, see e.g., Novak, Woźniakowski (2008), for a survey. It turns out that the results depend on the eigenpairs (λ_j, η_j) of the compact operator $W_1 := S_1^* S_1 : F_1 \to F_1$. We may assume that

$$W_1 \eta_j = \lambda_j \eta_j \quad \text{with} \quad \langle \eta_i, \eta_j \rangle_{F_1} = \delta_{i,j},$$

and that the eigenvalues are ordered,

$$\lambda_1 \geq \lambda_2 \geq \cdots \geq \lambda_j \geq \cdots \geq 0.$$

For $d \geq 1$, the eigenpairs of the operator $W_d = S_d^* S_d : F_d \to F_d$ have a particularly nice structure. Let us write these eigenpairs as $\{\lambda_{d,j}, \eta_{d,j}\}_{j \in \mathbb{N}^d}$. For $j = [j_1, j_2, \ldots, j_d] \in \mathbb{N}^d$ we have

$$\lambda_{d,j} = \prod_{i=1}^d \lambda_{j_i}$$

and

$$\eta_{d,j}(x) = \prod_{i=1}^d \eta_{j_i}(x_i) \quad \text{for} \quad x = [x_1, x_2, \ldots, x_d] \in D^d.$$

Let us order the eigenvalues $\{\lambda_{d,j}\}_{j\in\mathbb{N}^d}$. That is, let $\{\beta_{d,k}\}_{k\in\mathbb{N}} = \{\lambda_{d,j}\}_{j\in\mathbb{N}^d}$ with

$$\beta_{d,1} \geq \beta_{d,1} \geq \cdots \geq \beta_{d,k} \geq \cdots \geq 0.$$

Let $P : \mathbb{N} \to \mathbb{N}^d$ be a one-to-one and onto function such that $\beta_{d,k} = \lambda_{d,P(k)}$ for all $k \in \mathbb{N}$. Clearly, $\beta_{d,1} = \lambda_1^d$ with $P(1) = [1,1,\ldots,1]$, and $\beta_{d,2} = \lambda_1^{d-1}\lambda_2$ with $P(2) = [1,1,\ldots,1,2]$, etc. The linear algorithm

$$A_{n,d}(f) = \sum_{k=1}^{n} \langle f, \eta_{d,P(k)} \rangle_{H(K_d)} S_d \eta_{d,P(k)}$$

minimizes the worst case error among all algorithms that use n linear functionals from Λ^{all}. Furthermore,

$$e^{\text{wor}}(n,d) = e^{\text{wor}}(A_{n,d}) = \sqrt{\beta_{d,n+1}} = \sqrt{\lambda_{d,P(n+1)}}.$$

Hence the information complexity is given by

$$n^{\text{wor}}(\varepsilon,d) = \left|\{[j_1,j_2,\ldots,j_d] \in \mathbb{N}^d \mid \lambda_{j_1}\lambda_{j_2}\cdots\lambda_{j_d} > \varepsilon^2 \, \text{CRI}_d\}\right|.$$

If $\lambda_2 = 0$ then at most one $\lambda_{d,j}$ is positive. Then the problem is trivial since $n^{\text{wor}}(\varepsilon,d) \leq 1$. Hence, we shall assume that $\lambda_2 > 0$.

For the absolute error criterion, $\text{CRI}_d = 1$, Theorem 5.5 of Novak, Woźniakowski (2008) says the following:

- If $\lambda_1 > 1$ or $\lambda_1 = \lambda_2 = 1$ then S is intractable, and S suffers from the curse of dimensionality.
- If $\lambda_1 = 1$ and $\lambda_2 < 1$ then S is *not* polynomially tractable.
- Let $\lambda_1 = 1$. If S is weakly tractable then

$$\lambda_2 < 1 \quad \text{and} \quad \lambda_n = o\left([\ln n]^{-2}\right).$$

On the other hand, if

$$\lambda_2 < 1 \quad \text{and} \quad \lambda_n = o\left([\ln n]^{-2}[\ln\ln n]^{-2}\right)$$

then S is weakly tractable.

- Let $\lambda_1 < 1$. Then S is strongly polynomially tractable iff S is polynomially tractable iff there is a positive r such that

$$\lambda_n = \mathcal{O}(n^{-r}).$$

If so, the exponent of strong polynomial tractability is

$$p = \inf\left\{2\tau \;\middle|\; \sum_{j=1}^{\infty}\lambda_j^\tau \leq 1\right\}.$$

For the normalized error criterion $\mathrm{CRI}_d = \|S_d\| = \lambda_1^{d/2}$, Theorem 5.6 of Novak, Woźniakowski (2008) says that

- If $\lambda_1 = \lambda_2$ then S is intractable, and S suffers from the curse of dimensionality.
- If $\lambda_2 < \lambda_1$ then S is *not* polynomially tractable.
- If S is weakly tractable, then

$$\lambda_2 < \lambda_1 \quad \text{and} \quad \lambda_n = o([\ln n]^{-2}).$$

On the other hand, if

$$\lambda_2 < \lambda_1 \quad \text{and} \quad \lambda_n = o([\ln n]^{-2}[\ln \ln n]^{-2})$$

then S is weakly tractable.

In particular, if $\lambda_1 = \lambda_2 \geq 1$ then S suffers from the curse of dimensionality for both the absolute and normalized error criteria. Furthermore, if $\lambda_2 < \lambda_1$ then S is not polynomially tractable for the normalized case.

We hope to change these negative results when we switch to the weighted case. To derive weighted spaces, let $\{\eta_j\}_{j\in\mathbb{N}}$ be an orthonormal basis of F_1 consisting of the eigenfunctions of W_1. Then for $j = [j_1, j_2, \ldots, j_d] \in \mathbb{N}^d$, the functions $\eta_j(x) = \prod_{i=1}^{d} \eta_{j_i}(x_i)$ defined before, are the eigenfunctions of W_d and they are an orthonormal basis of F_d. Every $f \in F_d$ can be written as

$$f(x) = \sum_{j\in\mathbb{N}^d} \langle f, \eta_j \rangle_{F_d} \, \eta_j(x) \quad \text{for all} \quad x \in D^d.$$

For $j = [j_1, j_2, \ldots, j_d] \in \mathbb{N}^d$, define

$$\mathfrak{u}(j) = \{ k \mid j_k \geq 2 \}.$$

For any subset $\mathfrak{u}$ of $[d]$, let $f_\mathfrak{u} \colon D^d \to \mathbb{R}$ be defined by

$$f_\mathfrak{u}(x) = \sum_{j\in\mathbb{N}^d : \mathfrak{u}(j)=u} \langle f, \eta_j \rangle_{F_d} \, \eta_j(x) \quad \text{for all} \quad x \in D^d.$$

Assume for a moment that $\eta_1 \equiv 1$. Then it is easy to see that $f_\mathfrak{u}$ depends only on the variables in $x_\mathfrak{u}$, that is $f(x) = f(y)$ if $x_\mathfrak{u} = y_\mathfrak{u}$. If η_1 is not identically equal to 1, then $f_\mathfrak{u}$ depends on variables not in $\mathfrak{u}$ only through η_1. More precisely, in full generality, we have

$$f_\mathfrak{u}(x) = f_{\mathfrak{u},1}(x_\mathfrak{u}) \prod_{k\in[d]\setminus\mathfrak{u}} \eta_1(x_k)$$

with $f_{u,1}$ depending only on the variables in u. Furthermore, we have

$$\|f\|_{F_d}^2 = \sum_{u \subseteq [d]} \|f_u\|_{F_d}^2 .$$

This means that the contribution of each f_u is the same, so any group of variables is equally important.

We are ready to define weighted spaces. As with multivariate integration, assume that we have a sequence of weights

$$\gamma = \{\gamma_{d,u}\}_{u \subseteq [d], d \in \mathbb{N}} \quad \text{with} \quad \gamma_{d,u} \geq 0.$$

Then we define the Hilbert space $F_{d,\gamma}$ as a subset of F_d equipped with the norm

$$\|f\|_{F_{d,\gamma}}^2 = \sum_{u \subseteq [d]} \frac{1}{\gamma_{d,u}} \|f_u\|_{F_d}^2 .$$

Note that for positive weights $\gamma_{d,u}$, the norms of $F_{d,\gamma}$ and F_d are equivalent. If one of the weights $\gamma_{d,u} = 0$ then, as before, we assume that $f_u = 0$ and interpret $0/0 = 0$. In this case, $F_{d,\gamma}$ is a proper subspace of F_d. Again, by a proper choice of $\gamma_{d,u}$ we can control the contribution of each f_u, that is, each group of variables in u.

The weighted problem $S_\gamma = \{S_{d,\gamma}\}$ is defined by letting $S_{d,\gamma} : F_{d,\gamma} \to G_d$ with $S_{d,\gamma} f = S_d f$ for $f \in F_{d,\gamma}$. Polynomial and weak tractability of S_γ are defined as before, and depend on the behavior of the eigenvalues of the operator

$$W_{d,\gamma} := S_{d,\gamma}^* S_{d,\gamma} : F_{d,\gamma} \to F_{d,\gamma}.$$

Although $S_{d,\gamma}$ is equal to S_d, the operator $W_{d,\gamma}$ generally has different eigenvalues than W_d, since the change of the norm in $F_{d,\gamma}$ implies that the adjoint operator $S_{d,\gamma}^*$ is different than S_d^*. The eigenvalues of $W_{d,\gamma}$ are

$$\lambda_{d,\gamma,j} = \gamma_{d,u(j)} \lambda_{d,j} \quad \text{for all} \quad j \in \mathbb{N}^d.$$

We restrict ourselves to tractability for the normalized error criterion. Then the information complexity is

$$n_\gamma^{\mathrm{wor}}(\varepsilon, d) = \left| \{ [j_1, j_2, \ldots, j_d] \in \mathbb{N}^d \mid \lambda_{d,\gamma,j_1} \lambda_{j_2} \cdots \lambda_{d,\gamma,j_d} > \varepsilon^2 \|S_{d,\gamma}\| \} \right|.$$

Obviously,

$$\|S_{d,\gamma}\| = \max_{j \in \mathbb{N}^d} \left[\gamma_{d,u(j)} \lambda_{d,j} \right]^{1/2} = \max_{u \subseteq [d]} \left[\gamma_{d,u} \lambda_1^{d-|u|} \lambda_2^{|u|} \right]^{1/2} .$$

Polynomial and weak tractability of such weighted problems is characterized in Theorems 5.7 and 5.8 of Novak, Woźniakowski (2008). Here we only mention a few facts from these theorems.

We need to introduce the *normalized* weights

$$\gamma_{d,\mathfrak{u}}^* = \frac{\gamma_{d,\mathfrak{u}}(\lambda_2/\lambda_1)^{|\mathfrak{u}|}}{\max_{\mathfrak{v}\subseteq[d]} \gamma_{d,\mathfrak{v}}(\lambda_2/\lambda_1)^{|\mathfrak{v}|}},$$

and the *normalized* eigenvalues

$$\lambda_j^* = \frac{\lambda_{j+1}}{\lambda_2} \quad \text{for all} \quad j \in \mathbb{N}.$$

As in Wasilkowski, Woźniakowski (1999), define the exponent of summability for the normalized eigenvalues as

$$p_{\lambda^*} = \inf\left\{ \tau \geq 0 \ \Big| \ \sum_{j=1}^{\infty}[\lambda_j^*]^\tau < \infty\right\},$$

with the convention that the infimum of the empty set is infinity. Then

- S_γ is polynomially tractable iff $p_{\lambda^*} < \infty$ and there exist $q \geq 0$ and $\tau > p_{\lambda^*}$ such that

$$C := \sup_d \left[\sum_{\mathfrak{u}\subseteq[d]} [\gamma_{d,\mathfrak{u}}^*]^\tau \left(\sum_{j=1}^{\infty}[\lambda_j^*]^\tau\right)^{|\mathfrak{u}|}\right]^{1/\tau} d^{-q} < \infty. \qquad (8.6)$$

  If so, then

$$n_\gamma^{\mathrm{wor}}(\varepsilon, d) \leq C\, d^{q\tau}\, \varepsilon^{-2\tau}.$$

- S_γ is strongly polynomially tractable iff the condition (8.6) holds with $q = 0$, and then the exponent of strong polynomial tractability is given by

$$p = \inf\{2\tau \mid \tau > p_{\lambda^*} \text{ and } \tau \text{ satisfies (8.6) with } q = 0\}.$$

- For product weights $\gamma_{d,\mathfrak{u}} = \prod_{j\in\mathfrak{u}} \gamma_{d,j}$ with $\gamma_{d,j+1} \leq \gamma_{d,j}$ we have $\gamma_j^* = \gamma_{d,j}\lambda_2/\lambda_1$. For simplicity, assume that $\lambda_3 > 0$. Then

  - S_γ is polynomially tractable iff $p_{\lambda^*} < \infty$ and there exists $\tau > p_{\lambda^*}$ such that

$$\limsup_{d\to\infty} \frac{\sum_{j=1}^{d} \min(1, [\gamma_{d,j}^*]^\tau)}{\ln d} < \infty.$$

- S_γ is strongly polynomially tractable iff $p_{\lambda*} < \infty$ and $p_{\gamma**} < \infty$, where $\gamma^{**} = \{\min(1, \gamma_{d,j}^*)\}$ and $p_{\gamma**}$ is defined as $p_{\lambda*}$. If so, the exponent of strong polynomial tractability is

$$p = 2\max(p_{\lambda*}, p_{\gamma**}).$$

- For finite-order weights, $\gamma_{d,u} = 0$ for $|u| > \omega$, we have

  - S_γ is polynomially tractable iff $p_{\lambda*} < \infty$. If so then for any $\tau > p_{\lambda*}$

$$n_\gamma^{\text{wor}}(\varepsilon, d) \le 2\left(\sum_{j=1}^\infty [\lambda_j^*]^\tau\right)^\omega d^\omega \varepsilon^{-2\tau}.$$

  - S_γ is strongly polynomially tractable iff $p_{\lambda*} < \infty$ and there exists $\tau > p_{\lambda*}$ such that

$$\sup_d \left[\sum_{u \subseteq [d], |u| \le \omega} [\gamma_{d,u}]^\tau \left(\sum_{j=1}^\infty \lambda_j^*\right)^{|u|}\right]^{1/\tau} < \infty.$$

If so, then the exponent of strong tractability is

$$p = \inf\{2\tau \mid \tau > p_{\lambda*} \text{ and } \tau \text{ satisfies the condition above}\}.$$

- Let λ_1 be of multiplicity $p \ge 2$, i.e.,

$$\lambda_1 = \lambda_2 = \cdots = \lambda_p > \lambda_{p+1},$$

and let $\lambda_n = o([\ln n]^{-2}[\ln \ln n]^{-2})$. Define

$$m_p(\varepsilon, d) = \sum_{u \subseteq [d]:\ \gamma_{d,u} > \varepsilon^2} (p-1)^{|u|}.$$

Then

- S_γ is weakly tractable iff

$$\lim_{\varepsilon^{-1}+d \to \infty} \frac{\ln m_p(\varepsilon, d)}{\varepsilon^{-1} + d} = 0.$$

- For product weights, define $k(\varepsilon, d, \gamma) = k \in [d]$ such that

$$\prod_{j=1}^k \gamma_{d,j} > \varepsilon^2 \quad \text{and} \quad \prod_{j=1}^{k+1} \gamma_{d,j} \le \varepsilon^2.$$

If such an index k does not exist, set $k(\varepsilon, d, \gamma) = d$. Then S_γ is weakly tractable iff

$$\lim_{\varepsilon^{-1}+d \to \infty} \frac{k(\varepsilon, d, \gamma)}{\varepsilon^{-1} + d} = 0.$$

– For finite-order weights, S_γ is always weakly tractable.

The results presented above are typical for weighted spaces. They can be summarized by saying that for properly decaying weights, we can guarantee strong polynomial, polynomial or weak tractability. Furthermore, these conditions may not hold for the unweighted case, $\gamma_{d,\mathrm{u}} \equiv 1$. Indeed, this is the case even for a problem with only two positive eigenvalues, $\lambda_1 = \lambda_2 = 1 > \lambda_3 = 0$. Indeed, since S_d has 2^d eigenvalues equal to 1, we have $m_2(\varepsilon, d) = 2^d$, and so the unweighted problem is intractable.

We wish to stress that many other multivariate problems are at least polynomially tractable for finite-order weights. This holds for linear as well as a couple of nonlinear problems. The reader is referred to e.g., Wasilkowski, Woźniakowski (2004, 2005, 2008), Werschulz, Woźniakowski (2007a, 2007b).

Today, tractability of continuous problems is a popular research subject and tractability of many multivariate problems has been studied. I still believe that this only has been a beginning. Indeed, there are many open tractability problems. Maybe some readers of this paper will join the group of people interested in tractability.

8.10 Back to Numerical Tests

I hope that the reader still remembers the beginning of our story where we wanted to explain the success of numerical tests for finance computational problems. Do we really understand now why the numerical errors seem to be proportional to n^{-1} for large d and relatively small n?

From one point of view, we may be inclined to say yes. The reason is that probably the integrands of these tests belong to weighted spaces for which we have strong polynomial tractability with the exponent equal to 1. This is the case for:

- *Product weights* with $\gamma_{d,j} = j^{-\alpha}$ for some $\alpha > 2$. In fact, one can argue that for finance computations we have $\alpha = 2$. Indeed, since we often deal with approximation of path integrals equipped with the Wiener sheet measure (Brownian motion), the corresponding eigenvalues of the covariance operator behave like j^{-2}, and this may correspond to weights $\gamma_{d,j} = j^{-2}$. Although, the series $\sum_{j=1}^{\infty} \gamma_{d,j}^{1/2}$ does not converge, its divergence is quite weak since $\sum_{j=1}^{\infty} \gamma_{d,j}^{\tau} < \infty$ for all $\tau > 1/2$. In this case, it can be shown that the error is of order $n^{-1/(2\tau)}$, which is almost like n^{-1}.

- *Finite-order weights* with small ω. Some people claim that finance applications correspond to functions that can be well-approximated by sums of functions of one or two variables. Then $\omega = 1$ or $\omega = 2$. For general finite-order weights, as we know, the error may depend on $d^{\omega/2}$ or d^{ω}. But even for $d = 360$, the number $d^{\omega/2}$ for $\omega = 1$ is only about 19. Furthermore, we might have decaying finite-order weights, so that the condition of strong polynomial tractability may even be satisfied.

So what is wrong with this reasoning? The disappointing point is that functions from finance applications do *not* belong to the Sobolev space $H(K_d)$ or to the weighted Sobolev space $H(K_{d,\gamma})$ studied in Sections 3 and 5 for $d \geq 2$. The reason is simple. For finance integrands, we have many operations of min and max that correspond to decisions to buy or sell various options. For example, take

$$g(x) = \max(C, f(x))$$

for a smooth function f from $H(K_{d,\gamma})$ and a number C such that

$$\min_x f(x) < C < \max_x f(x).$$

For this C we have sometimes $g(x) = C$ and sometimes $g(x) = f(x)$, and so the function g is generally *not* smooth enough to belong to $H(K_{d,\gamma})$.

We have a really peculiar situation. The successful numerical tests were obtained by using algorithms whose error bounds can indeed be independent of d and of order n^{-1} for spaces $H(K_{d,\gamma})$ equipped with proper weights. But the finance integrands are *not* in these spaces. So the mystery is still present.

We believe that the QMC algorithms used in the numerical tests have one more useful property. Namely, these algorithms probably do not care much about the lack of smoothness introduced by operations of max or min, and so the error of at least some QMC is more or less the same with or without these operations. Note that for any real numbers a and b we have

$$\max(a, b) = \frac{a + b + |a - b|}{2} \quad \text{and} \quad \min(a, b) = \frac{a + b - |a - b|}{2}.$$

Therefore we can replace the operations of max and min by taking the absolute value. This lead us to the integration problem

$$\tilde{I}_d(f) = \int_{[0,1]^d} |f(t)| \, dt \quad \text{for all} \quad f \in H(K_{d,\gamma}).$$

We add in passing that for $d = 1$, if $f \in H(K_{1,\gamma})$ then also $|f| \in H(K_{1,\gamma})$. However, for $d \geq 2$ it is, in general, *not* true that $f \in H(K_{d,\gamma})$ implies $|f| \in H(K_{d,\gamma})$. Indeed, take $d = 2$, $\gamma_{d,\mathfrak{u}} = 1$ and a polynomial $f(x_1, x_2) = x_1 - x_2$. Assume that $|f| \in H(K_2)$. Then we would have

$$|x_1 - x_2| = \langle |f|, K_2(\cdot, x) \rangle_{H(K_2)} \quad \text{for all} \quad x = [x_1, x_2] \in [0,1]^2.$$

It is easy to compute the last inner product and check that it is $x_1 + x_2$, which contradicts the above formula.

Note that $\tilde{I}_d$ is a *nonlinear* functional and the existing theory for linear problems cannot be directly applied. Nevertheless, the nonlinearity of $\tilde{I}_d$ is very special and mild. There are preliminary arguments to support the claim that this nonlinear problem is roughly of the same difficulty as its linear counterpart I_d. We hope that some QMC algorithms are as good for $\tilde{I}_d$ as for the linear case I_d. But we still do not know it. So let me finish this paper with the following open problem whose solution seems necessary to close the loop and finally explain the success of numerical tests for finance applications.

Open Problem: Prove that the worst case errors of some QMC algorithms for multivariate integration are roughly the same for f as for $|f|$ if f belongs to $H(K_{d,\gamma})$, and tend to zero roughly like n^{-1} for properly chosen weights.

Acknowledgments

I am very grateful for useful and constructive comments from Erich Novak, Leszek Plaskota, Joseph Traub, Grzegorz Wasilkowski and Arthur Werschulz.

I also thank very much Frances Kuo and Christiane Lemieux for letting me know details of their tests for high dimensional integrals.

References

N. S. Bakhvalov (1971), 'On the optimality of linear methods for operator approximation in convex classes of functions' *USSR Comput. Maths. Math. Phys.* **11**, 244–249.

W. W. L. Chen (1985), 'On irregularities of distribution and approximate evaluation of certain functions', *Quarterly J. Math. Oxford* **36**, 173–182.

W. W. L. Chen (1987), 'On irregularities of distribution and approximate evaluation of certain functions, II', in *Analytic Number Theory and Diophantine Problems, Proceedings of a Conference at Oklahoma State University in 1984*, Birkhäuser, Basel, 75–86.

W. W. L. Chen and M. M. Skriganov (2002), 'Explicit constructions in the classical mean squares problem in irregularities of point distribution', *J. Reine Angew. Math. (Crelle)* **545**, 67–95.

J. Dick, I. H. Sloan, X. Wang and H. Woźniakowski (2006), 'Good lattice rules in weighted Korobov spaces with general weights', *Numer. Math.* **103**, 63–97.

K. K. Frolov (1980), 'Upper bounds on the discrepancy in L_p, $2 \leq p < \infty$', *Dokl. Akad. Nauk USSR* **252**, 805–807.

M. Giles, F. Y. Kuo, I. H. Sloan and B. Waterhouse (2008), 'Quasi-Monte Carlo for finance applications', submitted for publication.

J. Glimm and A. Jaffe (1987), *Quantum Physics*, Springer Verlag, New York.

M. Gnewuch and H. Woźniakowski (2007), 'Generalized tractability for multivariate problems, Part I: Linear tensor product problems and linear information', *J. Complexity* **23**, 262–295.

M. Gnewuch and H. Woźniakowski (2008), 'Generalized tractability for linear functionals', in: *Monte Carlo and Quasi-Monte Carlo Methods 2006*, A. Keller, S. Heinrich, H. Niederreiter (eds.), 359–381, Springer, Berlin.

A. Griewank and Ph. L. Toint (1982), 'Local convergence analysis for partitioned quasi-Newton updates', *Numer. Math.* **39**, 429–448.

S. Heinrich (2003), 'Some open problems concerning the star-discrepancy', *J. Complexity* **19**, 416–419.

S. Heinrich (2008a), 'Randomized approximation of Sobolev embeddings II', in preparation.

S. Heinrich (2008b), 'Randomized approximation of Sobolev embeddings III', in preparation.

F. J. Hickernell, G. W. Wasilkowski and H. Woźniakowski (2006), 'Tractability of linear multivariate problems in the average case setting', in: *Monte Carlo and Quasi-Monte Carlo Methods 2006*, A Keller, S. Heinrich, H. Niederreiter (eds.), 461–494, Springer, Berlin.

F. J. Hickernell and H. Woźniakowski (2000), 'Integration and approximation in arbitrary dimension', *Adv. Comput. Math.*, **12**, 25–58.

A. Hinrichs, E. Novak, and J. Vybiral (2008), 'Linear information versus function evaluations for L_2-approximation', to appear in *J. Approx. Th.*

Tak-San Ho and H. Rabitz (2003), 'Reproducing kernel Hilbert space interpolation methods as a paradigm of high-dimensional model representations: application to multidimensional potential energy surface construction', *J. Chem. Physics* **119**, 6433–6442.

F. Y. Kuo (2003), 'Component-by-component constructions achieve the optimal rate of convergence for multivariate integration in weighted Korobov and Sobolev spaces', *J. Complexity* **19**, 301–320.

F. Y. Kuo, I. H. Sloan, G. W. Wasilkowski and H. Woźniakowski (2008), 'On decompositions of multivariate functions', submitted for publication.

F. Y. Kuo, G. W. Wasilkowski and H. Woźniakowski (2008a), 'On the power of standard information for multivariate approximation in the worst case setting', to appear in *J. Approx. Th.*

F. Y. Kuo, G. W. Wasilkowski and H. Woźniakowski (2008b), 'On the Power of Standard Information for L_∞ Approximation in the Randomized Setting', submitted for publication.

Ch. Lemieux (2000), *L'utilisation de règles de réseau en simulation comme technique de réduction de la variance*, PhD, Université de Montréal.

C. Lemieux and P. L'Ecuyer (1999), 'Lattice Rules for the Simulation of Ruin

Problems', in *Proceedings of the 1999 European Simulation Multiconference*, **2**, 533–537, The Society for Computer Simulation, Ghent, Belgium.

H. Niederreiter (1992), *Random Number Generation and Quasi-Monte Carlo Methods*, SIAM, Philadelphia.

E. Novak and H. Woźniakowski (2001), 'Intractability results for integration and discrepancy', *J. Complexity* **17**, 388–441.

E. Novak and H. Woźniakowski (2008), *Tractability of Multivariate Problems*, Volume I, *European Mathematical Society*, Zürich.

E. Novak and H. Woźniakowski (2009), *Tractability of Multivariate Problems*, Volume II, to appear, *European Mathematical Society*, Zürich.

D. Nuyens and R. Cools (2006), 'Fast algorithms for component-by-component construction of rank-1 lattice rules in shift invariant reproducing kernel Hilbert spaces', *Math. Comp.* **75**, 903–920.

A. Papageorgiou and J. F. Traub (1996), 'Beating Monte Carlo', *Risk* **9**, 63-65.

S. Paskov and J. F. Traub (1995), 'Faster valuation of financial derivatives', *J. Portfolio Management* **22**, 113–120.

L. Plaskota, G. W. Wasilkowski and Y. Zhao (2008), 'New averaging technique for approximating weighted integrals', to appear in *J. Complexity*

H. Rabitz and Ö. F. Alis (1999), 'General foundations of high-dimensional model representations', *J. Math. Chemistry* **25** 197–233.

K. F. Roth (1954), 'On irregularities of distributions', *Mathematika* **1**, 73–79.

K. F. Roth (1980), 'On irregularities of distributions IV', *Acta Arith.* **37**, 67–75.

J. Rust, J. F. Traub and H. Woźniakowski (2002), 'Is there a curse of dimensionality for contraction fixed points in the worst case?', *Econometrica* **70**, 285–329.

I. H. Sloan, F. Y. Kuo and S. Joe (2002), 'Constructing randomly shifted lattice rules in weighted Sobolev spaces', *SIAM J. Numer. Anal.* **40**, 1650-1665.

I. H. Sloan, X. Wang and H. Woźniakowski (2004), 'Finite-order weights imply tractability of multivariate integration', *J. Complexity* **20**, 46–74.

I. H. Sloan and H. Woźniakowski (1998), 'When are quasi-Monte Carlo algorithms efficient for high dimensional integrals?' *J. Complexity* **14**, 1–33.

I. H. Sloan and H. Woźniakowski (2001), 'Tractability of integration for weighted Korobov spaces', *J. Complexity* **17**, 697–721.

J. F. Traub, G. W. Wasilkowski and H. Woźniakowski (1988), *Information-Based Complexity*, Academic Press, New York.

J. F. Traub and A. G. Werschulz (1998), *Complexity and Information*, Cambridge University Press, Cambridge.

G. W. Wasilkowski and H. Woźniakowski (1999), 'Weighted tensor-product algorithms for linear multivariate problems', *J. Complexity* **15**, 402–447.

G. W. Wasilkowski and H. Woźniakowski (2001), 'On the power of standard information for weighted approximation', *Found. Comput. Math.* **1**, 417–434.

G. W. Wasilkowski and H. Woźniakowski (2004), 'Finite-order weights imply tractability of linear multivariate problems', *J. Approx. Th.* **130**, 57–77.

G. W. Wasilkowski and H. Woźniakowski (2005), 'Polynomial-time algorithms for multivariate problems with finite-order weights: worst case setting', *Found. Comput. Math.* **5**, 451–491.

G. W. Wasilkowski and H. Woźniakowski (2006), 'The power of standard information for multivariate approximation in the randomized setting', *Math. Comp.* 76, 965-988.

G. W. Wasilkowski and H. Woźniakowski (2008), 'Polynomial-time algorithms for multivariate linear problems with finite-order weights: average case setting', to appear in *Found. Comput. Math.*

A. G. Werschulz and H. Woźniakowski (2007a), 'Tractability of quasilinear problems I: general results', *J. Approx. Th.* **145**, 266–285.

A. G. Werschulz and H. Woźniakowski (2007b), 'Tractability of quasilinear problems II: second-order elliptic problems', *Math. Comp.* **258**, 745–776.

H. Woźniakowski (1991), 'Average case complexity of multivariate integration', *Bull. AMS* **24**, 185–191.

H. Woźniakowski (1994a), 'Tractability and strong tractability of linear multivariate problems', *J. Complexity* **10**, 96–128.

H. Woźniakowski (1994b), 'Tractability and strong tractability of multivariate tensor product problems', *J. of Computing and Information* **4**, 1–19.

Printed in the United States
By Bookmasters